NITROGEN CYCLING IN THE NORTH ATLANTIC OCEAN AND ITS WATERSHEDS

NITROGEN CYCLING IN THE NORTH ATLANTIC OCEAN AND ITS WATERSHEDS

Edited by
ROBERT W. HOWARTH

Reprinted from Biogeochemistry 35(1)

KLUWER ACADEMIC PUBLISHERS
DORDRECHT / BOSTON / LONDON

A C.I.P. Catalogue record for this book is available from the Library of Congress.

ISBN 0-7923-4281-X

Published by Kluwer Academic Publishers,
P.O. Box 17, 3300 AA Dordrecht, The Netherlands.

Kluwer Academic Publishers incorporates
the publishing programmes of
D. Reidel, Martinus Nijhoff, Dr W. Junk and MTP Press.

Sold and distributed in the U.S.A. and Canada
by Kluwer Academic Publishers,
101 Philip Drive, Norwell, MA 02061, U.S.A.

In all other countries, sold and distributed
by Kluwer Academic Publishers Group,
P.O. Box 322, 3300 AH Dordrecht, The Netherlands.

Printed on acid-free paper

Printed in the Netherlands

CONTENTS

Biogeochemistry **35:** 1, 1996.

Foreword

The papers in this special issue of *Biogeochemistry* summarize the results of a workshop held on Block Island, Rhode Island, USA, in May 1994. This was the first of several workshops planned as part of the International SCOPE Project on Nitrogen Transport and Transformations: A Regional and Global Analysis. This project is sponsored by SCOPE, the Scientific Committee on Problems of the Environment, International Council of Scientific Unions, because of the recognized need to better understand how humans have altered nitrogen cycling at large regional scales, and what the consequences of this alteration are.

This first workshop assessed what is known about nitrogen cycling in the North Atlantic Ocean and its watersheds, with a particular emphasis on how human activity has altered nitrogen fluxes. Over 40 scientists from 13 nations contributed to the great success of this workshop. The resulting papers consist of a summary overview, 4 core chapters which reflect the concensus viewpoints of the 4 working groups of the meeting, and a few notes which elaborate on particular aspects of nitrogen cycling in the North Atlantic Basin. I hope that the readers of *Biogeochemistry* share my enthusiasm and that of the meeting participants for the many exciting findings presented in this issue.

The Block Island workshop was sponsored by the United Nations Environment Program, the World Meteorological Organization, and the Mellon Foundation. The Mellon Foundation also provides additional financial support for the SCOPE Nitrogen Project. James Galloway and I served as co-chairs for the meeting. Other members of the organizing committee included Ragnar Elmgren, Joseph Prospero, Anthony Knap, Anthony Michaels, Jennie Moody, and Scott Nixon. Scott Nixon also served as local host for the workshop. Support for the meeting was provided by Ellen Yoder of the University of Rhode Island, Mary Scott Kaiser of the University of Virginia, and Roxanne Marino, Emily Ehrenfeld, and Dennis Swaney of Cornell University. I am grateful to all.

Robert W. Howarth
Editor-in-Chief

International Scope Project on Nitrogen Transport and Transformations: A Regional and Global Analysis

Objectives and Activities

- to foster the necessary synergism between scientist of many disciplines (marine ecologists, forest ecologists, agricultural scientists, microbiologists, atmospheric chemists, oceanographers, hydrologists) in order to help develop new approaches for the study of nitrogen cycling;
- to refine the global nitrogen budget and develop regional budgets for selected key and contrasting regions of the world;
- to more fully understand the problems stemming from accelerated nitrogen cycling, and the inter-relationships among these problems.

Scientific Advisory Committee

Co-chairs:
Robert Howarth (USA)
John Freney (Australia)

Members:
Frank Berendse (The Netherlands)
Pornpimol Chaiwanakupt (Thailand)
Valery Kudeyarov (Russia)
Scott Nixon (USA)
Peter Vitousek (USA)
Zhu Zhao-liang (People's Republic of China)

Consultants:
Ragner Elmgren (Sweden)
James Galloway (USA)

SCOPE
SCIENTIFIC
COMMITTEE
ON PROBLEMS
OF THE ENVIRONMENT

UNEP
United Nations Environment Programme

Biogeochemistry **35:** 3–25, 1996.

Nitrogen and phosphorus budgets of the North Atlantic Ocean and its watershed

J. N. GALLOWAY[1], R. W. HOWARTH[2], A. F. MICHAELS[3], S. W. NIXON[4], J. M. PROSPERO[5] & F. J. DENTENER[6]

[1]*Environmental Sciences, University of Virginia, Charlottesville, VA 22903 USA;* [2]*Ecology and Systematics, Cornell University, Ithaca, NY 14853 USA;* [3]*Bermuda Biological Station for Research, St. Georges GE-01 Bermuda;* [4]*Graduate School of Oceanography, University of Rhode Island, Naragansett RI 02882 USA;* [5]*RSMAS, University of Miami, Miami, FL 33149 USA;* [6]*Department of Airquality, Wageningen, NL 6700 EV Wageningen, The Netherlands*

Received 22 March 1996; accepted 22 March 1996

Abstract. Anthropogenic food and energy production extensively mobilize reactive nitrogen (N) in the watershed of the North Atlantic Ocean (NAO). There is wide spread N distribution by both hydrologic and atmospheric processes within the watershed of the NAO, resulting in reactive N accumulation in terrestrial systems. Net denitrification in most estuaries and continental shelves exceeds the amount of N supplied to the shelves by rivers and requires a supply of nitrate from the open ocean. Thus riverine N is only transported to the open ocean in a few areas with the flow from a few major rivers (e.g., Amazon). Atmospheric N deposition to the open ocean has increased and may increase the productivity of the surface ocean. In addition, as a consequence of increased Fe deposition to the open ocean (due in part to anthropogenic processes), the rate of biological N-fixation may have increased resulting in N accumulation in the ocean. Phosphorus (P) is also mobilized by anthropogenic processes (primarily food production). Relative to N, more of the P is transported across the shelf to the open ocean from both estuaries and major rivers. There are several consequences of the increased availability of N and P that are unique to each element. However, the control on primary productivity in both coastal and open ocean ecosystems is dependent on a complex and poorly understood interaction between N and P mobilization and availability.

Introduction

Nitrogen (N) is a key element of many biogeochemical processes and can be a limiting element of aquatic and terrestrial ecosystem processes (Schlesinger 1991; Vitousek & Howarth 1991). However, about 99% of global N exists as stable atmospheric N_2 (Mackenzie et al. 1993) and thus is unavailable to ecosystems unless it is converted into a reactive N species (reactive N = NH_3, NH_4^+, organic N, NO, NO_2, HNO_3, NO_3, N_2O_5, HNO_4, HNO_2, NO_2^-, and NO_3^-). Once created, one species of reactive N can be converted into other species of reactive N by a variety of chemical and microbial processes. In addition, reactive N species are very mobile via atmospheric and hydrologic pathways. After formation, reactive N can only be converted back to unreactive N_2 by denitrification, an anaerobic process which only occurs at significant rates in specific types of ecosystems. Reactive N accumulates in the environment if the denitrification rate is less than the rate of N-fixation.

In the absence of humans, natural processes create reactive N by biological N-fixation and lightning. Biological N-fixation occurs in specific microbes when atmospheric N_2 is converted to NH_3 by the enzyme nitrogenase. Lightning produces NO by the reaction of N_2 and O_2 at high temperatures. The former process is about two orders of magnitude greater than the latter on a global basis (Galloway et al. 1995). Human intervention in the N cycle has increased the formation rate of reactive N by fertilizer production, legume and rice cultivation, and combustion of fossil fuels. Results from several recent analyses of the global N cycle (Mackenzie et al. 1993; Ayres et al. 1994; Galloway et al. 1995) generally agree that anthropogenic activities mobilize about 10 Tmol N yr^{-1} (Tmol = 10^{12} moles) and that human activities mobilize N at rates equal to natural terrestrial processes.

There is significant distribution of anthropogenic N by hydrologic and atmospheric transport. Combustion of fossil fuels injects reactive N (NO) directly into the atmosphere. Fertilizer and cultivation N increase the productivity of agricultural landscapes. However, on average, no more than 50% of the applied N fertilizer is removed by crop harvest; the remainder is lost to the atmosphere or hydrosphere, or stored in the soil (Howarth et al. this volume). In addition, N fixed in crops has a short residence time; it is quickly transformed into human or animal waste, which also results in significant inputs to the atmosphere and hydrosphere (Howarth et al. this volume). If the N introduced into the environment by fossil fuel combustion, fertilizer production and cultivation is not denitrified, then reactive N accumulates in downwind or downstream ecosystems.

The mobilization, distribution and accumulation of anthropogenic N impacts a number of physical and ecosystem processes. As discussed in Howarth et al. (this volume), Nixon et al. (this volume), and Michaels et al. (this volume), increased availability of reactive N increases forest productivity, and as a consequence stores atmospheric CO_2, contributes to forest decline (if soil is N-saturated) and climate change, and results in shifts in community structure and ecosystem function. It also increases coastal eutrophication in estuaries (including their wetlands), and results in increased N supply to oligotrophic mid-ocean gyres with concomitant affects on the ecology of the upper ocean. In addition to being important to ecosystems, reactive N also affects atmospheric chemistry (Prospero et al. this volume). High levels of NO_x ($NO + NO_2$) play an important role in the photochemical production of O_3 (Moxim et al. 1994). NH_3 is a major source of alkalinity in the atmosphere and a source of acidity in soils (Schlesinger 1991). Although N_2O is not viewed as a reactive form of N in the troposphere, it adsorbs IR

Figure 1. The North Atlantic Ocean and its watershed. The watershed is subdivided into its drainage basins. The boundary between the shelf and the open ocean is 200m depth.

radiation and acts as a greenhouse gas. In the stratosphere, N_2O impacts O_3 concentrations (Warneck 1988). Thus, any change in the rate of formation of reactive N (or N_2O), its global distribution, or its accumulation rate can have a fundamental impact on many environmental processes.

Our understanding of the global distribution, fate and impacts of anthropogenic N is insufficient because of a lack of data. To partially alleviate this problem, it is necessary to examine the N cycle on a smaller scale where data are more available, while still keeping the salient features of the N cycle – natural and anthropogenic sources, atmospheric and hydrologic transport processes, and agricultural, forested, freshwater, coastal and marine ecosystems. Hence we focus on a watershed analysis in general, and the North Atlantic Ocean and its watershed (NAO&W) specifically (Figure 1). We selected the watershed scale because its physiographic and hydrological features shape a biogeochemical system; these same features shape systems of human interaction. We selected the NAO&W because, compared to other ocean basins, it contains a rich data record of the atmosphere:watershed:ocean units on N cycling and because of the high degree of N disturbance by human activities.

Our analysis of the NAO&W N budget was centered about a workshop, held in May 1994 at Block Island, RI, USA. The central scientific question of the workshop was:

What are the current sources and sinks of nitrogen in the North Atlantic Ocean and its watershed? How might these have been changed over natural background levels as a result of human activity?

Since the impact of anthropogenic N on ecosystems is determined in part by the relative availability of N and P, our investigation of the N budget of the NAO&W includes an analysis of the magnitude and fate of P transferred from continents to the ocean.

To address the question of N cycling on the scale of an entire ocean basin and its associated watershed, a unique group of individuals assembled, each with expertise in physical, chemical and biological aspects of marine, coastal, freshwater and terrestrial ecosystems, the atmosphere, and the connections among the components. Working groups were formed on the basis of reservoirs: atmosphere, watershed, coastal and shelf region, and open ocean. Each working group developed an annual N cycle for their respective reservoir. The watershed of the NAO and associated coastal segments were further divided (Figure 1) to investigate spatial variability in N and P fluxes. The task of each group was to characterize the internal N fluxes and the hydrologic/atmospheric export losses for their reservoir; input fluxes to the reservoir were the domain of the upstream or upwind group. To enhance integration, each group was allowed 'to borrow' members from other groups to address specific issues of N exchange.

The primary products of the workshop are the working group papers on the N and P fluxes of each sub-reservoir: atmosphere – deposition to watershed and oceans (Prospero et al. this volume); watershed – N inputs, fates, and riverine losses to the coastal margins, and riverine P fluxes to the coastal ocean (Howarth et al. this volume); coastal and shelf region–fate of riverine and atmospheric inputs (Nixon et al. this volume); and the open ocean (Michaels et al. this volume). In addition, in the process of developing these papers, it became apparent that several topics warranted special attention. Thus accompanying these papers are four notes: NH_3 exchange between the atmosphere and the NAO (Quinn et al. this volume), N-fixation in the NAO (Lipschultz and Owens, this volume); denitrification in the ocean boundaries (Seitzinger and Giblin, this volume); and shelf areas of the North Atlantic Ocean (Pilson and Seitzinger, this volume).

This paper is a synthesis of the workshop findings. It presents the N and P budgets for the entire NAO&W system by addressing the following specific topics:

- Mobilization of reactive N in the NAO&W
- Re-distribution of reactive N among the subsystems of the NAO&W
- Removal of reactive N from the NAO&W relative to sources
- P distribution patterns in the NAO&W

• Anthropogenic N and P impacts on the NAO&W

An advantage of examining the NAO&W as one system is that it provides an overview; a disadvantage is that it masks the extensive spatial variability in the N and P budgets of the portions of four continents, and associated coastal margins, that constitute the watershed of the NAO. This variability is extensive. Europe and North America have significantly greater increases in the mobilization of reactive N than do Africa and South America. We highlight some of the detailed regional findings from the 14 sub-watershed portions of the watershed of the NAO and adjacent coastal regions and refer the readers to the appropriate papers for the rest of the detail.

The results

We illustrate the overall behavior of N and P in the NAO&W by dividing the NAO&W into five systems: atmosphere, watershed (26.9×10^6 km^2), estuaries, shelf (to 200m depth, 5.7×10^6 km^2), and open ocean (100×10^6 km^2) (e.g., Figure 2). Shaded arrows in Figure 2 represent reactive N creation; open arrows represent reactive N transfers between the NAO&W sub-systems. The arrow width is sized to the median flux magnitude. The ranges about the median fluxes are found in the text. Units are 10^9 moles yr^{-1} (Gmol yr^{-1}). For consistency, the fluxes in this paper have been rounded to two significant figures from the values reported in the accompanying papers.

Mobilization of reactive N in the NAO&W

There are two sources of reactive N – creation within, and transport to, the NAO&W. Reactive N is created and mobilized by biological N-fixation, lightning, commercial N fertilizer use, fossil fuel combustion and legume cultivation. The watershed and open ocean are the most important source regions for reactive N in the NAO&W; per area, the land masses of the North Atlantic are a far greater source of reactive N, reflecting the greater human disturbance on land (Figure 2).

Watershed. Fertilizer use introduces 1,600 Gmol yr^{-1} to the watershed of the NAO and is the most important anthropogenic N source, followed by the deposition of NO_y ($NO_y = NO_x + HNO_3 + NO_3^-$ aerosol + other oxygen-containing N species except N_2O) from fossil fuel combustion (590 Gmol yr^{-1}), and legume cultivation (330 Gmol yr^{-1}). The total anthropogenic N addition to the NAO watershed is 2,500 Gmol yr^{-1} (Howarth et al. this volume). Figure 2 shows a watershed input of anthropogenic NH_x ($NH_3 + NH_4^+$) from atmospheric deposition (290 Gmol yr^{-1}). We view this input as

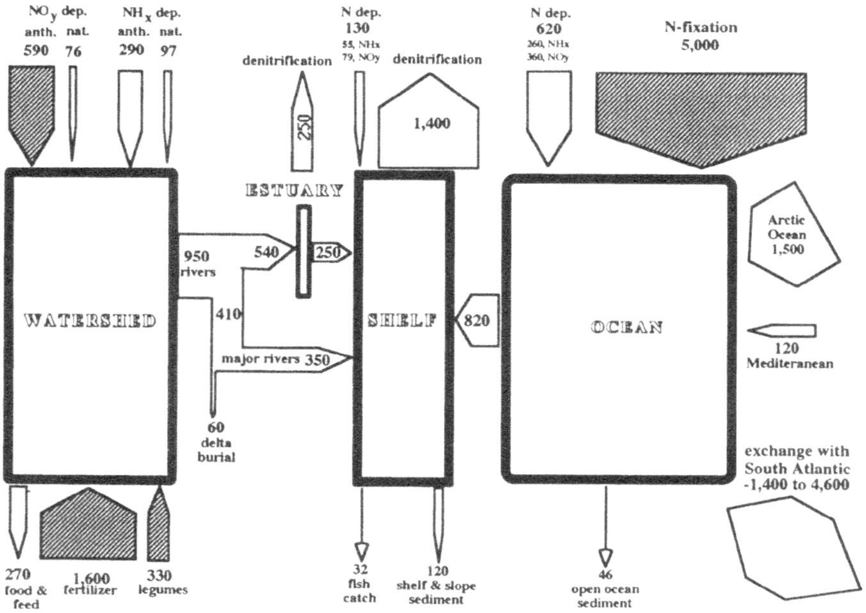

Figure 2. The N budget for the North Atlantic Ocean and its watershed (Gmol yr^{-1}). The stippled arrows represent the introduction of new sources of reactive N into the NAO&W. The open arrows represent exchange of reactive N between the subsystems of the NAO&W.

an internal recycling of NH_x following its introduction by fertilizer use and subsequent NH_3 volatilization from soils and animal waste (Howarth et al. this volume).

There are two natural sources of reactive N: lightning and terrestrial biological N-fixation. Lightning sources are small relative to fertilizer sources – the global total is only about 210 Gmol yr^{-1} (Levy et al. 1996). Natural terrestrial biological N-fixation for the watershed of the NAO may be important relative to anthropogenic N sources. Globally, the range of fifteen estimates is 5,700 to 13,600 Gmol yr^{-1} (as reviewed in Jaffe 1992; Mackenzie et al. 1993). Thus, natural biological N-fixation in the watershed of the NAO, with 18% of the earth's land area, could be significant. However, given the probable strong latitudinal dependence on the N-fixation rate, the dearth of data for many tropical and temperate ecosystems, and our focus at examining the fate of anthropogenic N mobilized in the watershed of the NAO, we choose not to estimate natural terrestrial biological N-fixation rate. We do point out however, that studies more specifically focused on developing an integrated N cycle for watershed units need to include natural biological N-fixation.

The relative importance of anthropogenic N sources varies by sub-watershed unit (as defined in Figure 1). Agricultural N sources dominate inputs in many regions, particularly the Mississippi and the North Sea sub-watersheds. In the NE United States, NO_y deposition as a consequence of fossil fuel combustion is the major control over river N. In the temperate region, there is a strong correlation of total N flux with population density. This feature is manifested in a strong linear correlation between river fluxes of total N and the sum of anthropogenic N inputs (Howarth et al. this volume). Although on the average, N in food and feed is exported from the watershed of the NAO (Figure 2), for some sub-watersheds it can be a significant source. For example, in the NE coast region, about 1/3 of the N sources are from food and feed import (Howarth et al. this volume).

Estuary and shelf. Biological N-fixation certainly occurs in estuarine ecosystems but its magnitude is small relative to riverine input of reactive N previously created in the NAO watershed. In the shelf region biological N-fixation is estimated to also be small, on the order of 20 Gmol yr^{-1} (Nixon et al. this volume).

Ocean. Biological N-fixation in the open ocean is the most important N source for the NAO&W. Michaels et al. (this volume) identify a source of nitrate of 3,700 to 6,400 Gmol yr^{-1} (median, 5000 Gmol yr^{-1}, Figure 2) within the main thermocline of the Sargasso Sea that they infer is caused by N-fixation. This N-fixation may be regulated by the supply of Fe in dust and thus may have occurred at a lower rate before the expansion of the Sahara in the early 1970s. The most conspicuous N-fixing organisms in the North Atlantic Ocean are the filamentous cyanobacteria *Trichodesmium sp.* Lipschultz and Owens (this volume) conclude that the measurements of N-fixation by *Trichodesmium sp.* imply a rate of only 1,100 Gmol N yr^{-1} for this taxa alone. The difference between Lipschultz and Owens and Michaels et al. may come from the scarcity of the data, questions about the accuracy of the methods and from the potential contributions of other organisms than *Trichodesmium sp.*

The second general source of N to the NAO is the transport into the region of reactive N created elsewhere. The large scale currents in the NAO result in significant horizontal advection of nutrients (Michaels et al. this volume). There is strong northerly surface transport in the Gulf Stream. There is a net southerly flow in the North Atlantic Deep Water, and net northerly transport in the Antarctic Bottom and Intermediate Waters. Michaels et al. summarized a number of techniques to estimate meridional transport in the NAO. From the north they estimate that 300 to 2,800 Gmol yr^{-1} (median: 1,500 Gmol

yr^{-1}, Figure 2) are advected into the NAO from the Arctic Ocean. In addition, Michaels et al. (this volume) estimate that 50 to 200 Gmol yr^{-1} (median: 120 Gmol yr^{-1}, Figure 2) are transported from the Mediterranean to the NAO. To the south, the uncertainty of the calculation is such that both the magnitude and direction of transport are unknown. The range is – 1,400 Gmol yr^{-1} out of the NAO to the South Atlantic Ocean (SAO), to 4,600 Gmol yr^{-1} from the SAO to the NAO. The implications for the overall balance of N in the NAO are discussed in a following section.

In summary, the watershed of the NAO receives its anthropogenic reactive N primarily from fertilizer use, with smaller contributions from combustion of fossil fuels and legume cultivation. Some regions within the watershed of the NAO, especially within the temperate zone, are heavily impacted by not only fossil fuel combustion and fertilizer application, but also N importation in foods and feeds. The NAO receives its reactive N primarily from N-fixation, although transport from the SAO may also be important. As discussed in the next two sections, additional N sources are atmospheric and hydrologic transport of reactive N created in the watershed.

Re-distribution of reactive N among the subsystems of the NAO&W

Via atmospheric transport. NO_x and NH_x are emitted into the atmosphere by natural and anthropogenic processes. The most important anthropogenic processes are energy production (NO_x emissions from fossil fuel combustion) and food production (NH_3 emissions from fertilizer use and animal waste) (Prospero et al. this volume). Once emitted, NO_x and NH_3 are transformed, transported downwind and deposited as NO_y and NH_x, respectively. Unlike ecosystems, there are no important atmospheric reactions that convert NH_x to NO_y and visa-versa. Thus deposition of anthropogenic NO_y and NH_x can be generally linked to energy and food production, respectively.

Total (present-day natural and anthropogenic) NO_y atmospheric deposition to the watershed, shelf and open ocean regions of the NAO&W is 670 Gmol yr^{-1}, 79 Gmol yr^{-1} and 360 Gmol yr^{-1}, respectively (Figure 2) (Prospero et al. this volume; Howarth et al. this volume; Nixon et al. this volume; Michaels et al. this volume). Anthropogenic NO_y deposition (590 Gmol yr^{-1}) to the entire watershed of the NAO is about seven-fold greater than deposition of naturally produced NO_y (76 Gmol yr^{-1}) (Figure 2). In regions heavily impacted by fossil fuel combustion (i.e., NE coast of US), NO_y deposition is estimated to have increased by a factor of 30 relative to natural rates. For more remote portions of the watershed of the NAO (i.e., Amazon and Tocantins basin) NO_y deposition has increased about 2–3 times over natural values (Howarth et al. this volume).

A similar situation exists for NH_x deposition. The watershed, shelf and open ocean regions receive total deposition of 390 Gmol yr^{-1}, 55 Gmol yr^{-1} and 260 Gmol yr^{-1} respectively (Figure 2). For the watershed of the NAO, current NH_x deposition is about 3 times greater than natural rates for the entire watershed and up to 12 times greater for regions of intense agricultural activity (i.e., NW European coast) and about two times for more remote regions (i.e., Amazon and Tocantins basin) (Howarth et al. this volume). Relative to riverine sources, direct N atmospheric deposition to estuaries is small (Nixon et al. this volume).

While we have not estimated the natural and anthropogenic components of N deposition to the shelf and open ocean portions of the NAO, there is likely to be a strong human influence, especially for NO_y, given the disparity between natural NO atmospheric sources and anthropogenic. As discussed in Prospero et al. (this volume) and Galloway et al. (1995), the present-day global deposition rate of NO_y is about five times greater than pre-industrial times largely due to emissions from energy production and biomass burning. The relative increase will be even greater for the NAO because it is adjacent to region of intense fossil fuel combustion. While there is no doubt that NH_x deposition to the NAO has increased (Prospero et al. this volume), the magnitude of the increase is more difficult to estimate because of a potentially large natural emission of NH_3 from the ocean (Quinn et al. 1988). At this stage, the data on NH_3 emissions for the NAO are too sparse to make an estimate of an ocean wide flux (Quinn et al. this volume). It should be noted that our estimates of total N deposition to the NAO&W may be underestimates for a growing literature suggests that atmospheric deposition of organic N species may be equivalent to deposition of NO_y and NH_x, especially in marine regions (Prospero et al. this volume).

There is significant spatial variability in the NO_y and NH_x deposition rates. By way of illustration, we examine NO_y (Dentener & Crutzen 1993) and NH_x (Dentener & Crutzen 1994) deposition across transects of 45° N and 5° N at ten degree blocks of latitude and longitude (Figure 3). To show how N deposition to the NAO compares to the North Pacific Ocean, we begin the transects at 175° W. To show how European and African emissions impact deposition to the east of the study area, we extend the transect to 35° E.

At 45° N, N deposition in North America and Europe is about 20 and 30 times, respectively, greater than the Pacific Ocean (Figure 3). Downwind of North America there is a rapid reduction in NO_y and NH_x deposition to the NAO. However, even in the middle of the NAO region, N deposition is still about double that of the North Pacific Ocean. These model results support previous statements that N deposition to the entire NAO has increased as a consequence of human activity (Whelpdale & Galloway 1994; Prospero et

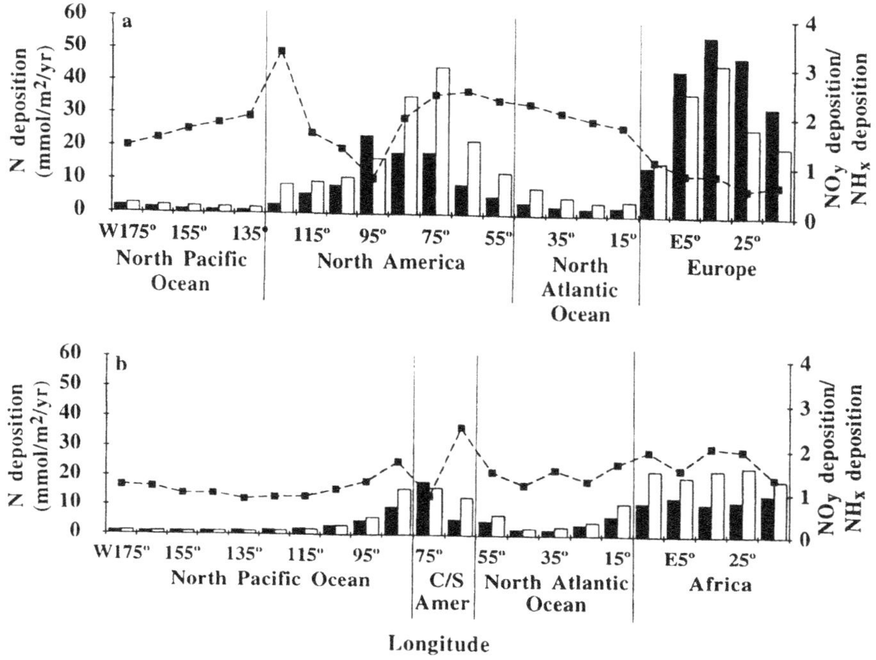

Figure 3. NO_y (open bars) and NH_x (solid bars) deposition (mmol/m2/yr) from 175° W eastward to 35° E at (a) 45° N latitude and (b) 5° N latitude. Ratio of NO_y/NH_x deposition shown by dotted line.

al. this volume). At the western boundary of Europe, N deposition increases significantly and remains high well outside of our study region due to atmospheric transport of European N emissions. Within the NAO&W study region, the relative importance of NO_y to NH_x deposition varies by about a factor of five. At the western boundary, NO_y/NH_x is about 0.7 reflecting the predominance of agricultural sources of NH_3. To the east, the ratio increases to about 2 due to the relative increase in energy over food production. The ratio gradually decreases over the NAO and is about 1 at the western boundary of Europe. It then decreases to about 0.5, again due to the strength of the agriculture sector relative to the energy sector.

For the 5° N transect, total deposition is significantly less than at 45° N (Figure 3) due to the reduced impact of NO_x and NH_3 sources. Nonetheless, N deposition to the NAO at 5° N is still about 2–3 times greater than to the North Pacific Ocean. In addition, because of the predominant easterly winds, atmospheric transport of N is from east to west, accounting for the gradual reduction in deposition to the west of Africa and Central and South America. With the exception of the eastern portion of the Central and South America, the NO_y/NH_x ratio varies between about 1 and 2, reflecting the importance of combustion over agriculture as N sources to the atmosphere.

In summary, there is a net transfer of NO_y and NH_x from the continental to the marine atmosphere. A substantial fraction of the ~2,500 Gmol yr^{-1} of reactive N mobilized in the watershed of the NAO by human activities is deposited into the marine environment following atmospheric emission and transport.

Via hydrologic transport. N is hydrologically exchanged between the subsystems of the NAO&W by riverine, estuarine and cross shelf processes. 2,500 Gmol yr^{-1} of anthropogenic N are introduced into the watershed of the NAO. 950 Gmol yr^{-1} of natural and anthropogenic N are lost via rivers to the marine environment; 530 Gmol yr^{-1} from the temperate portion of the watershed and 420 Gmol yr^{-1} from the tropical (Howarth et al. this volume). Since we do not have estimates of the natural N-fixation rate in the watershed, or precise estimates of the anthropogenic N in rivers, we can only say that the maximum loss of anthropogenic N via rivers in the temperate portion is 35%. Any amount of natural N-fixation within the watershed will reduce this percentage. The data are too sparse to make an equivalent estimate for the tropical portion.

Of the 950 Gmol yr^{-1} lost via rivers from the entire watershed of the NAO, about 410 Gmol yr^{-1} are from large rivers that debouch directly into the NAO shelf region; the remainder (540 Gmol yr^{-1}) flow into estuaries. Ultimately, large rivers and estuaries deliver about 350 Gmol yr^{-1} (290 from major rivers to shelf waters and ~61 Gmol yr^{-1} to slope sediments) and 170–340 Gmol yr^{-1} to the NAO shelf region, respectively (250 Gmol yr^{-1}, Figure 2) (Nixon et al. this volume). From the ocean side of the shelf region, about 820 Gmol yr^{-1} are transported to the shelf from the open ocean, primarily as nitrate (Nixon et al. this volume; Michaels et al. this volume). Thus the shelf region, via hydrologic transport, receives about 1,400 Gmol N yr^{-1} of N from major rivers, estuaries and the open ocean (Figure 2). It also receives ~20 Gmol yr^{-1} by N-fixation (not shown on Figure 2). This amount, combined with the atmospheric flux of 130 Gmol yr^{-1}, results in an input of ~1,600 Gmol yr^{-1} to the shelf. As discussed in the following section, most of the N delivered to the shelf region is denitrified.

There is significant spatial variability in the degree of hydrologic N transport. About two-thirds of the riverine N is injected from the western portion of the watershed of the NAO, about equally divided between North America and Central and South America. 25% of the N comes from one basin – the Amazon (Howarth et al. this volume). Similarly, all the major rivers that inject N directly to the shelf region are in the western portion of the watershed of the

NAO and again, the Amazon exerts a large influence. It accounts for about half of the 600 Gmol yr^{-1} of riverine N that are injected on to the NAO shelf (Nixon et al. this volume).

There is also spatial variability in the impact that human activities have had on hydrologic transfer of N between the NAO and its watershed. Using data from relatively pristine areas as an index of change, Howarth et al. (this volume) estimate that riverine N fluxes in many of the temperate regions have increased from pre-industrial times by 2 to 20 fold, although some regions such as northern Canada are relatively unchanged. Fluxes from the most disturbed region, the North Sea drainages, have increased by 6 to 20 fold. Fluxes from the Amazon basin are also at least 2 to 5 fold greater than estimated fluxes from undisturbed temperate-zone regions. However, we do not know the relative contribution of natural vs. anthropogenic activities. Deforestation may be contributing to the Amazon N flux, but tropical regions may also have naturally greater riverine fluxes due to higher amounts of N-fixation and phosphorus rather than N limitation of net primary productivity in forests (Howarth et al. this volume).

For the NAO shelf as a whole, N fluxes from rivers and estuaries (600 Gmol yr^{-1}) exceed atmospheric deposition (130 Gmol yr^{-1}) by about 4-fold (Figure 2), but this varies widely among regions of the shelf. For example, on the U. S. Atlantic shelf and on the NW European shelf, atmospheric N deposition may exceed estuarine exports.

Removal of reactive N from the NAO&W relative to sources

This section discusses the removal of reactive N from the NAO&W by denitrification, storage in long term reservoirs (e.g., forests, sediments, groundwater), transport out of the NAO&W via the atmosphere, the hydrosphere and food and feed. It concludes with a summary of the N budget for the NAO&W (Table 1).

Denitrification within the NAO&W. The conversion of reactive N to N_2 by biological denitrification decreases the reactive N stock in the NAO&W and can be viewed as a permanent sink (atmospheric N_2 residence time is $\sim 10^6$ yrs). Important locations of watershed denitrification are wetlands and the sediments of freshwater aquatic ecosystems. Howarth et al. (this volume) estimate that about >920 Gmol yr^{-1} are denitrified or stored in wetlands, lakes, streams and rivers of the temperate portions. The equivalent value for tropical portions is unknown. As pointed out in Howarth et al. (this volume), there is significant spatial variability in the amount of denitrification and storage over the watershed of the NAO.

Table 1. Summary of Nitrogen Fluxes for the Watershed, Estuary, Shelf and Open Ocean Regions of the North Atlantic Ocean and its Watershed, Gmol yr^{-1}*

INPUTS		OUTPUTS	
• *Temperate Watershed*		• *Temperate Watershed*	
Fertilizer (2)	1460	River discharge (2)	520
Combustion (2)	510	Food/feed (2)	230
N-fixation by legumes (2)	330	Groundwater storage (2)	<60
Natural biological N-fixation	??	Forest storage (2), [when nat. biolog N-fix is low]	<530
N-fixation by lightning	small	Denitrification & storage in wetlands, streams, rivers (2)	>920
Total input	2300	Total output	2300
(*w/o natural biological N-fixation*)		(*Plus whatever is fixed in natural biological N-fixation*)	
• *Tropical Watershed*		• *Tropical Watershed*	
Fertilizer, excl. Africa (2)	85	River discharge (2)	420
Combustion, excl. Africa (2)	74	Food/feed (2)	??
N-fixation by legumes (2)	??	Groundwater storage (2)	??
Food & feedstock imports	??	Forest storage (2)	??
Natural biological N-fixation	??	Denitrification & storage in	
N-fixation by lightning	??	wetlands, streams, rivers (2)	??
Total input	??	Total output	??
• *Estuary*		• *Estuary*	
Rivers (2)	540	Transport to shelf (3)	250
Atmospheric deposition (1)	small	Storage in sediment(3)	small
Nat. biological N-fixation (3)	small	Denitrification (3, 5)	250
Total input	540	Total output	500
• *Shelf*		• *Shelf*	
From estuaries (3)	250	Denitrification (3, 5)	1400
Major rivers (2)	410	Delta, shelf, slope burial (3)	180
Atmospheric depositon (1)	130	Fish catch (3)	32
Nat. biological N-fixation (3)	20		
From ocean (3,4)	820		
Total input	1630	Total output	1612
• *Ocean*		• *Ocean*	
Nat. biological N-fixation (4,6)	5000	Denitrification (4)	small
Atmospheric depositon (1)	620	Storage in sediment (4)	46
From Arctic Ocean (4)	1500	To SAO (4)	??
From Mediterranean Sea (4)	120	Transport to shelf (3, 4)	820
From SAO (4)	??		
Total input	7240	Total output	966

* The values in this table are averages and are presented to give an appreciation of the magnitude of the fluxes. For some parameters, the range around the average can be substantial. For all values, see paper for details. The numbers in parentheses refer to the original paper that discusses the flux: (1) Prospero; (2) Howarth; (3) Nixon; (4) Michaels; (5) Seitzinger and Giblin; (6) Lipschultz.

There are three other locations where denitrification may be important: estuaries, shelf and open ocean. Estuaries receive about 540 Gmol N yr^{-1} via rivers, of which about half is buried or denitrified, primarily the latter mechanism (Nixon et al. this volume). The shelf region receives about 1,600 Gmol yr^{-1} from major rivers, estuaries, atmospheric deposition, N-fixation, and the open ocean. It is estimated that ~90% is denitrified in the shelf sediments and the water column; the remainder is buried in shelf and slope sediments or removed in fisheries harvest (Figure 2) (Nixon et al. this volume; Seitzinger & Giblin, this volume). Denitrification in the open ocean can occur in the sediments and the water column. Most of the sediment denitrification occurs in the shelf, which we have already accounted for. Open ocean denitrification is uncertain, but thought to be small because it requires areas with strong sub-surface oxygen minimum zones and anoxic basins. Anoxic conditions only occur in a few marginal seas like the Cariaco trench. Even if there is active denitrification in anoxic microenvironments within the water column (e.g. inside particles) then oceanic rates of denitrification must be small relative to the N-fixation rate. Denitrification removes nitrate and lowers the nitrate:phosphate ratios below the nominal Redfield stoichiometries of 16N:1P. In most of the NAO, nitrate:phosphate ratios are anomalously high, relative to the Redfield ratio (Fanning 1989; 1992; Michaels et al. 1994) indicating net N-fixation (Michaels et al. this volume).

Storage within the NAO&W. Storage in the watershed of the NAO occurs in biomass and groundwater. The temperate watershed of the NAO stores up to 530 Gmol yr^{-1} of the 2,500 Gmol yr^{-1} in forest biomass in an unknown mixture of anthropogenic and natural N; storage in groundwater of the temperate portion is less than a few percent (<60 Gmol yr^{-1}) (Howarth et al. this volume). Equivalent values for the tropical forest and groundwater are unknown (Howarth et al. this volume). Storage of reactive N in deltas and estuarine shelf and open ocean sediments is thought to be small; in all cases estimated N storage is $<10\%$ of the N inputs (Figure 2) (Nixon et al. this volume; Michaels et al. this volume). The overall balance of fluxes in the open NAO indicates that it may be accumulating N at a rate of 0.06–0.4% per year, even given the considerable uncertainty in the transport estimates (Michaels et al. this volume). This apparent accumulation is largely a result of the large inputs from the estimated nitrogen-fixation and a modest atmospheric deposition, both of which may vary with time and are likely larger today because of human activity. There is a net flux of nitrate to the shelves to meet the shelf denitrification demand that exceeds the nitrogen supply from the rivers. In the pre-industrial world, these riverine N fluxes would also have been lower and the denitrification demands on oceanic nitrate larger. Unfortunately, the large size of the total nitrate pool in the oceans and our short observational record preclude the direct observation of changes in the total stock of reactive N.

Transport out of the NAO&W. There are three transport mechanisms that distribute anthropogenic N beyond the borders of the NAO&W: atmospheric, hydrologic and food/feed exports. About 10% of the anthropogenic N formed in the watershed of the NAO is exported from the NAO&W as food and feed (270 Gmol yr^{-1}, Figure 2) (Howarth et al. this volume). Hydrologic transport of N out of the region can occur via deep water flow to the South Atlantic Ocean. There is a potential for transport of up to 20 to 40% of the N fixed by biological activity in the NAO. The uncertainty in this transport term is such that there is also the potential for the SAO to be a source for N that is equivalent to N-fixation in the NAO (Figure 2) (Michaels et al. this volume). Atmospheric transport of N out of the region does occur; however, the magnitude is uncertain. Regions of significant transport are to the east of Europe and the west of Central and South America (Figure 3).

N Budget of the NAO&W. In summary, there are three important N sources to the NAO&W: anthropogenic N introduction to the watershed, N-fixation in the open ocean and transport of N into the NAO from adjacent water bodies. Natural biological N-fixation in the watershed of the NAO is probably an important, but unknown N source. Important N sinks are storage within the NAO&W subsystems, denitrification, and possibly transport to the SAO.

Significant uncertainties in some of the sinks and sources precludes an overall budget for the NAO&W. However, we can compare N sources and sinks for the NAO subsystems (Table 1). For the NAO temperate watershed, 2,300 Gmol N yr^{-1} are introduced by fertilizer, fossil fuel combustion and legume cultivation. Natural biological N-fixation is unknown but probably significant. Of this N, 520 Gmol yr^{-1} are lost via rivers; a small amount ($<$60 Gmol yr^{-1}) is stored in groundwater. Food and feed exports account for another 230 Gmol yr^{-1}. Storage in temperate forests is less than 530 Gmol yr^{-1}, storage and denitrification in temperate wetlands, streams and rivers is greater than 920 Gmol yr^{-1}. We are not able to differentiate between storage and denitrification in wetlands, streams and rivers. For tropical regions we have uncertain estimates of anthropogenic inputs and thus we are not able to estimate either forest storage or storage and denitrification in freshwater ecosystems (Table 1). It is clear, nonetheless, that the tropics export more N in rivers as a consequence of anthropogenic activities (Howarth et al. this volume).

The estuaries of the NAO&W receive most of their N from riverine discharge (540 Gmol yr^{-1}); atmospheric deposition and N-fixation are less important. About half of the N delivered to estuaries is denitrified and half is transported to the shelf waters (Table 1).

The NAO shelf receives about 1,600 Gmol N yr^{-1} from: the ocean (820 Gmol yr^{-1}), major rivers (410 Gmol yr^{-1}), estuaries (250 Gmol yr^{-1}), atmospheric deposition (130 Gmol yr^{-1}), and N-fixation (20 Gmol yr^{-1}). Most of this N is denitrified, and only a small amount (<10%) is lost to delta, shelf and slope sediments (180 Gmol yr^{-1}) or as fish catch (32 Gmol yr^{-1}) (Table 1).

The major N source to the NAO is biological fixation (5,000 Gmol yr^{-1}) (Table 1). Hydrologic transport from the SAO could be very important, but both the sign and the magnitude of the transport is very uncertain. Estimates range from a loss of 1,400 Gmol yr^{-1} to the SAO to a gain of 4,600 Gmol yr^{-1} from the SAO. In spite of the large uncertainties in all the fluxes, the NAO appears to be accumulating N in today's world. Even if the estimate of transport across the equator is significantly wrong, the other sub-regions of the NAO are also accumulating N (Michaels et al. this volume). The injection of anthropogenic N to the NAO by cross shelf transport from major rivers and atmospheric deposition is small relative to natural N sources in the NAO. However, if we consider as anthropogenic the possible increase in N-fixation in the Sargasso Sea from increased Saharan dust (75% higher than before the drought) this would add another 2,800 to 4,800 Gmol yr^{-1} to the anthropogenic flux (see later discussion) (Michaels et al. this volume).

Overall, the greatest areas of uncertainty in the N budget for the NAO&W are (1) storage vs. denitrification losses in the watershed, (2) natural N-fixation in the watershed, (3) N-fixation in the open ocean, and (4) the exchange of N between the NAO and the SAO. It is important to gain additional information on all of these fluxes. Until we learn the balance between N storage vs. denitrification in the NAO watershed we are unable to assess to what degree anthropogenic N is accumulating in continental ecosystems. In addition, until we have better information on natural terrestrial N-fixation we are unable to estimate the rate at which human activities are fixing N relative to natural processes. Our uncertainty about N-fixation rates in the ocean and the degree that the NAO is a N source or a sink relative to the SAO make it difficult to determine not only the natural N cycle for the NAO but also the possible impact of Fe input on N-fixation in the NAO.

P distribution patterns in the NAO&W

The distribution patterns of N and P through the environment are quite different. N has a significant gas phase, and is distributed by atmospheric as well as hydrologic processes. P has no stable gas phase in the atmosphere and is only transported on particles, limiting its atmospheric lifetime and hence its

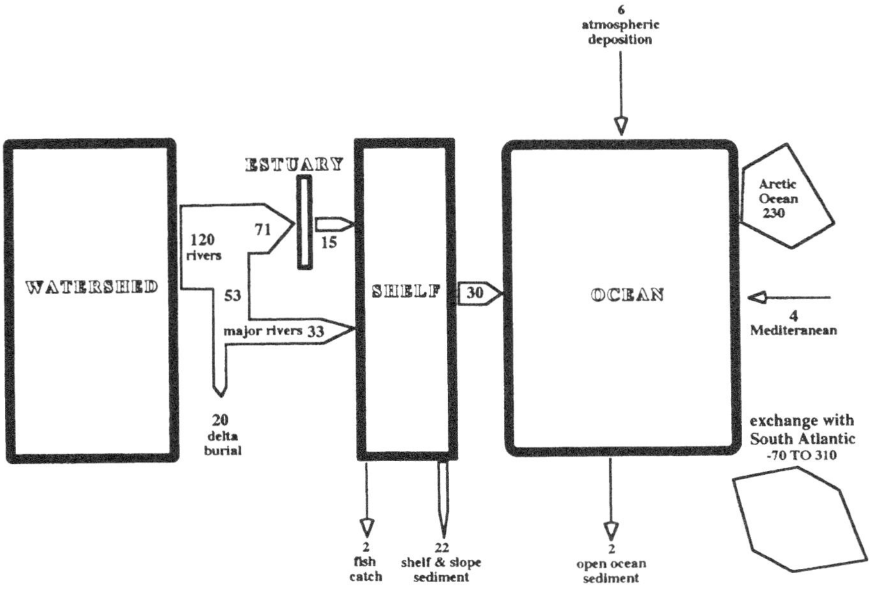

Figure 4. The P budget for the North Atlantic Ocean (Gmol yr^{-1}).

global dispersion. Hence atmospheric and riverine N inputs into the marine environment are both important (Figure 2), while P inputs from rivers are substantially greater than those from the atmosphere (Figure 4). There is another critical difference between N and P. N mobilized by human activities can be "unmobilized" by both long term storage (e.g., sediments) and by denitrification, which returns N to an inactive state (N_2). For P, there is no process equivalent to denitrification. Once mobilized by human activity, it accumulates in either a short term reservoir (biomass) or a long term reservoir (sediments). Given the importance of hydrologic transport for P, and its potential for storage in sediments, we focus our effort on the coastal, shelf and open ocean subsystems of the NAO&W.

Annually, about 120 Gmol yr^{-1} P are lost from the watershed via rivers of which about 53 Gmol P are from large rivers (mostly the Amazon) and the remainder from smaller rivers that flow into estuaries (Figure 4). Most of the riverine P is deposited to delta, estuarine, shelf and slope sediments (~100 Gmol yr^{-1}, Figure 4). Only about 30 Gmol yr^{-1} are injected into the open ocean from rivers. Another 160–300 Gmol yr^{-1} are injected to the NAO from the Arctic Ocean (230 Gmol yr^{-1}, Figure 4). It appears that the NAO may be accumulating P at a rate of about 0.1 % per year (Michael et al. this volume). However, given the relatively large uncertainties around many of the estimates of P fluxes and stocks, this small difference may not be separable from a balance.

Impact of anthropogenic N and P on the NAO&W

Terrestrial systems. NO_x emissions from fossil fuel combustion have increased the oxidizing potential of the atmosphere on a regional basis in the watershed of the NAO with measurable impacts on human health. The same emissions have also increased precipitation acidity in wide spread portions of the watershed of the NAO, with concomitant impacts on terrestrial and aquatic ecosystems (e.g., Grennfelt et al. 1995). The addition of fertilizer N to the watershed of the NAO has purposefully increased productivity of agricultural ecosystems, resulting in the fixation of more carbon and other elements associated with plant and animal growth. There are also inadvertent increases in productivity in other ecosystems which result in alterations to other elemental cycles. Persistent atmospheric deposition of fixed N to forest ecosystems may eventually lead to retrogressive changes in ecosystem function. Aber et al. (1989) suggested that forests in the northeastern United States are becoming N sources rather than N sinks because of extensive N deposition resulting in N supply in excess of demand. Not only are the forests potentially impacted by the high rate of atmospheric deposition but they are also N sources to downstream ecosystems and to the atmosphere (N_2O emissions). The emission and subsequent deposition of N results in increased growth of forests and sequestering of C and other nutrients (Schindler & Bayley 1993; Schimel et al. 1995).

Coastal systems. Riverine injection of anthropogenic N and P to coastal ecosystems has the potential to increase productivity (Howarth 1988; Turner & Rabelais 1991; Vitousek & Howarth 1991; NRC 1993; NRC 1994; Nixon 1995; Howarth et al. 1995). The importance of the N versus P contribution to increased productivity depends on their relative injection rates. If the N:P ratio is <16 ("Redfield" ratio), theoretically N is the limiting nutrient; if the ratio is >16, P is the limiting nutrient. The N:P ratios vary by region. In the Amazon, it is 4.5:1; waters flowing into the western Gulf of Mexico have a 271:1 ratio. However, for most of the riverine fluxes to the NAO coastal region, the N:P ratios are within a factor of 2 or 3 of the "Redfield" ratio, suggesting that estuaries in most of these regions have the potential to be either N or P limited, depending on the bioavailability of the N and P, and the estuarine processes that alter the ratio (Howarth et al. this volume).

Anthropogenic N in rivers probably does not directly impact the open ocean. Riverine N is primarily denitrified in the shelf region (except for the Amazon), and thus never reaches the open ocean. However, an indirect impact is that anthropogenic N in rivers results in a decrease in the open ocean N export to the shelf. In the pre-industrial world, if shelf denitrification was unchanged from today, than supply of N from the open ocean was perhaps about double of what it is today. If this is correct, then the oceanic accumulation may have an additional anthropogenic cause.

Open ocean. Anthropogenic N can be deposited to the open ocean, but at rates that are generally small compared to other sources (Michaels et al. 1993). While anthropogenic N probably does not have a major impact on the overall biogeochemistry of the NAO, there may be episodic impacts on productivity in stratified regions of mid-ocean gyre, where upwelling supply of N is limited (Owens et al. 1992; Michaels et al. 1993). In these situations, P limitation may be a confounding factor. The N:P ratio of the deposition of reactive N and P from the atmosphere to the ocean varies systematically across the basin. Reactive N is primarily from pollution sources and is maximal near the U.S. coastline and decreases rapidly to the east. P deposition is primarily with dust and thus has a maximum near the coast of Africa and declines to the west. As a result, the ratio of reactive N to P varies from values of 10–30 near the African coast to values of 500–1500 near the U.S. coastline (calculated from Prospero et al. this volume). This analysis does not include organic N or P; however, since levels of organic phosphorous are low, the N:P ratio is dominated by the substantial pools of organic nitrogen. Given that the N:P ratios are always greater than the Redfield ratio, it is possible that the P deposition may regulate the impact on surface biology. However, there are confounding factors; e.g., much of the deposited P is tightly bound in insoluble compounds such as apatite and, consequently, the P is not readily available to organisms.

Phytoplankton in the surface waters of the Sargasso Sea are exposed to low concentrations of both reactive N and phosphate; in situ concentrations are below the detection limits of standard analysis for both nutrients. Thus, the instantaneous growth rates of most phytoplankton could be co-limited by both nutrients. Some organisms may have the ability to fix atmospheric N and these would obviously be limited by the availability of phosphate. The atmospheric deposition patterns modify this picture. In nearly all cases the deposition ratio of N and P is higher than the traditional stoichiometry of phytoplankton growth, the Redfield ratio of 16N:1P. Thus, a deposition event, even if it adds considerable reactive N to the surface mixed layer, will rarely add enough phosphate to allow complete utilization of the N. Thus, in these deposition events, surface populations are likely phosphate limited. Since the required phosphate supply to sustain the N-fixation postulated by Michaels et al. this volume, is very large compared to the atmospheric deposition, the phosphate to support N-fixation must come from the stocks in the thermocline. They suggest that the migration behavior of the large and colonial organism responsible for the nitrogen-fixation allow them access to these deeper phosphate stocks. Consequently, the nutrient limitation patterns for the surface community are probably determined by the dynamics of the phosphate transport from below and not the atmospheric deposition.

There may be an indirect impact of human activities on the open ocean N cycle. There is an interesting and potentially very important interaction between the N cycle and Fe. It is proposed that Fe deposition to the NAO by atmospheric transport from Africa is supporting N-fixation rates in the NAO (Prospero et al. this volume; Michaels et al. this volume). A substantial fraction of the mineral dust carried to the NAO appears to be mobilized as a consequence of human activities in Africa compounded by rainfall variability (Prospero et al. this volume); thus, the impact of dust-borne Fe on N-fixation rates in the NAO could be regarded as being largely an anthropogenic effect. Also, because dust transport is highly variable from year-to-year due to changes in weather and climate (Prospero & Nees 1986) the impact of Fe deposition to the ocean could also vary greatly. This linkage of emission processes in Africa to ecosystem productivity in far downwind ecosystems is not new – it has also been proposed for the supply of P to portions of the Amazon forest (Swap et al. 1992). Further, the nitrogen-fixation that occurs in the Sargasso Sea would require an additional uptake of dissolved CO_2 of approximately 0.3–1.2 Gt yr^{-1} (Michaels, this volume). This carbon is remineralized at depths with residence times of decades, so the process does not result in a long-term sequestration of surface or atmospheric carbon. However, if the rate of nitrogen-fixation is regulated by the highly variable dust deposition, then the resulting variations in N-fixation rate could yield significant interannual fluctuations in the carbon drawdown and perhaps affect the interannual variations in atmospheric CO_2 as well.

Acknowledgments

We wish to acknowledge and express our appreciation for the funding of this project from the UNEP and the Mellon Foundation to support the efforts of the SCOPE Nitrogen Project, and from the World Meteorological Organization to investigate the interaction between the N cycles of the atmosphere and the NAO&W. The Rhode Island Sea Grant Program contributed logistical support for the meeting. We appreciate the dedicated and cheerful support services provided at the meeting by Roxanne Marino (Cornell University), Emily Ehrenfeld (Cornell University), Dennis Swaney (Cornell University), and Ellen Yoder (University of Rhode Island). We are also thankful for the efforts of Mary Scott Kaiser (University of Virginia) in making arrangements prior to the meeting and to Judith Peatross (University of Virginia) in assisting with manuscript editing and preparation. James Galloway is grateful to The Ecosystems Center of the Marine Biological Station, the Woods Hole Oceanographic Institute, and the International Institute for Applied Systems Analysis for hosting him during his sabbatical year. In addition, partial support from NOAA (Global Precipitation Chemistry Project) and the NSF (Atmosphere-Ocean Chemistry Experiment) is gratefully acknowledged in the preparation of this manuscript. This paper is a contribution to the Bermuda Biological Station for Research, Inc.

References

Aber JD Nadelhoffer KJ Steudler P & Melillo JM (1989) Nitrogen saturation in northern forest ecosystems. BioScience 39: 378–386

Ayres RU, Schlesinger WH & Socolow RH (1994) Human impacts on the carbon and nitrogen cycles. In: Socolow RH, Andrews C, Berkhout R & Thomas V (Eds) Industrial Ecology and Global Change (pp 121–155). Cambridge University Press, Cambridge Mass

Dentener FJ & Crutzen PJ (1993) Reaction of N_2O_5 on tropospheric aerosols: Impact on the global distributions of NO_x O_3 and OH. J. Geophys. Res. 98: 7149–7163

Dentener FJ & Crutzen PJ (1994) A three-dimensional model of the global ammonia cycle. J. Atmos. Chem. 19: 331–369

Fanning KA (1989) Influence of atmospheric pollution on nutrient limitation in the ocean. Nature 339: 460–463

Fanning KA (1992) Nutrient provinces in the sea: Concentration rations reaction rate ratios and ioodeal covariation. J. Geophys. Res. C Oceans 97: 5693–5712

Galloway JN, Schlesinger WH, Levy II H, Michaels A & Schnoor JL (1995) Nitrogen-fixation: anthropogenic enhancement–environmental response. Global Biogeochemical Cycles 9: 235–252

Grennfelt P, Rodhe H, Thörnelöf E & Wisniewski J (1995) Acid Reign '95? Proceedings from the 5th International Conference on Acidic Deposition Goteborg Sweden 26–30 June. Water, Air and Soil Pollut. 85, Nos. 1, 2, 3, & 4

Howarth RW (1988) Nutrient limitation of net primary production in marine ecosystems. Ann. Rev. Ecol. & Syst. 19: 89–110

Howarth RW, Jensen H, Marino R & Postma H (1995) Transport to and processing of P in near-shore and oceanic waters. In: Tiessen H (Ed) Phosphorus in the Global Environment: Transfers Cycles and Management SCOPE 54 (pp 323–345). John Wiley & Sons, Chichester UK

Howarth RW, Billen G, Swaney D, Townsend A, Jaworski N, Lajtha K, Downing JA, Elmgren R, Caraco N, Jordan T, Berendse F, Freney J, Kudeyarov V, Murdoch P & Zhu Zhao-liang (1996) Regional nitrogen budgets and riverine N and P fluxes for the drainages to the North Atlantic Ocean: natural and human influences. Biogeochemistry, this volume

Jaffee DA (1992) The Nitrogen Cycle. In: Butcher SS, Charlson RJ, Orians GH & GV Wolfe (Eds) Global Biogeochemical Cycles (pp 263–284). Pub Academic Press

Levy II H, Moxim WJ & Kasibhatla PS (1996) A global 3-dimensional time dependent lightning source of tropospheric NO_x. J. Geophys. Res. in press

Lipschultz F & Owens NJP (1996). An assessment of nitrogen-fixation as a source of nitrogen to the North Atlantic Ocean. Biogeochemistry, this volume

Mackenzie FT, Ver LM, Sabine C, Lane M & Lerman A (1993) C N P S global biogeochemical cycles and modeling of global change NATO ASI Series Vol 14. In: Wollast R, Mackenzie FT & Chou L (Eds) Interactions of C N P and S Biogeochemical Cycles and Global Change (pp 1–61). Springer-Verlag Berlin Heidelberg
Michaels AF, Knap AH, Dow RL Gundersen K, Johnson RJ, Sorensen J, Close A, Knauer GA, Lohrenz SE, Asper VA Tuel M & Bidigare RR (1994) Seasonal patterns of ocean biogeochemistry at the U S JGOFS Bermuda Atlantic Time-series Study site. Deep-Sea Res. 41: 1013–1038
Michaels AF, Siegel DA, Johnson RJ, Knap AH & Galloway JN (1993) Episodic inputs of atmospheric nitrogen to the Sargasso Sea: Contributions to new production and phytoplankton blooms. Global Biogeochem. Cycles 7: 339–351
Michaels AF, Olson D, Ammerman JW, Fanning K, Jahnke R, Knap AH, Lipschultz F & Prospero JM (1996) Inputs losses and transformations of nitrogen and phosphorus in the pelagic North Atlantic Ocean. Biogeochemistry, this volume
Moxim WJ, Levy II H, Kasibhatla PS & Chameides WL (1994) The impact of anthropogenic and natural NOx sources on the global distribution of tropospheric reactive nitrogen and the net chemical production of ozone. Eos Trans AGU 75 Fall Meet Supple A41E
National Research Council (1993) Managing Wastewater in Coastal Urban Areas Committee on Wastewater Management for Coastal Urban Areas. National Academy Press. Washington DC
Nixon SW (1995) Coastal marine eutrophication: A definition social causes and future concerns. Ophelia 41: 199–219
Nixon SW, Ammerman JW, Atkinson LP, Berounsky VM, Billen G, Boicourt WC, Boynton WR, Church TM, DiToro DM, Elmgren R, Garber JH, Giblin AE, Jahnke RA, Owens NJP, Pilson MEQ & Seitzinger SP (1996) The fate of nitrogen and phosphorus at the land-sea margin of the North Atlantic Ocean. Biogeochemistry, this volume
Owens NJP, Galloway JN & Duce RA (1992) Episodic atmospheric nitrogen deposition to oligotrophic oceans. Nature 357: 397–399
Pilson MEQ & Seitzinger SP (1996) Areas of shallow water in the North Atlantic. Biogeochemistry, this volume
Prospero JM & Nees RT (1986) Impact of North African drought and El Nino on mineral dust in the Barbados trade winds. Nature 320: 735–738
Prospero JM, Barrett K, Church T, Dentener F, Duce RA, Galloway JN, Levy II H, Moody J & Quinn P (1996) Atmospheric deposition of nutrients to the North Atlantic basin. Biogeochemistry, this volume
Quinn PK, Charlson RJ & Bates TS (1988) Simultaneous observations of ammonia in the atmosphere and ocean. Nature 335: 336–338
Quinn PK, Barrett KJ, Dentener FJ, Lipschultz F & Kurz KD (1996) Estimation of the air/sea exchange of ammonia for the North Atlantic Basin. Biogeochemistry, this volume
Schimel D, Enting I, Heimann M, Wigley TML, Raynaud D, Alves D & Siegenthaler U (1995) CO_2 and the carbon cycle. In: Houghton J et al. (Eds) Climate Change, 1994, Radiative Forcing of Climate Change. Cambridge University Press, England, 339 p
Schlesinger WH (1991) Biogeochemistry: An Analysis of Global Change. Academic Press, San Diego, 443 pp
Schindler DW & Bayley SE (1993) The biosphere as an increasing sink for atmospheric carbon: Estimates from increased nitrogen deposition. Global Biogeochem Cycles 7: 717–725
Seitzinger SP & Giblin AE (1996) Estimating denitrification in the North Atlantic continental shelf sediments. Biogeochemistry, this volume
Swap R, Garstang M, Talbot R & Kakaallberg P (1992) Saharan dust in the Amazon basin. Tellus 44b: 133–149
Turner RE & Rabalais NN (1991) Changes in Mississippi River water quality this century: Implications for coastal food webs. Bioscience 41: 140–147
Vitousek PM & Howarth RW (1991) Nitrogen limitation on land and in the sea: How can it occur? Biogeochemistry 13: 87–115

Warneck P (1988) Chemistry of the Natural Atmosphere. Academic Press, London
Whelpdale DM & Galloway JN (1994) Sulfur and nitrogen oxide fluxes in the North Atlantic atmosphere. Global Biogeochem. Cycles 8: 481–493

Biogeochemistry **35**: 27–73, 1996.

Atmospheric deposition of nutrients to the North Atlantic Basin

J.M. PROSPERO[1], K. BARRETT[2], T. CHURCH[3], F. DENTENER[4], R.A. DUCE[5], J.N. GALLOWAY[6], H. LEVY II[7], J. MOODY[8] & P. QUINN[9]
[1]*University of Miami, RSMAS, Miami, FL;* [2]*Norwegian Meteorological Institute, Oslo, Norway;* [3]*University of Delaware, College of Marine Studies, Newark, DE;* [4]*Wageningen University, Department of Air Quality, Wageningen, The Netherlands;* [5]*Texas A&M Univ., College of Geosciences & Maritime Studies, College Station, TX;* [6]*University of Virginia, Dept. of Environmental Sciences, Charlottesville, VA;* [7]*NOAA/GFDL, Princeton University, Princeton, NJ;* [8]*University of Virginia, Dept. of Environmental Sciences, Charlottesville, VA;* [9]*NOAA, Pacific Marine Environmental Laboratory, Seattle, WA*

Received 21 September 1995; accepted 6 December 1995

Key words: dust, deposition, iron, models, NH_x, NO_y, nutrients

Abstract. Atmospheric chemical models are used to estimate the deposition rate of various inorganic oxides of nitrogen (NO_y), reduced nitrogen species (NH_x) and mineral dust to the North Atlantic Ocean (NAO). The estimated deposition of NO_y to the NAO (excluding the coastal ocean) and the Caribbean is 360×10^9 Moles-N m^{-2} yr^{-1} (5.0 Tg N); this is equivalent to about 13% of the estimated global emission rate (natural and anthropogenic) and a quarter of the emission rate from sources in North America and Europe. In the case of NH_x, 258 Moles-N m^{-2} yr^{-1} (3.6 Tg N) are deposited to the NAO and the Caribbean; this is about 6% of the global continental emissions. There is relatively little data on the deposition rate of organic nitrogen species; nonetheless, this evidence suggests that concentrations and deposition rates are comparable to those for inorganic nitrogen.

Because of anthropogenic emissions, the present-day deposition rate of NO_y to the NAO is about five times greater than pre-industrial times largely due to emissions from energy production and biomass burning. The present-day emissions of NH_x from continental anthropogenic sources are about four-to-five times greater than natural sources, mostly due to the impact of emissions from animal wastes associated with food production. Indeed, present-day emissions of NH_x from animal waste are estimated to be about 10 times greater than the pre-human era. The deposition rate of mineral dust to the NAO is about 170 Tg yr^{-1}; deposited with the dust (assuming average crustal abundances) is about 6 Tg yr^{-1} of Fe and 0.2 Tg yr^{-1} of P. Dust deposition in the NAO is almost completely attributable to transport from North African sources; a substantial fraction of the dust over the NAO is probably mobilized as a consequence of land use practices in arid regions and, consequently, it should be regarded as a pollutant.

Abbreviations: AEROCE – Atmosphere/Ocean Chemistry Experiment; AOT – aerosol optical thickness; EMEP – Cooperative Program for Monitoring and Evaluation of Long Range; – Transmission of Air Pollution in Europe; GESAMP – A model used by the Group of Experts on the Scientific Aspects of Marine Pollution; GCM – General Circulation Model; GCTM – Global Chemical Transport Model; GFDL – Geophysical Fluid Dynamic Laboratory; MOGUNTIA – Model of the Global Universal Tracer Transport in the Atmosphere; MSC – Meteorological Synthesizing Centre; NAB – North Atlantic Basin; NAO – North Atlantic Ocean; NAS – National Academy of Sciences

Introduction

The atmosphere is an important pathway by which many gas and aerosol species are carried by winds from their sources to the larger world where they are eventually deposited. There is now ample evidence that a substantial fraction of the global cycles of many chemical species, both natural and anthropogenic, involves transport through the atmosphere. Although the fluxes of some species have been characterized for some regional waters and basins, they are not accurately known for the open ocean. Nonetheless, it is clear that some fluxes could be large enough to have a significant impact on the chemical and biological processes in natural waters and soils.

The North Atlantic Ocean (NAO) is a relatively small ocean basin that is closely bounded by four large continental masses. Consequently there is a relatively high probability that continental emissions, whether natural or anthropogenic, will be carried to the NAO by winds. Pollutant emissions are of particular interest because of the high level of human activities in this region. For example, sources of anthropogenic NO and NO_2 (Levy & Moxim 1989; Logan 1983; Hameed & Dignon 1992) are highly concentrated in northeastern North America and western Europe which together emit about half of the global total. Also, large quantities of soil dust are transported out of North Africa all year long, affecting much of the tropical NAO and Caribbean; dust concentrations are probably greater over the NAO than over any other large ocean region. Biomass burning products are also transported out of North Africa and over the NAO especially during the winter and spring. Thus, if wind-transported pollutants and other continental emissions are having an impact on nutrient cycling in the oceans, then the impacts should be most readily observable in the NAO and associated waters (North Sea, Caribbean, etc.).

In this paper we assess the atmospheric transport and deposition fluxes of a number of nitrogen species that could play a role in the nutrient cycle of natural waters. The primary productivity in surface waters of the ocean depends on the availability of nutrients (see for example, Eppley 1980; McCarthy & Carpenter 1983; Levitus et al. 1993), the most important being phosphate and nitrate. Historically, the major sources of nutrients in the photic zone were considered to be: the fixation of nitrogen in surface waters; the upward transport of nutrients from deeper waters; the recycling of nutrients in the surface waters. However, Duce (1986) showed that in some oligotrophic regions the atmospheric input of NO_3^- and Fe could play a significant role in primary production under certain conditions. A number of studies have supported this conclusion although the magnitude and importance of the effect is still debated (Fanning 1989; Michaels et al. 1993; Owens et al. 1992; Willey & Paerl 1993). Certain trace metals such as Fe (Martin & Fitzwater

1988; Donaghay et al. 1991; DiTullio et al. 1993) could also be important because of their role as essential micronutrients. Wind transported dust could be the dominant source of Fe for many ocean regions (Young et al. 1991; Morel et al. 1991; Miller et al. 1991; Duce 1986; Duce & Tindale 1991).

We present estimates of the atmospheric input to the oceans of several classes of nitrogen species that could play a role in the nutrient cycle. One group consists of nitrogen species that fall primarily into two classes: oxidized nitrogen species including aerosol NO_3^- and gas-phase oxides of nitrogen (NO, NO_2, HNO_3 and related species); reduced nitrogen species including aerosol NH_4^- and gaseous NH_3 (and related organic nitrogen species). There is evidence that organic nitrogen species could also play an important role. However, there are relatively little data for organic nitrogen compounds; most efforts has focused on dissolved organic nitrogen species (DON). We will discuss DON deposition but quantitative estimates are precluded.

The second group of nutrient-related species consists of Fe and P. These elements are primarily associated with mineral dust particles and, hence, the assessment of the fluxes of these elements will depend on estimates of dust transport to the oceans. To this end in this paper we discuss dust transport to the NAO and then use the model estimates of dust deposition to estimate the deposition of Fe and P.

The impact of atmospheric deposition to the NAO depends not only on the absolute quantities of material but also on the spatial and temporal variability of the deposition. Obviously, the deposition of nutrients into oligotrophic waters will have a much greater impact than on nutrient-rich waters (e.g., in a region of upwelling). Also we know that atmospheric transport and deposition processes are highly episodic; thus, nutrient species will often be deposited in short periods of time thereby increasing the nutrient enhancement process, albeit for a brief time. Nonetheless, these short bursts of enhanced productivity could have a number of important implications. For this reason, we devote a section of this paper to the discussion of temporal variability on time scales ranging from seasons to those of individual deposition events which typically extend over a period of hours to days.

In our review of the emissions rates of NO_x and NH_x to the atmosphere, we distinguish between the present-day fluxes due to natural and anthropogenic processes; the "natural" component of the present day emission rates serves as an estimate of "pre-industrial" emissions. A longer-term picture of emission trends is obtained from studies of chemical species in ice cores. We also make projections into the future based on population growth estimates and various scenarios for energy use. Although the projections are somewhat uncertain, they serve to demonstrate the magnitude of possible future impacts and changes in spatial deposition patterns.

The subject of long range atmospheric transport to the oceans has been addressed in a number of workshops (NAS 1978; Duce et al. 1991). A number of papers and reports have summarized various aspects of this subject (Hasse 1983; Merrill 1986; Moody & Galloway 1988; Rodhe 1985; Whelpdale & Moody 1990; Levy 1989; Whelpdale & Galloway 1994; Knap 1990; Riley et al. 1989; Merrill 1989; Whelpdale & Kaiser 1996). This current effort differs from the earlier ones in a number of respects, most notably in the use of a wide range of models to estimate transport and deposition to the North Atlantic.

Scientific background

In this section we review some aspects of the chemistry of the subject species and provide an overview of the larger issues regarding fluxes to the atmosphere. In preparing this section we do not attempt to provide an exhaustive literature review; rather we cite only representative papers and we make use of some comprehensive reviews. The source emissions discussed in the following section are summarized in Table 1. The data for NO_x species in Table 1 are taken largely from Galloway et al. (1995, 1996); the data for NH_x species are taken largely from Schlesinger & Hartley (1992). The purpose of this table is to provide the reader with a sense of the magnitude of the emissions to the atmosphere; the fact that we do not cite other literature sources more extensively in Table 1 (although others are cited in the discussion section) is not intended to imply that the values in Table 1 are necessarily the preferred ones. Nonetheless, as indicated in the text below, there is fairly good agreement for many species.

Nitrogen species

The important oxidized nitrogen species from the standpoint of transport and deposition are NO, NO_2 (collectively referred to as NO_x) and NO_3^- (both in the particulate form and as gas phase HNO_3). There are a number of organic nitrates that may play a significant role, in particular peroxyacetyl nitrate (PAN) which is believed to serve as a substantial reservoir for NO_3^- in the upper troposphere; however, PAN is unstable at typical lower troposphere temperatures, where it decomposes to yield HNO_3. The entire ensemble of reactive gas phase N-species is generally referred to as NO_y which is defined as the sum of NO_x, HNO_3, PAN and other related minor (mostly organic) species. The following discussion of sources and budgets is based largely on Galloway et al. 1995, 1996.

The dominant reduced nitrogen species are NH_4^+ and NH_3 (which, as a pair, are typically referred to as NH_x). It is important to note that in the atmosphere,

Table 1. Source emission estimates for NO_x and for NH_x. Estimates are shown for Pre-industrial missions and Present-day emissions; the difference between the two yields the present-day Anthropogenic emission rate. a) Source emissions of NO_y based on data taken largely from Galloway et al. (1995). b) Source emissions for NH_x. Data are largely from Schlesinger & Hartley (1992) except for pre-industrial biomass burning and animal excreta (Dentener & Crutzen 1994) and ocean sources (Quinn et al., this volume).

Source	Pre-industrial	Present day	Anthropogenic
	NO_x (Tg N yr^{-1})*		
Lightning	3	3	–
Soils (and crops)	3.6	5.5	1.9
Biomass burning	0.8	8.5	7.7
Stratospheric injection	0.6	0.6	–
Energy production	0	21.3	21.3
Aircraft	0	0.5	0.5
Total	8.0	39.4	31.4
	NH_x (Tg N yr^{-1})+		
Ocean	3■–13	3■–13	–
Soils (and fertilizer)	10	20	10
Biomass burning	0.5◊	5	4.5
Animal excretia	2.5◊	32	29.5
Total	16–26	60–70	44

* Levy et al. 1995; Galloway et al. 1995.
\+ Schlesinger & Hartley (1992) except: ◊ Dentener & Crutzen 1994; ■ Quinn et al., this volume.

there is relatively little conversion or exchange between the NO_x/NO_y and the NH_x forms of nitrogen; in particular, NH_3 is quite resistant to oxidation under normal circumstances although in tropical regions model results suggest that significant amounts of NH_3 may be oxidized due to high OH and low sulfate concentrations (Dentener & Crutzen 1994). Similarly, N_2O is relatively stable in the troposphere and does not participate to a significant degree in NO_x/NO_y chemistry nor that of NH_x.

Natural sources

The primary natural sources of NO_x are biological fixation and lightning. Lightning is a relatively minor source of NO, about 3 Tg N yr^{-1} (Lawrence et al. 1995; Levy et al. 1995); while there have been much higher estimates in the past, they are no longer tenable based on wide-spread measurements of consistently low values of NO_x and NO_3^- in remote regions.

Pre-industrial biotic N fixation in terrestrial ecosystems is estimated to be about 90–130 Tg N yr^{-1} (Stedman & Shetter 1983); fixed N is converted to organic N or lost from the organism as reduced inorganic N (NH_4^+). N will remain in reduced form until an element other than N becomes limiting or until there is a reduction of biotic mass (e.g., as a result of fire, seasons with limited growth, widespread vegetation decay, etc.). Of this total, only about 4.4 Tg N yr^{-1} is emitted to the atmosphere as NO_x; 3.6 Tg N yr^{-1} is emitted directly from soils (Yienger & Levy 1995) and 0.8 Tg N yr^{-1} from burning (Galloway et al. 1994). There are no major oceanic sources of NO_x or NO_y. Logan (1983) estimates that less than 1 Tg N yr^{-1} is emitted from the oceans as NO each year. As shown in Table 1, the total pre-industrial emissions of NO_x are about 8 Tg N yr^{-1}.

Major natural sources of NH_x include excreta from wild animals (2.5 Tg N yr^{-1}, Dentener & Crutzen (1994), 4 Tg N yr^{-1}, Warneck (1988)), and emissions from soils and vegetation (5.1 Tg N yr^{-1}, Dentener & Crutzen (1994); 10 Tg N yr^{-1}, Schlesinger & Hartley (1992)). The oceans, especially regions of the Pacific Ocean that are not heavily impacted by anthropogenic activities, are believed to be a source of NH_3 (Quinn et al. 1988, 1990; Quinn et al., this volume). Estimates of a global source strength are highly uncertain, however, due to a lack of confirming atmospheric and seawater measurements. Schlesinger & Hartley (1992) estimate the oceanic source of NH_3 at 13 Tg N yr^{-1}; Dentener & Crutzen (1994) estimate 7 Tg N yr^{-1}. In more recent work (Quinn et al., this volume) an even lower emission rate of about 3 Tg N yr^{-1} has been estimated. Emissions from pre-industrial biomass burning are very difficult to estimate but they are believed to be small (e.g., 0.5 Tg N yr^{-1}; Dentener & Crutzen 1994). The total pre-industrial emissions of NH_x (Table 1) is about 16–26 Tg N yr^{-1}, based largely on the estimates of Schlesinger & Hartley (1992). The lower value reflects the very low emission rate obtained for the oceans by Quinn et al. (this volume). Dentener & Crutzen (1994) also obtain a lower total, 14.6 Tg N yr^{-1}, largely because of their lower estimates for the oceanic source and for emissions from soils and vegetation.

Although inorganic nitrogen has been studied in aerosols and precipitation for some time, only recently has it been recognized that organic nitrogen can be of comparable importance. Organic nitrogen in aerosols and precipitation can result from reactions of gas phase species and by mechanical processes that release particles directly (e.g., plants fragments, sea spray droplets). The protein in pollens or oceanic surface films may degrade in the atmosphere to lower molecular weight organic species such as amino acids or primary amines. Because many different techniques have been used to study organic nitrogen, it is often difficult to make a direct comparison of results. Most research has focused on the DON fraction. The composition, sources and sinks of DON are so poorly characterized that it is impossible to make detailed model estimates of global budgets as we do for NO_x and NH_x species (see below).

Anthropogenic sources and sinks of NH_x and NO_y

There are three types of anthropogenic activities that are major producers of reactive N: fuel combustion; fertilizers; and the cultivation of certain crops, especially legumes. It is generally agreed that the present day utilization of energy sources (combustion of coal, petroleum products, natural gas) produces about 20 Tg N yr^{-1} as NO_x (Levy & Moxim 1989; Hameed & Dignon 1988, 1992; Benkovitz et al. 1995). This source is growing at a steady rate – from 1960 to 1986, at about 2.7% per year. There is also reasonable agreement that the production of fertilizer converts about 78 Tg N yr^{-1} from N_2 to NH_3 with an annual rate of increase of 5.3% per year (see Galloway et al. 1994). Because of the increased N fixed in soils by crops, there is an additional release of 1.9 Tg N yr^{-1} of NO from soils (Yienger & Levy 1995) and biomass burning produces 7.7 Tg N yr^{-1} (Levy et al. 1991) yielding a total of 9.6 Tg N yr^{-1}.

Large quantities of NH_3 are emitted directly from fields as a result of added fertilizer: 10 Tg N yr^{-1} (Schlesinger & Hartley 1992); 6.4 Tg N yr^{-1} (Dentener & Crutzen 1994). Excreta from domesticated animals yield: 32 Tg N yr^{-1} NH_3 (Schlesinger & Hartley 1992); 22 Tg N yr^{-1} (Dentener & Crutzen 1994). Biomass burning yields: 5 Tg N yr^{-1} (Schlesinger & Hartley 1992); 2 Tg N yr^{-1} (Dentener & Crutzen 1994). Coal combustion is believed to be a negligible source (Dentener & Crutzen 1994); Warneck (1988) estimates less than 2 Tg N yr^{-1}. The present day emissions of NH_x (including both natural and anthropogenic sources) is about 60–70 Tg N yr^{-1} (Table 1), based largely on the estimates of Schlesinger & Hartley (1992); Dentener & Crutzen (1994) obtain a substantially lower value, 45 Tg N yr^{-1}.

Once in the atmosphere, NH_3 can react with acidic non-seasalt sulfate and nitrate aerosol particles to form partially-to-fully neutralized ammoniated salts. Wet deposition is the major removal pathway of this particulate NH_4^+ from the atmosphere. NH_3 also can be dissolved directly into cloud water where, if the cloud does not evaporate, NH_3 will be removed through wet deposition. The lifetime of NH_3 with respect to reaction with sulfate aerosol is very short compared to the lifetime with respect to oxidation by OH (Quinn et al. 1992; DeMore et al. 1992) although in the tropics with high OH and low sulfate concentrations some oxidation of NH_3 may occur (Dentener & Crutzen 1994).

Conclusions

The present day rate of emissions of NO_x and related species is about five times greater than the emissions in pre-industrial times largely due to emissions from energy production and biomass burning. The present day emissions of NH_x are about three times greater mostly due to the impact of emissions from animal wastes associated with food production. Indeed, emissions of NH_x from animal waste have increased by about a factor of 10. From the standpoint of oceanic impacts, the effective increase of NH_x emissions is much greater than the gross numbers would indicate. The oceanic source of NH_x accounts for as much as half of the total pre-industrial emissions. If we compare only the continental components of the NH_x budgets (because these are the only components that could lead to a net increase in oceanic deposition), the present day NH_x emissions are about four-to-five times greater than the pre-industrial rates.

The total present-day emissions for both NO_x and NH_x is about 100–110 Tg N yr^{-1} (Table 1); total pre-industrial emission rates were about 24–34 Tg N yr^{-1}.

A substantial fraction of the continental emissions, whether natural or anthropogenic, are deposited relatively close to the source regions. These deposits may impact processes in watersheds and in coastal waters. However a substantial fraction is carried to the open oceans by winds. In subsequent sections we estimate the total deposition to these various regions. Although we do not attempt to assess the deposition fluxes for anthropogenic and natural sources independently, it can be assumed that the relative deposition rates for these two classes of substances will be approximately the same as those for the gross budgets as shown in Table 1.

Mineral dust

Huge amounts of soil dust are mobilized by winds, mostly in arid regions, and substantial quantities can be carried great distances. The importance of eolian transport to oceanic processes was first suggested by studies of the mineral distributions in pelagic ocean sediments. The concentration patterns of certain minerals (e.g., quartz, kaolinite, illite) in sediments off the coasts of some continents (e.g., the west coast of North Africa and North America, the east coast of Asia) were not related to fluvial sources but rather to the pattern of large scale wind fields (Prospero 1981). The subject of dust mobilization and transport has been extensively studied during the past two decades (for reviews, see for example: Prospero 1981, 1990, 1996; Pye 1987; Middleton 1990; Duce 1995; Goudie & Middleton 1992). A recent issue of Atmospheric Environment is devoted to the results of a large Russian/American field study of dust mobilization in central Asia (Golitsyn & Gillette 1993). Many aspects of dust transport to the ocean are summarized in Duce et al. 1991.

It is difficult to quantitatively estimate dust transport amounts because of the difficulty in defining what is meant by "transport". Vast quantities of soil material are moved by winds each year especially in regions of sand dunes and in eroding agricultural areas (Pye 1987). However, much of the transported mass in the source regions consists of large particles, that is, particles with diameters greater than about ten microns (d'Almeida and Schütz 1983). Such large particles have a very short residence time in the atmosphere because of their high settling velocity and will not be carried far by winds. For this reason, in assessments of long-range transport it is customary to base flux estimates on that portion of the particle mass that is below about 10–20 μm diameter. Because of the rapid fallout of large particles during transport, the size distribution rapidly shifts to small particles; at distances of about 1000 km or more from the source, the dust attains a relatively stable size distribution with a mass median diameter of several μm (Duce 1995) although some very large particles can be carried great distances (Betzer et al. 1988). Because of these variables, estimates of the source strength of dust are probably the poorest of any major aerosol species and estimates of deposition are also highly uncertain. Nonetheless, on a global scale, dust production rates are estimated to be in the range of about 1000–3000 Tg yr^{-1} (Duce et al. 1991; Duce 1995; Andreae 1995).

North Africa is one of the world's largest sources of dust that can be transported great distances. Large amounts of dust are carried to the north across the Mediterranean to Europe (Bergametti et al. 1989a) and to the west (Bergametti et al. 1989b) across the North Atlantic to the Caribbean, the east coast of North America (Prospero & Nees 1986; Prospero et al. 1987) and to South America (Prospero et al. 1986; Swapp et al. 1992). Satellite imagery of aerosol optical thickness (AOT) (Ignatov et al. 1995; Rao et al. 1989) shows that the highest values of AOT and the largest areal coverage over the oceans is clearly related to dust sources (Prospero 1996). Over the North Atlantic, the pollution plumes (as manifested by AOT distributions) that emerge from the east coast of the US and west coast of Europe are relatively small and weak in comparison to the African dust plume. Furthermore dense African dust plumes are highly visible all year long while the European and North American pollution plumes are prominent only in spring and summer.

The chemical and mineralogical composition of mineral dust over the North Atlantic has been extensively studied (Prospero 1981; Schütz et al. 1990). The elemental composition of the dust is similar to that of average crustal material (Zhuang et al. 1992; Uematsu et al. 1985; Arimoto et al. 1989, 1990, 1992; Prospero 1981; Prospero et al. 1987, 1989; Lepple 1975; Talbot et al. 1986; Graham & Duce 1979; Spokes et al. 1994). Although there have been few studies of the organic composition of the dust, total organic carbon concentrations are on the order of a few percent (Lepple & Brine 1976; Delaney & Zenchelsky 1976; Duce et al. 1983).

Stragety

Introduction

Our strategy is to estimate deposition to the NAO by means of transport/deposition models and to assess the relative performance of the models by comparing the calculated deposition to the NAO. We also compare the model results with the few deposition measurements that are available. The detailed description of the models and the comparisons of results are found in Prospero et al. (1995). Because the models are very different (they use different "meteorology", source configurations, and parameterizations of the transformation and removal processes), the degree of agreement (or disagreement) will give us some idea of the validity of the computed deposition over the entire NAO. The comparison showed that the models produced surprisingly concordant results for most regions. Consequently, our estimates of deposition will be based on the means of all available model results for specific regions. In the following sections we describe the models and present the estimated deposition rates for NO_y and NH_x. We then compare the model estimates with the deposition rates measured at a few island and coastal sites. Finally we present the deposition estimates for the North Atlantic Basin (NAB), i.e., the entire NAO and the land areas that drain into it.

Model descriptions

Five models are used: Global Chemical Transport Model (GCTM) of the NOAA Geophysical Fluid Dynamics Laboratory (GFDL); EMEP (Cooperative Program for Monitoring and Evaluation of Long Range Transmission of Air Pollution in Europe); MOGUNTIA (Model of the Global Universal Tracer Transport In the Atmosphere); and two versions of GESAMP (a model used by the Group of Experts on the Scientific Aspects of Marine Pollution. Duce et al. 1991). These are briefly described here. Three models (GCTM, EMEP and MOGUNTIA) are based on meteorological modules (either generated by the model or based on real meteorological data). Emissions (natural and pollutant) from sources are injected into the models' atmosphere where they are transported according to the meteorological wind fields. The chemical transformation and removal processes are parameterized (in varying degrees of detail, depending on the complexity of the model and the computer resources).

The GFDL/GCTM uses transport fields that are derived from a general circulation model (GCM); the model yields twelve months of 6-hour time-averaged wind, temperature, and precipitation fields (Mahlman & Moxim 1978). In common with all GCMs, the GCTM generates a detailed "meteorology" for an average year. While the model may successfully produce the "meteorology" and "climatology" that one might expect for a typical time period (e.g., a typical day in July or a typical July), there is no correspondence to the weather or climate of any specific period (e.g., 4 July 1994). The GCTM has a horizontal resolution of 265 km (2.4° × 2.4°) (Levy et al. 1982; Kasibhatla et al. 1993).

The MSC (Meteorological Synthesizing Centre) EMEP model is a high resolution 3-D model that covers Europe and adjacent seas including the northeast Atlantic, the European Arctic Ocean and the adjacent seas: North, Norwegian, Baltic, Black and Mediterranean. The model has a resolution of 150 km and incorporates real meteorological data. Because of its high resolution and limited spatial coverage, EMEP does not provide estimates for deposition to the open ocean in the North Atlantic.

MOGUNTIA (Zimmermann 1988) is a low-resolution (10° × 10°) 3-D model based on average winds obtained from meteorological data (Oort 1983). Transport by deep cumulus convection is parameterized according to Feichter & Crutzen (1990). Horizontal transport is described by monthly average winds and an eddy diffusion parameterization based on the standard deviation of the winds from the means. Transport by deep cumulus convection is described through parameterizations.

The GESAMP (Duce et al. 1991) models are low resolution (10° × 10°) 2-D empirical models. Wet deposition is calculated using climatological precipitation fields which are combined with existing atmospheric chemical concentration data. The GESAMP-I model uses gas and aerosol concentration fields over the oceans to calculate wet and dry deposition. GESAMP-II (Whelpdale & Galloway 1994) is basically the same as the GESAMP-I model. However, instead of calculating the wet deposition on the basis of gas and aerosol concentration fields, it uses existing data on precipitation concentrations. These are compiled on the map grid and interpolated using physically reasonable approximations; GESAMP-II deposition rates are obtained using the same rainfall amounts used in GESAMP-I.

The comparison of model performance (Prospero et al. 1995) showed that there was general good agreement amongst the various model estimates for the deposition of NO_y and NH_x to the NAO and to the coastal ocean and continental watershed. Indeed, the agreement was surprisingly good when one considers the very different rationales employed in the models, the different meteorologies upon which they are based and the different data sets that are used for source emissions in the models. For the most part the mid-ocean deposition estimates (using GCTM, MOGUNTIA, GESAMP-I and GESAMP-II) for NO_y agree to within ±50% of the mean. Only two models (MOGUNTIA and GESAMP-I) yield estimates for NH_x; even here the agreement is remarkably good for the mid-ocean regions.

Only the MOGUNTIA and EMEP provide estimates of NH_x deposition to the coastal ocean and the continental watersheds. Even these models yield surprising agreement in their estimates except for NH_x in the western NAO.

Comparison of model estimates with measured deposition rates

There is very little long-term (a year or longer) deposition data with which to compare the model estimates. In Table 2 we summarize wet deposition data (Galloway, pers. comm.) for NO_3^- and NH_4^+ from five stations in the NAB and compare them with the values obtained with the models. For this comparison, it was not convenient to obtain the modeled wet deposition values separate from the dry deposition amounts; instead we show the total deposition (wet and dry) rates obtained from the models and compare them with the measured values for wet deposition only. However, we would expect that the dry deposition fraction would constitute only a small fraction of the total flux at these sites.

Mace Head, Ireland. The averaged models results are close to the measured values for both NO_3^- and NH_4^+ (four years of data).

Lewes, Delaware. Although there is a considerable spread among the model values for NO_y, the mean is close to the measured values (ten year means) for both NO_3^- and NH_4^+.

Bermuda. The Bermuda comparison is the poorest of the entire set. All models substantially overestimate the deposition rate to this region. Table 2 shows the measured means for two periods: 1981–1988, when the sampling site was located at Harbor Tower, near the eastern end of Bermuda; 1989–1993, when the sampling was carried out at the AEROCE (Atmosphere/Ocean Chemistry Experiment) site at Tudor Hill, at the extreme western end of the island. The deposition rate at the eastern site was markedly higher for both species, 31% greater for NO_3^- and 75% greater for NH_4^+ (Figure 1). The relatively high deposition rate at the eastern site appears to be largely due to differences in rainfall amount; the annual mean precipitation at the western site during 1989–1993 was 100.0 cm while at the eastern site during 1981–1988 the mean was 146.3 cm; i.e., 46% greater. In the case of Bermuda, we must ask which deposition rate is most appropriate for use in the comparison with the models. In this regard, it should be noted that the distance between these two sites is only about 30 km, much less than the resolution of any of the models. The large difference in deposition amounts demonstrates the close

Table 2. Comparison of model estimates of deposition with measured values. (Units: mMoles-N m^{-2} yr^{-1}.)

	NO_3^-	NH_4^+
Mace Head		
MOG	6.1	3.5
GCTM	4.9	–
GES-I	–	3.9
GES-II	7.1	–
Model avg.	6.0	3.7
Meas. (1990–93)	5.2	4.4
Lewes, DE		
MOG	29.8	16.3
GCTM	52.9	–
GES-I	12.7	8.0
GES-II	21.4	–
Model avg.	29.2	12.1
Meas. (1981–90)	22.5	15.5
Bermuda		
MOG	14.7	7.6
GCTM	14.1	–
GES-I	7.9	7.8
GES-II	14.3	–
Model avg.	12.8	7.7
Meas. (1981–88)	6.4	4.2
(1989–93)	4.9	2.4
Everglades, FL		
MOG	7.6	10.0
GCTM	21.4	–
GES-I	5.4	8.8
GES-II	–	–
Model avg.	12.5	9.4
Meas. (1990)	12.1	9.7
Barbados		
MOG	1.6	1.1
GCTM	1.4	–
GES-I	4.1	1.2
GES-II	6.4	–
Model avg.	3.4	1.1
Meas. (1990–93)	2.0	2.0

Coordinates

Mace Head, Ireland	53.32N 9.85W
Lewes, DE	38.75N 75.14W
Bermuda	32.27N 64.87W
Everglades Park, FL	25.10N 80.53W
Barbados	13.17N 59.43W

linkage between scavenging rates and precipitation and shows how the model deposition rates will depend to a considerable degree on the accuracy of the precipitation modeling. In the case of Bermuda, it could be argued that the increased rainfall at the eastern site is due to topographic effects and that the rainfall amounts at the western end should be more representative of oceanic rainfall amounts.

Thus, the models as a group appear to seriously overestimate the eastward transport of materials from North America. However the overestimate does not appear to extend to the NAO on a broader scale. The average model deposition rates for the grid box immediately to the east of the Bermuda box (6.4 mMoles-N m^{-2} yr^{-1} for NO_y, 3.8 mMoles-N m^{-2} yr^{-1} for NH_x) are close to the mean values measured on Bermuda; similarly, for the box to the south of Bermuda the average model values are 5.6 mMoles-N m^{-2} yr^{-1} and 3.0 mMoles-N m^{-2} yr^{-1} for NO_y and NH_x respectively. The mismatch between the model results and the measured values may be due in part to the very coarse grid size that is used in this comparison.

Everglades, Florida. There is a considerable spread among the model values for NO_3^- but the mean is essentially identical to the measured value. The modeled NH_4^+ values are also quite close to the measured values.

Barbados, West Indies. The model estimate for NH_4^+ is about half the measured values. In contrast the model average for NO_y is about 70% greater than the measured value. The model average is shifted strongly to the high side by the results from the two GESAMP models; the MOGUNTIA and GCTM values are quite close to the measured mean (four years). The deposition results for NO_y for the Barbados grid cell are somewhat anomalous in that the agreement amongst the models is much better for the adjacent grid cells on all sides.

In summary, the models do a reasonable job of estimating the deposition at the stations. The model estimates for three stations (Mace Head, Lewes and Everglades) were extremely close to the measured values. The Bermuda results were quite poor but they could be rationalized into better agreement. On the basis of these limited comparisons, it is difficult to estimate the overall reliability of the model estimates. Nonetheless, it should be noted that the poorest estimates were off at most by a factor of about two (allowing for a rationalization of the Bermuda results). Thus, for the NAB as a whole, we would expect the model estimates to be within a factor of two of the true value.

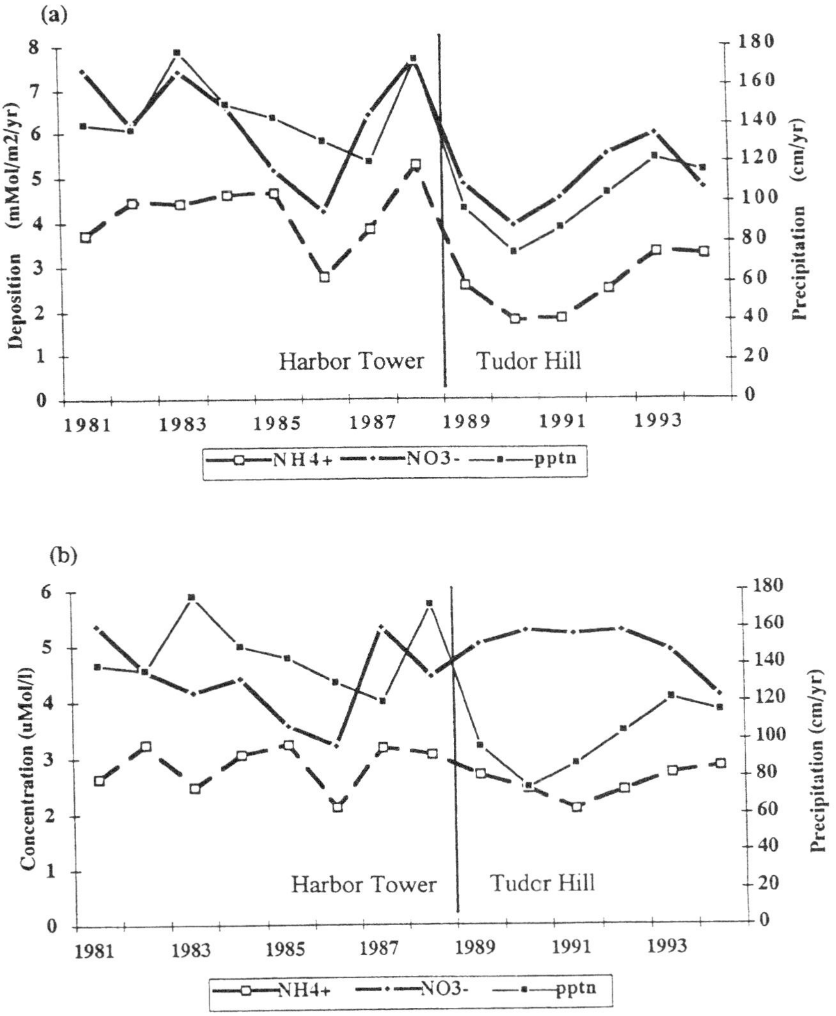

Figure 1. a.) Annual precipitation amounts (cm yr^{-1} and annual volume weighted deposition rates for nitrate-N and ammonium-N (mMoles-N m^{-2} yr^{-1}) for Bermuda for 1981–1993. b.) Annual mean concentration (μMol l^{-1}) for nitrate-N and ammonium-N. The vertical line indicates the change in location of the sampling station from Harbor Tower located near the eastern end of the island to Tudor Hill on the western end (Galloway, pers. comm.).

Nitrogen species

NO_y

The annual mean deposition rates obtained from the four models for the NAO and the Caribbean are shown in Table 3. These rates are computed for a 10° ×

10° grid and they include both the wet and the dry deposition fluxes. Table 3 presents both the deposition rate per unit area (mMoles-N m^{-2} yr^{-1}) and the total annual deposition per grid square (10^9 Moles-N). The table also shows the sums of the total annual deposition in 10° latitude bands.

The model-estimated deposition rates of NO_y and NH_x (mMoles-N m^{-2} yr^{-1}) to continental watersheds and to coastal waters are shown in Table 4. In those areas where the coastal shelves are very narrow (e.g., West Africa, Caribbean Islands, etc.), the model resolution was too coarse to provide an estimate. For the same reason, we do not attempt to assess the total deposition to the watersheds and coastal regions, but provide only the rate per unit area.

Howarth et al. (this volume.) and Nixon et al. (this volume) use the values in Table 4 (deposition rate per unit area) to estimate the total deposition to the shelves and the watersheds.

The impact of continental sources is quite obvious. The deposition rates are much higher in the grid cells adjacent to North America and Europe and they decrease rapidly with distance to the ocean. The NO_y deposition rates along the east coast of North America and the Gulf of Mexico (Table 4) range from about 20 to 40 mMoles-N m^{-2} yr^{-1}; for western European waters, the rates are somewhat lower, about 20 to 25 mMoles-N m^{-2} yr^{-1}. In contrast, the central NAO deposition rates tend to fall in the range of 2 to 4 mMoles-N m^{-2} yr^{-1}, about a factor of ten lower. Looking at the total deposition of NO_y to the NAO as a function of latitude (Table 3), the deposition from 30° N to 60° N is somewhat less than twice as great as that the 10° N to 30° N band. But note that the flux in the band from 0° N to 10° N is comparable to that in the mid-latitudes. Finally, the deposition to the 60°–70° N band is quite low, about a quarter of that in the 30°–60° N band.

NH_x

The mean deposition rates of NH_x species to the open NAO are shown in Table 5. Only two models provided NH_x estimates, MOGUNTIA and GESAMP-I; nonetheless, the agreement is quite good. The rates are computed for a 10° × 10° grid and they include both the wet and the dry deposition fluxes. Table 5 presents both the deposition rate per unit area (mMoles-N m^{-2} yr^{-1}) and the total annual deposition per grid square (10^9 Moles-N). The table also shows the sums of the total annual deposition in 10° latitude and longitude bands. The model-estimated deposition rates of NH_x (mMoles-N m^{-2} yr^{-1}) to continental watershed and to coastal waters are shown in Table 4.

As was the case of NO_y, the deposition rates are quite high in the coastal waters and decrease rapidly with distance to the oceans. Values along the east coast of North America and the Gulf of Mexico fall in the range of about 10–20 mMoles-N m^{-2} yr^{-1} while those for European coastal waters are in

Table 3. Annual average NO_y deposition to the open ocean. Total (wet plus dry) per 10° square. Also shown are the sums in 10° latitude and longitude bands. The sums do not include deposition to the Caribbean or the Mediterranean.

Lat.	Longitude											Lat. sum
	100–90	90–80	80–70	70–60	60–50	50–40	40–30	30–20	20–10	10–00	00–10	
					a: Total NO_y deposition: mMoles N m^{-2} yr^{-1}							
60–70					1.9		2.1	2.7	3.9	6.7	7.1	
50–60					6.4	4.3	3.6	3.8	6.0	50.4		
40–50				17.5	9.5	6.1	4.3	3.3	4.6	13.4		
30–40			29.0	12.8	6.4	4.3	3.1	2.8	4.1			
20–30	18.4	14.9	8.7	5.6	3.5	2.5	2.1	2.6	3.1			
		10.2	6.8	5.6	3.4	2.6	2.4	3.1	8.5			
00–10					7.3	4.4	4.2	5.0	8.1	8.8	12.6	
					b: Total NO_y deposition: 10^9 moles N 10°-box^{-1} yr^{-1}							
60–70					1.0		1.1	1.4	2.0	3.5	3.7	12.8
50–60					4.5	3.0	2.6	2.7	4.3	35.6		52.7
40–50				15.2	8.3	5.3	3.8	2.9	4.0	11.6		51.2
30–40			29.2	12.9	6.5	4.3	3.2	2.8	4.1			62.9
20–30	20.6	16.6	9.7	6.3	3.9	2.7	2.3	2.9	3.4			31.3
10–20		12.1	8.1	6.7	4.0	3.1	2.9	3.7	10.1			38.6
00–10					9.0	5.4	5.2	6.1	10.0	10.7	15.4	61.8
									N. Atlantic total			311.2
									Carib. total			49.3

Table 4. NO_y and NH_x Deposition Rates to Continental Watersheds and to Coastal Ocean Regions: Average of Models (mMoles-N m^{-2} yr^{-1}). For total deposition amounts to the watershed and coastal regions, see Howarth et al. (this volume) and Nixon et al. (this volume).

Region	Code	NO_y cont	NO_y coast	NH_x cont	NH_x coast	Ratio: NO_y/NH_x Cont	Ratio: NO_y/NH_x Coast
North Canadian rivers	NCR	6	5	5	2	1.2	2.0
St. Lawrence Basin	SLB	40	15	19	6	2.1	2.5
North-East USA	NEU	61	38	17	11	3.6	3.5
South-East USA	SEU	66	22	25	10	2.6	2.3
Eastern Gulf of Mexico	EGM	60	20				
Mississippi	MIS	38	30	19	18	2.0	1.6
Western Gulf of Mexico	WGM	20	18	16	14	1.2	1.3
Cent & No. S. Amer. (to Carib)	CNS	19	5	13	7	1.5	0.6
Amazon	AMA	16	5	10	10	1.6	0.5
Mediterranean	MED		16	11	10		1.6
Spain and Portugal	SP	24	22	23	14	1.1	1.6
W. Europe, Grt Brit. (to N. Sea)	WEB	55	25	53	22	1.0	1.2
Baltic	BAL	33	24	40	28	0.8	0.9

the range of 10–30 mMoles-N m^{-2} yr^{-1}. Central ocean values range from about 1–3 mMoles-N m^{-2} yr^{-1}. In the case NH_x high deposition rates are also found in the low latitudes, from 0° N to 20° N; indeed, total deposition in the 0°–10° N band is about twice as high as that in any other latitude band, largely reflecting the impact of sources in North Africa. However, it should be emphasized that our knowledge of NO_y and (especially) NH_x sources in Africa is poor compared to those for Europe and North America.

Organic nitrogen

None of the models provide estimates of organic nitrogen transport and deposition. Because most research has focused on DON and because there is very little data on particulate organic nitrogen (PON), we simply survey the available data on DON. We then use the relative concentrations of DON and dissolved inorganic nitrogen (DIN, largely NO_3^- and NH_4^+) to extrapolate to the entire NAO using the modeled deposition rate of inorganic nitrogen to the region. This estimate assumes that the DON-to-DIN ratio is generally applicable to the region; obviously, this must be regarded as a very rough estimate of DON deposition.

The first reported measurements of DON in rain were made off the coast of California by Williams (1967), (Table 6); values were about 1.7 μmolar or 49 to 65% of the total nitrogen. More recently, in the North Atlantic,

Table 5. Annual average NH_x deposition to the open ocean. Total per 10° square. Also shown are the sums in 10° latitude and longitude bands. The sums do not include deposition to the Caribbean or the Mediterranean.

Lat.	Longitude											Lat. sum
	100–90	90–80	80–70	70–60	60–50	50–40	40–30	30–20	20–10	10–0	0–10	
	a: Total NH_x deposition: mMoles N m^{-2} yr^{-1}											
60–70					2.4		3.6	3.9	3.4	4.4	7.6	
50–60					4.9	3.1	3.1	3.1	3.7	4.8		
40–50				8.5	6.7	3.8	2.0	1.7	2.7	10.3		
30–40			12.1	7.7	3.8	1.8	1.4	1.9	3.0			
20–30	12.5	9.4	7.4	3.0	1.1	0.8	0.7	1.9	1.5			
10–20		13.7	9.1	6.2	1.1	0.9	1.3	3.0	7.1			
00–10					13.1	3.8	2.1	3.2	11.0	13.4	12.6	
	b: Total NH_x deposition: 10^9 moles N 10°-box^{-1} yr^{-1}											
60–70					1.3		1.9	2.0	1.8	2.3	4.0	13.2
50–60					3.5	2.2	2.2	2.2	2.6	3.4		16.0
40–50				7.4	5.8	3.3	1.8	1.5	2.3	9.06		31.1
30–40			12.2	7.8	3.8	1.9	1.4	2.0	3.1			32.1
20–30	14.0	10.5	8.3	3.3	1.3	0.9	0.8	2.1	1.7			18.3
10–20		16.2	10.8	7.4	1.4	1.1	1.6	3.5	8.4			34.2
00–10					16.0	4.7	2.5	4.0	13.5	16.5	15.5	72.7
									N. Atlantic total			217.6
									Carib. total			40.7

Table 6. Measured concentrations of DON in precipitation.

Site	Rainwater conc. (μMole l^{-1})	Reference
33° N, 117° W	1.73	Williams 1967
6° S, 174° W	1.65	Williams 1967
Bermuda	8	Knap et al. 1986
	16	Cornell et al. 1995
NE Atlantic	9	Jickells et al. 1990
North Sea	6	Rendell et al. 1993
Chesapeake Bay	21	Jordan & Correll 1995
Coastal Delaware	<3 to 30	Scudlark et al., pers. comm.
Coastal Rhode Is.	5 to 10	Nixon et al., pers. comm.
Philadelphia (urban)	15 to 35	Seitzinger, pers. comm.
Texas (rural)	30	Cifuentes, pers. comm.

concentrations have been measured at Bermuda (Knap et al. 1986; Cornell et al. 1995), the northeast Atlantic (Jickells et al. 1990), and the North Sea (Rendell et al. 1993). The measurements yield remarkably similar values, 6 to 16 μmolar. Nearer to the Atlantic coast, reported concentrations are higher (10–25 μmolar) with an average of 16 μmolar (Jordan et al. 1995; Cornell et al. 1995). Extensive measurements of DON and DIN in precipitation from various continental and marine locations including the North Atlantic (Cornell et al. 1995) show that at marine locations the concentration of DON is comparable to, or somewhat greater than, that of DIN.

Concentrations of 3 to 30 μmolar DON have been measured at coastal sites in Delaware (Scudlark et al., pers. comm.), the Carolinas (Paerl et al., pers. comm.), and Rhode Island (Nixon et al., pers. comm.). A similar concentration range has been found for urban Philadelphia (Seitzinger, pers. comm.) and rural Texas (Cifuentes et al., pers. comm.). In addition, there appears to be a distinct seasonality with peak DON precipitation concentrations in late spring or early summer and a correlation between DON and NO_3^- deposition.

There are only a few measurements of the speciation of DON. Mopper & Zika (1987) measured concentrations of 7 μmolar for disassociated free amino acids and aliphatic amines in the western Atlantic and Gulf of Mexico. Measurements in the North Atlantic have indicated that total amino acids make up only a few percent of the DON (Gorzelska & Galloway 1990; Gorzelska et al. 1995).

From the data discussed above, DON wet deposition fluxes were estimated for a few coastal and mid-Atlantic areas (Table 7). The mid-Atlantic value is near 3 mMoles-N m^{-2} yr^{-1}. Mean values off the northeast and southeast coast of the U.S. are 20 and 8 mMoles-N m^{-2} yr^{-1}, respectively. Fluxes at

Table 7. Wet deposition fluxes of DON (mMoles-N m^{-2} yr^{-1}) in selected regions, estimated from data presented in Table 6.

Site	Wet deposition (mMol N m^{-2} yr^{-1})
Northeast coast of U.S.	mean = 20, winter = 5, summer = 50
Southeast coast of U.S.	6 to 10
Texas (rural)	20
North Sea	4
NE Atlantic	8
N. Atlantic background	3

Bermuda are around 10 mMoles-N m^{-2} yr^{-1} while those in the North Sea and NE Atlantic are 4 and 8 mMoles-N m^{-2} yr^{-1}, respectively.

There are not enough data to make estimates of PON deposition. However, the deposition of pollen alone may be seasonally important in watersheds and coastal areas; the fact that summer-time DON deposition rates are 2.5 times the winter values may be due in part to DON extracted from pollen and other forms of PON brought down in precipitation.

Summary discussion: NO_y, NH_x, DON

Europe and North America appear to have distinctly different emission ratios for NO_y and NH_x. As shown in Table 4, the NO_y/NH_x ratio in deposition along the east coast of North America is about 2–3 whereas for European waters it is close to one. This may reflect the relatively higher concentration of domestic animals in western Europe compared to North America where animals tend to be concentrated in the central and western regions of the continent.

As shown in Table 3, the estimated deposition of NO_y to the NAO (excluding the coastal regions) and the Caribbean is 360×10^9 Moles-N m^{-2} yr^{-1} (5.0 Tg N); this is about 13% of the estimated global emission rate (natural and anthropogenic) for NO_y. We would expect that most of the deposited NO_y in this region is derived from sources in North America and Europe which jointly emit about half of the global total. Our estimated deposition rate comprises about a quarter of the total from these sources. The impact of the continental sources would be greatest in coastal waters. Nixon et al. (this volume) present the total deposition amounts to the continental shelf regions in the NAB. For the western NAO self areas extending to the 200m depth line, the total NO_y deposition is 48.5×10^9 Moles-N yr^{-1} while along the European coast it is 30.1×10^9 Moles-N yr^{-1}; the total, 78.6×10^9 Moles-N yr^{-1}, is about 22% of the total deposited to the open ocean regions of the NAO.

In the case of NH_x, 258 Moles-N m^{-2} yr^{-1} (3.6 Tg N) are deposited to the North Atlantic and the Caribbean (Table 5); this is about 6% of the global continental emissions (Table 1).

Table 8 presents the sum of the NO_y and NH_x depositions to each of the NAO 10° grid cells. Note that the highest deposition occurs in the latitude band 0°–10° N. This is due in part to the relatively high NO_y deposition rates and the very high NH_x deposition rates in this region. The total of NO_y and NH_x is 530×10^9 Moles-N yr^{-1} (7.4 Tg N yr^{-1}). This is about 7% of the total global emissions of NO_x and NH_x. Nixon et al. (this volume) obtain total deposition amounts for NO_y and NH_x of 74.9×10^9 Moles-N yr^{-1} on shelf areas in the western NAO and 58.2×10^9 Moles-N yr^{-1} for the eastern NAO. The total shelf deposition (133.1×10^9 Moles-N yr^{-1}) is about 25% of the amount deposited to the open ocean.

Because of the dearth of organic nitrogen data, it is not possible to make an accurate assessment of total nitrogen deposition to the NAO. Nonetheless, it should be noted that the DON concentrations in precipitation at ocean sites are comparable to those of DIN (Cornell et al. 1995). Also the measurements of DON deposition rates for the few sites where there are data (Table 7) are comparable to those estimated for NO_y and NH_x from the models (Table 8). This suggests that the total deposition amounts for all nitrogen species (NO_y, NH_x and DON) may be as much as twice that which we estimate for NO_y and NH_x alone (530×10^9 Moles-N, Table 8), yielding a total of about 1×10^{12} mMoles-N yr^{-1} for the entire NAO, excluding the coastal ocean regions.

Cornell et al. (1995), assessed the relative importance of DON and DIN in precipitation from various sites around the world. On the basis of various lines of evidence (principally isotopic data and spatial trends in the concentration of DON and DIN), they conclude that a large fraction of the DON is derived from continental sources. However, they could not assess the relative importance of anthropogenic versus natural sources of the DON (or its precursors). Regardless of the source, they conclude that on a global scale the deposition rate of DON is about equal to that of DIN which is consistent with our conclusion for deposition to the NAO.

Mineral dust

The mineral aerosol deposition rates (units 10^9 gm mineral dust) for each 10° grid box are shown in Table 9. For estimating the deposition rates of P and Fe as shown in Table 9, we assume that these elements are present in dust in concentrations that are typical of average crustal material. We use values of upper crustal abundances (Taylor & McLennan 1985): Fe, 3.5%; P, 0.105%. Measurements of P in aerosols often yield P/Al ratios that are higher than that

Table 8. Sum of NO_y and NH_x deposition to 10° squares in the NAO (10^9 mMoles-N yr^{-1}). Also shown are the sums in 10° latitude and longitude bands; these do not include deposition to the Caribbean or the Mediterranean.

Lat.	Longitude											Lat. sum
	100–90	90–80	80–70	70–60	60–50	50–40	40–30	30–20	20–10	10–00	00–10	
60–70					2.2		3.0	3.4	3.8	5.8	7.7	26.0
50–60					8.0	5.2	4.7	4.9	6.9	39.0		68.7
40–50				22.7	14.1	8.6	5.5	4.3	6.4	20.6		82.3
30–40			41.5	20.6	10.3	6.2	4.6	4.7	7.2			95.0
20–30	34.5	27.1	18.0	9.6	5.2	3.6	3.1	5.0	5.1			49.6
10–20		28.4	18.9	14.1	5.4	4.2	4.5	7.2	18.5			72.8
00–10					25.0	10.1	7.7	10.1	23.5	27.2	30.9	134.6
									No. Atlantic total			528.9
									Carib. total			90.0

of crustal materials but this is believed to the presence of P from sources other than mineral dust (Graham & Duce 1979). In cases where mineral dust is the dominant component in the aerosol (Bergametti et al. 1992), P/Al ratios are close to those of average crustal material. Similarly, Fe/Al ratios are similar to average crustal material (Zhuang et al. 1992).

The GESAMP-I model yields a total mineral dust deposition to the NAO of about 170 Tg yr^{-1}. The equivalent Fe deposition is about 6 Tg yr^{-1} while that for P is about 0.2 Tg yr^{-1}. Dust deposition in the NAO is dominated by transport from African sources. Dust deposition rates in the central and western tropical NAO are relatively high (on the order of several mg m^{-2} yr^{-1}) but they are consistent with the measured accumulation rates of non-biogenic sediment accumulation rates in this regions (Prospero et al. 1987).

There are very few models with which we can compare our data. A mechanistic model of African dust transport (Schütz et al. 1981) based on measured particle size distributions and estimated transport and loss rates yields an annual transport of 260 Tg yr^{-1} about half of which is deposited within the first 500 km of the coast of Africa. This leaves about 130 Tg for transport to the remainder of the NAO, a value that is comparable to our model results.

There have been a number of efforts to model the transport of dust to the oceans using various meteorological models (Joussaume 1990; Tegen & Fung 1994, 1995); however, these models are in their early stages of development and they do not seem to be capable of properly simulating the distribution of dust sources or dust transport – for example, as evidenced by satellite imagery (Tegen & Fung 1995). Another model of Saharan dust transport is based on the inversion of AOT measurements (d'Almeida 1986) and, consequently, it is very sensitive to assumptions about the aerosol size distribution; furthermore, these estimates can not be adequately extended to the open ocean.

The dust deposition measurements (and the corresponding estimates of Fe and P) are subject to a great deal of uncertainty. It is unlikely that we will soon resolve the critical issues that must be addressed to obtain good estimates of dust deposition (more concentration data, better data on size distributions, better understanding of the processes affecting removal from the atmosphere). Nonetheless, it is clear that Africa dust transport is a major phenomenon that affects a large area of the NAO.

An additional factor concerns the biological availability of the Fe and P in the dust to marine organisms. There have been a number of studies of the solubility of P species in aerosols. In some cases solubilities can be quite high. For example, 36% of the total P present in aerosols collected on the U.S. east coast was soluble in water (Graham & Duce 1982); the high solubility is possibly a consequence of anthropogenic effects on aerosol composition, perhaps

Table 9. The mineral aerosol deposition rates (units 10^9 gm mineral dust) for each 10° grid box.

Lat.	Longitude W90–100	80–90	70–80	60–70	50–60	40–50	30–40	20–30	10–20	10–00	00–10	10–20	20–30E	Lat. sum
Mineral dust deposition (wet plus dry) to 10° square (10^9 gm yr^{-1})														
60–70					73		28	28	74	72	412			686
50–60					110	43	44	45	110	100	1089			1541
40–50				523	274	137	79	74	119	665				1871
30–40			955	1575	868	262	244	465	1458		3457	6979	14250	5826
20–30	886	1514	2254	1536	912	604	936	1734	17419					25395
10–20		2846	2430	4004	4016	3466	4206	6915	36594					61631
00–10					4681	4840	4882	7765	24727	17764	11083			75742
											No. Atlantic total			172693
											Medit. total			24685
											Carib. total			5246
											Grand total			202624
Fe deposition (wet plus dry) to 10° square (10^9 gm yr^{-1})														
60–70					3		1	1	3	3	14			24
50–60					4	2	2	2	4	4	38			54
40–50				18	10	5	3	3	4	23				65
30–40			33	55	30	9	9	16	51		121	244	499	204
20–30	31	53	79	54	32	21	33	61	610					889

Table 9. Continued.

Lat.	Longitude W90–100	80–90	70–80	60–70	50–60	40–50	30–40	20–30	10–20	10–00	00–10	10–20	20–30E	Lat. sum
10–20		100	85	140	141	121	147	242	1281					2157
00–10					164	169	171	272	865	622	388			2651
											No. Atlantic total			6044
											Medit. total			864
											Carib. total			184
											Grand total			7092
					P deposition (wet plus dry) to 10° square (10^9 gm yr^{-1})									
60–70					0.1		0.0	0.0	0.1	0.1	0.4			1
50–60					0.1	0.0	0.0	0.0	0.1	0.1	1.1			2
40–50				0.5	0.3	0.1	0.1	0.1	0.1	0.7				2
30–40			1.0	1.7	0.9	0.3	0.3	0.5	1.5		3.6	7.3	15.0	6
20–30	0.9	1.6	2.4	1.6	1.0	0.6	1.0	1.8	18.3					27
10–20		3.0	2.6	4.2	4.2	3.6	4.4	7.3	38.4					65
00–10					4.9	5.1	5.1	8.2	26.0	18.7	11.6			80
											No. Atlantic total			181
											Medit. total			26
											Carib. total			6
											Grand total			213

organic materials, although no studies have been made of the speciation. In contrast, in regions dominated by mineral dust, P solubility is considerably lower. Bergametti et al. (1992) find that in precipitation samples collected in the Mediterranean in the presence of high concentrations of Saharan dust, the soluble P is almost totally explainable by the presence of organic P. We would expect that the P in mineral dust would be relatively insoluble; the most abundant mineral form of P is apatite which accounts for 95% of all P in the Earth's crust (Jahnke 1992). Laboratory studies of P solubility in Saharan dust filter samples collected over the North Atlantic indicate that P solubility in aqueous solutions is at most a few percent (J. M. Prospero, unpublished data). Thus, of the estimated 0.2 Tg P yr^{-1} that is deposited in the NAO with African dust, only a small percentage would be readily available to marine organisms. Even if all the P in dust were to be available, it would be unimportant to processes in surface waters compared to other oceanic sources (e.g., nutrient-rich upwelling waters). Duce (1986) estimated that in the Sargasso Sea, atmospheric deposition of P accounted for as much as 5% and as little as 0.2% of the new phosphorus in the photic zone. The relative impact of atmospheric P would be even less in other regions where upwelling sources are stronger.

There have been a number of studies of the solubility of Fe in aerosols (Hardy & Crecelius 1981; Moore et al. 1984; Spokes et al. 1994; Zhuang et al. 1990; Zhu et al. 1992). As much as half of the Fe can be readily extracted from some marine aerosols (Zhuang et al. 1992). However, studies of Fe solubility in Saharan dust collected at Barbados and in laboratory studies (Zhu et al. 1992, 1993; Spokes et al. 1994) show that, in the mean, about 7–8% of the total Fe is soluble in moderately acid saline solutions having a composition similar to that which might be expected to exist on sea-salt-coated mineral dust particles suspended in the marine atmosphere (i.e., 1 *M* NaCl with the pH adjusted to 1.0 with HCl). We would expect that this Fe fraction (i.e., 7–8%) would be rapidly liberated from the dust particle once it is deposited in the ocean. Other processes in the ocean (including those associated with biological activity) could conceivably mobilize more Fe from the dust matrix (Jickells 1995).

Temporal variability in deposition

Annual budgets are important from the standpoint of understanding the regional fluxes of nitrogen-nutrients throughout the North Atlantic basin. However, from the standpoint of the impact of atmospheric inputs to oceanic productivity, it is necessary to understand the temporal variability of the deposition to the ocean surface. In general, we would expect that the impact of deposition to be greatest during the season when insolation is greatest and nutrient requirements are highest (i.e., spring and summer, Owens et al. 1991; Michaels et al. 1993). The effects of the nutrient deposition will be enhanced at those times when the ocean surface mixed layer is relatively shallow (i.e., generally in the summer) because the incremental increase in concentration over ambient ocean levels will be greatest (Michaels et al. 1993). The effects will be further enhanced if the deposition takes place primarily in a few large flux events rather than in a large number of low flux events (e.g., Paerl 1985).

The variability in deposition rates depends on a number of factors: the uneven geographical distribution of sources; the seasonal variability of anthropogenic and natural emissions and of the meteorological conditions that transport and deposit the material to the earth's surface. In this section we examine some factors that could affect the temporal variability of precipitation deposition to a few island sampling stations in the North Atlantic basin.

NO_y and NH_x

Seasonal emission patterns for nitrogen species

Some fuel combustion sources have a clear seasonal pattern (e.g., space heating and cooling) while others (e.g., industrial activity or vehicles) do not. For example in Europe (for 1985) the combined industrial and heating sources made up 51% of total anthropogenic nitrogen emissions (Pacyna et al. 1989). European energy generation peaks in winter, with emissions in the maximum month estimated at a factor of two higher than the minimum (OECD 1977). As a measure of the importance of this variability for nitrogen emissions, the peak month would account for around 10% of total yearly emissions, falling only slightly to around 7% in the minimum. Hence the seasonal variability for nitrogen emissions is relatively small.

Reduced nitrogen emissions comprise ~54% of the total anthropogenic nitrogen emissions in Europe (Sandnes 1993). There is a strong spring/summer maximum and an autumn/winter minimum with the emission rates in March/April being 1.4–2.3 times greater than in July, and 3–4 times greater than in late autumn. The seasonal maximum is largely the result of agricultural activities and temperature. Estimates (e.g., Tuovinen et al. 1994) suggest that maximum emissions (35% of annual emissions) occur in March–May, and minimum emissions (approximately 20%) in October–December.

In summary, the combination of the oxidized and reduced nitrogen emissions yield a maximum in March–May, with 30% of the annual total, and a minimum in October to December, with 24%. Thus, from the standpoint of total oxidized and reduced nitrogen emissions, the seasonal variability is quite small compared to other factors discussed below.

Meteorological influences

Given the relative constancy of emissions, the temporal and spatial variability of deposition of nitrogen species will be largely controlled by meteorological factors. The local meteorological conditions in the source regions will determine whether the emissions are confined to a relatively small area (as would be the case during a period of stagnation, typically associated with a stable high pressure system) or whether the emissions are carried to the middle and upper troposphere (as would occur during the passage of a frontal system) where winds could rapidly carry the emissions over great distances.

A second major consideration is the occurrence of clouds and precipitation. Because the secondary oxidized nitrogen products and reduced nitrogen species are generally soluble, clouds scavenge them from the atmosphere with very high efficiency. Over the oceans, precipitation is the dominant removal mechanism for most species. Thus the total deposition rates for many species will be largely determined by the occurrence of precipitation. Hence the deposition to regional seas and the shelf will usually be strongly controlled by mesoscale meteorology. The meteorological factors controlling wet removal rates have been studied in a few field studies; in the western NAO: Galloway et al. (1983); Galloway (1985); Galloway & Whelpdale (1987); Moody & Galloway (1988); Galloway et al. (1989, 1993). Studies in the eastern NAO include: Buijsman et al. (1991); Lacaux et al. (1992); Hanseen et al. (1990); Risbo et al. (1981); Mayewski et al. (1990).

Temporal variability of deposition

In this section we present wet-deposition data for relatively remote ocean stations. For this comparison we use data from three AEROCE stations in the North Atlantic where precipitation data has been gathered for at least several years. These are Bermuda, Barbados and Mace Head, located on the west coast of Ireland. The AEROCE data are collected on a daily schedule using automatic (wet-only) precipitation collectors. For this comparison we use the data for a three year period, 1989 to 1991.

Variability of annual deposition rates

Figure 1a shows the annual (volume weighted) deposition rates for NO_3^- and NH_4^+ (mMoles N m^{-2} yr^{-1}) for Bermuda which has the longest continuous deposition record for an open-ocean NAO site, 1980–1993. As discussed in an earlier section, there was a marked change in deposition rates in 1989 when the sampling site was moved from Harbor Tower to the AEROCE site on the western end of the island, Tudor Hill. There is considerably less rainfall at Tudor Hill compared to Harbor Tower. It is clear that the deposition rates track the precipitation rates across the entire record. This relationship is clearer in Figure 1b which shows the annual mean concentration of NO_3^- and NH_4^+

along with precipitation amount; the concentrations are relatively uniform across the entire record despite the very large changes in rainfall amounts.

Frequency distributions of deposition events and episodicity

At the AEROCE sites, the number of deposition events ranged from about 50 to 100 days depending on site and year. The average for each site for the three years combined (1989–1991) was about 80 events per year. Figure 2a shows for each site the cumulative percentile plots of daily NO_3^- deposition amounts for the years combined, 1989–1991. This figure shows that there can be considerable spatial variability in deposition amounts; note, in particular the large difference between Barbados and sites in the higher latitudes, Bermuda and Mace Head. In contrast, the year-to-year variability of the cumulative frequency distribution is not very great. As an example, Figure 2b shows the cumulative frequency distribution plots for Bermuda for each of the three years, 1989–1991. Note that the number of events per year ranged from 66 to 100.

When we look at the actual number of events that deliver relatively large amounts of deposition, we see that about 30 percent of the daily NO_3^- deposition values measured at Bermuda (23–28 events per year, depending on the year) deposited more than 50 μmoles-N m^{-2}; the deposition rates in individual events ranged from 65 to 450 μmoles-N m^{-2} (see Figure 2a) and accounted for 67% of the total annual NO_3-N deposition. At Mace Head, 15–26 events per year (28% of the total) were greater than 50 μmoles-N m^{-2} during the period 1989–1991; these accounted for 80% of the NO_3-N deposition. In contrast, at Barbados, only 5–9 events per year (10%) yielded nitrate deposition rates greater than 50 μmoles-N m^{-2} and these large events deposited only 37% of the NO_3-N deposition. The lower frequency of high-deposition events at Barbados reflects the lower impact of pollution sources because of the distance from the sources (mainly in Europe and North Africa) and the scavenging that has occurred during transport over the tropical NAO.

Seasonal deposition trends

The average seasonal deposition of NO_3-N measured at three AEROCE sites for a three-year period (1989–1991) is shown in Table 10 along with annual averages. We also show for each season the percentage of the deposition that occurred in events which deposited more than 50 μmoles-N m^{-2} . These values provide an indication of spatial and temporal variation in the episodic N-deposition. At all sites there is a well-defined seasonality in deposition amounts; the ratio of the maximum season to the minimum is about a factor of two for Bermuda and Mace Head and about a factor of four for Barbados. It is also clear that large events (defined here as those depositing more than

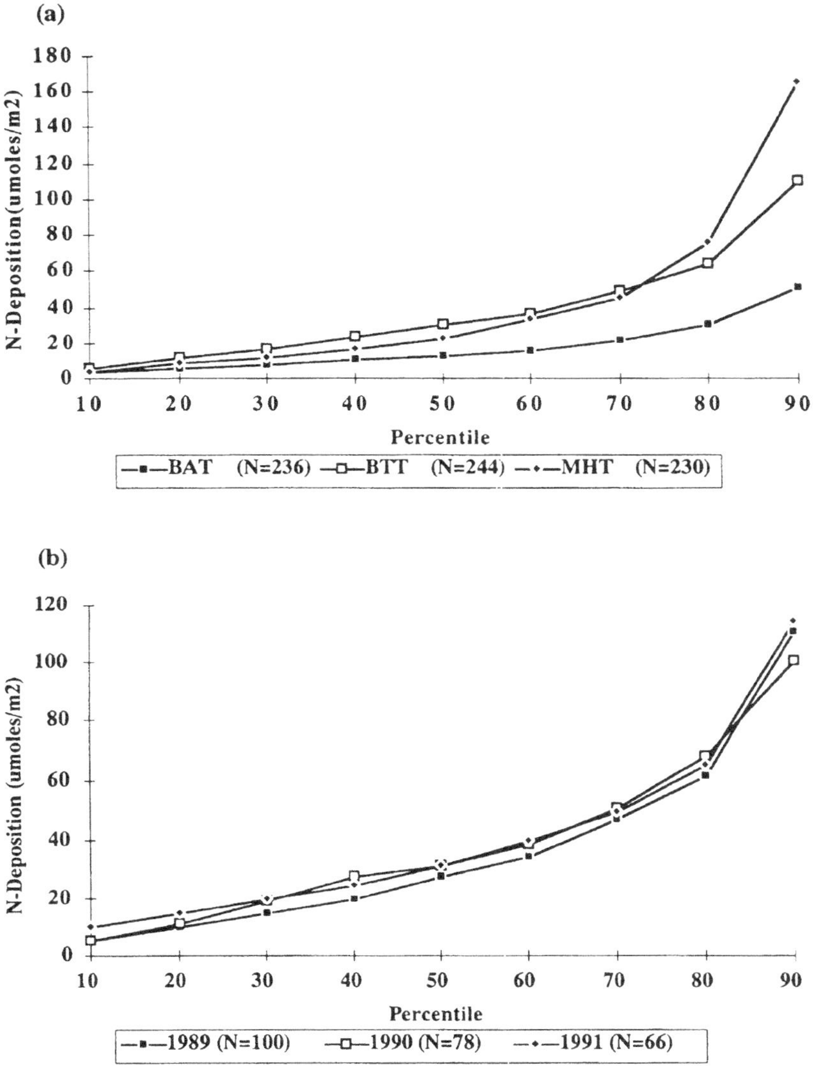

Figure 2. a.) The cumulative frequency distribution of daily nitrate deposition values for each of the AEROCE sites (using data from the years 1989–1991) for Bermuda (BTT), Barbados (BAT) and Mace Head (MHT) (N: number of deposition events for the three year period). b.) The cumulative frequency distribution for nitrate-N deposition at Bermuda for each of the three years, 1989–1991. (N: number of deposition events per year.)

50 μmoles-N m^{-2}) account for a large fraction of the deposition, indeed, the major fraction at Bermuda and Mace Head.

Table 10. Seasonal deposition of NO_3^--N to AEROCE sites: a) measurements; b) GFDL GCTM model.

	Bermuda	% events	Barbados	% events	Mace Head	% events
a)						
Jan–Mar	793	48%	280	2%	1211	79%
Apr–Jun	892	63%	589	40%	916	78%
Jul–Sep	1610	77%	649	34%	896	71%
Oct–Dec	1144	72%	699	49%	1953	85%
Annual avg:	3772		1532		4976	
b)						
Jan–Mar	2160	19%	240	0%	733	7%
Apr–Jun	2147	45%	189	0%	1602	42%
Jul–Sep	1146	20%	303	0%	1582	54%
Oct–Dec	1928	40%	481	0%	761	16%
Annual avg:	7650		1212		4679	

a: Seasonal deposition of NO_3^- N (μMoles-N m^{-2}) at AEROCE sites for the period 1989–1991. The "% events" category shows the percentage of the total seasonal deposition that occurred in episodes with amounts $\geq$50 μMoles-N m^{-2}.
b: GFDL GCTM model NO_3^-N (μMoles-N m^{-2}) in the model grid cells of AEROCE precipitation collection sites. The "% events" category shows the percentage of the total seasonal deposition that ccurred in daily episodes with amounts $\geq$50 μMoles-N m^{-2}.

To more explicitly compare the occurrence of significant deposition events, the data were divided into four categories: 50< NO_3^- <100 μmoles-N m^{-2}, 100< NO_3^- <150 μmoles-N m^{-2}, 150< NO_3^- <200 μmoles-N m^{-2}, and NO_3^- >200 μmoles-N m^{-2}). For each site observations were sorted for these deposition intervals and plotted by season (Figure 3).

The measured deposition distributions show clear evidence of seasonality. At Bermuda (Figure 3a) there is a greater frequency of high deposition events in the warm season quarters of the year (April–June and July–September); this is especially noticeable for events in the range 50< NO_3^- <100 μmoles-N m^{-2}. In contrast at Mace Head (Figure 3b), there is a no consistent seasonal difference except for very large deposition events (i.e., those greater than 200 μmoles-N m^{-2}) which are much more frequent in January–March. Also at Mace Head there is a higher frequency of very large deposition events compared to Bermuda and (especially) Barbados (Figure 3c). In the three years of data, at Mace Head there were 19 daily-deposition events with deposition greater than 200 μmoles-N m^{-2}; in contrast, at Bermuda there

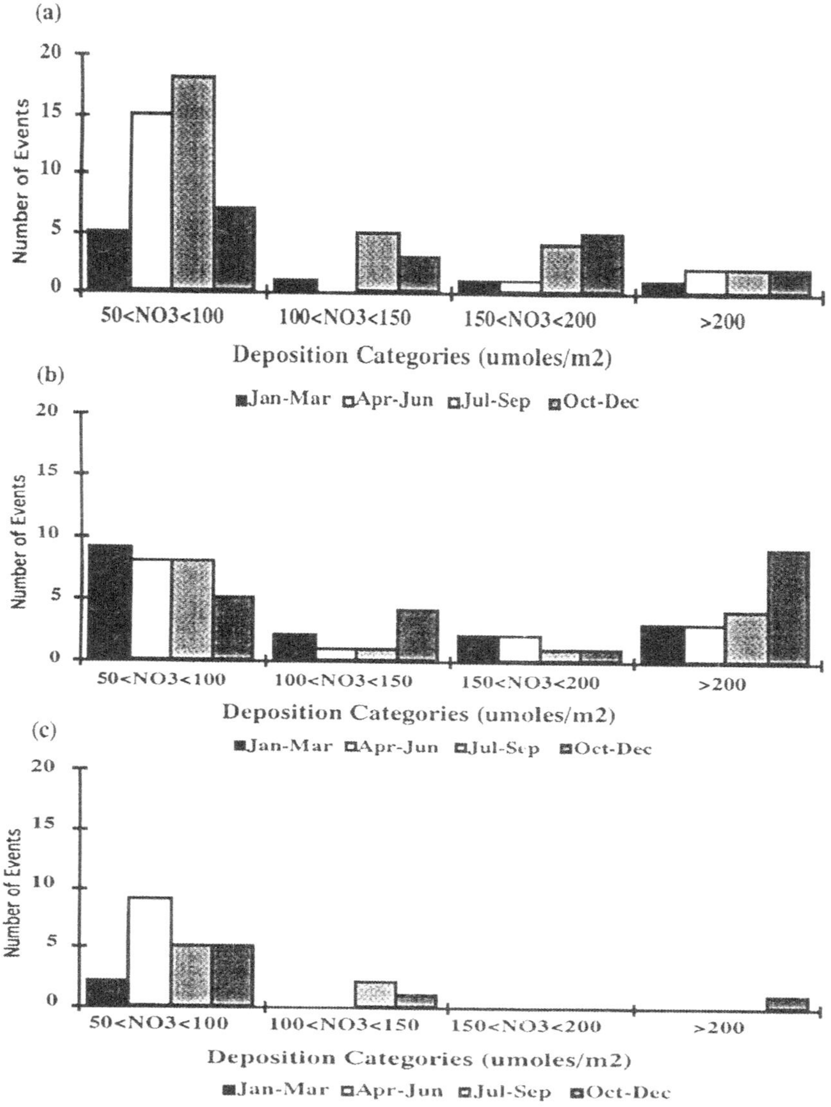

Figure 3. Seasonal distribution of large nitrate-N deposition events. Measured values for the years 1989–1991 at Bermuda (a), Mace Head (b) and Barbados (c).

were 7 events of this magnitude while at Barbados there was only 1 event. Thus of the total of 27 such very large events at the three sites, about 70% of them occurred at Mace Head. At Bermuda, in contrast to Mace Head, there were many more events with intermediate deposition rates (i.e., in the range 50 to 200 μmoles-N m^{-2}).

As an exercise we have compared the spatial and seasonal variability of the GCTM model with our observations (Table 10). This is the only model for which we could obtain daily values with which to compare our observations. The GCTM model accurately predicts the annual NO3-N deposition at Barbados and Mace Head, but it over predicts the amount at Bermuda by a factor of two. The model does capture some of the variability in deposition; however, it under-predicts large deposition events, producing a large percentage of the deposition in very small events (i.e., in the model the precipitation occurs mainly as drizzle).

While the GCTM accurately predicts the total annual NO3-N deposition amounts for Mace Head and Barbados, it does poorly in predicting the seasonal cycle. The model under-predicts spring and summer deposition at Barbados, and over-predicts spring and summer deposition at Mace Head while under-predicting the fall and winter deposition at Mace Head. These comparisons suggest that this model could not be used to estimate the seasonal variability of deposition.

Conclusions

Field measurements suggest that the temporal and spatial character of NO_y and NH_x deposition episodes is variable over the North Atlantic and that this variability is related to meteorological factors. The data from Mace Head show that the greatest deposition fluxes and the most extreme deposition events occur under conditions of transport of European emissions over the NE North Atlantic during the cold half of the year. During winter the surface waters are well-mixed, the input of atmospheric nitrogen is not likely to be significant. Biological activity is at a maximum during the spring and summer; thus it is during the summer, when the surface waters are highly stratified, that contributions from the atmosphere may play a significant role as a nutrient (Michaels et al. 1993). Bermuda and Barbados both show a seasonal pattern with a maximum atmospheric contribution during the summer, although atmospheric inputs are a factor of two lower at Barbados relative to Bermuda. It may be significant that the deposition of atmospheric nitrogen at Bermuda is highly episodic during the summer, with over three quarters of the total deposition delivered in large events. However, it is clear that at all three sites there is the potential for episodic inputs from the atmosphere to briefly impact the ocean.

Mineral dust and related species

The temporal and spatial variability of dust deposition is expected to be large relative to that for nitrogen aerosols. First of all, the emission rate of dust in the source regions varies strongly with season (Prospero & Nees 1986), much more so than for NO_x, NO_y and NH_x sources. Secondly, the dust particle size distribution is shifted to relatively large sizes compared to that of N-bearing aerosols. Because dust has a mass median diameter of several microns or greater (Duce 1995), sedimentation is an important (often dominant) removal mechanism, especially in coastal regions near the source.

There is relatively little data on the temporal and spatial variability of dust deposition. Nonetheless, it is clear that a large fraction of the annual deposition takes place in a very small fraction of the precipitation events. In a one-year study of Saharan dust deposition in Miami (Prospero et al. 1987), 22% of the annual deposition occurred in once day and 68% in rain events that occurred during two dust episodes spread over a total of four days. In a study at Midway Island (Uematsu et al. 1985), about half of the annual deposition of dust occurred during a two week period. Because of the highly sporadic nature of these events, the concentration of mineral dust in ocean surface waters will be highly variable in time and space. Thus, in ocean regions where there is active dust transport, the mean concentration of dust in the underlying ocean could be relatively high but the concentration distribution could be highly non-uniform, much more so than for the deposition of nitrogen species.

On an annual basis, there is a large seasonal variability in dust concentration which is reflected in the modeled dust deposition rates (Prospero et al. 1995). Over much of the tropical and subtropical NAO, dust concentrations are at a seasonal minimum during the winter; they increase sharply during the spring and attain a maximum during the summer when monthly means concentrations are typically more than an order of magnitude higher than in the winter (Prospero & Nees 1986; Savoie et al. 1992).

The emission and transport of soil dust is highly sensitive to climate variables such as rainfall and wind velocity. African dust concentrations in the trade winds have been studied essentially continuously since 1965 at Barbados, West Indies (Prospero & Nees 1986). The mineral dust measurements at Barbados for the period 1965 to 1992 (Figure 4) show that concentrations increased sharply beginning in 1970 when a persistent drought began in the sub-Saharan (Sahel) regions of Africa. Prior to 1970, the annual average dust concentration was 3.9 $\mu g\ m^{-3}$ (range 3.2–4.5 $\mu g\ m^{-3}$); from 1970 to 1992, the average was 11.0 $\mu g\ m^{-3}$ (5.4–18.7 $\mu g\ m^{-3}$), roughly three times greater than the pre-drought mean. The range of annual means is greater, about a factor of six from the lowest annual mean (in the mid-1960's prior to the drought) to the maximum. The highest dust concentrations at Barbados

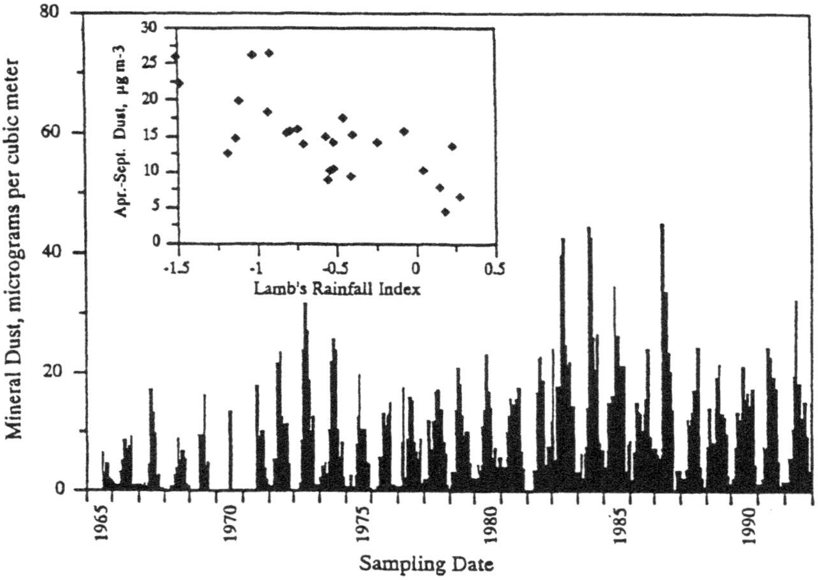

Figure 4. Monthly mean mineral dust concentration (μg m^{-3} at Barbados, 1965–1992. Inset is a scatter plot of the mean annual August–September dust concentration at Barbados vs. the rainfall in the Sahel (see text).

occurred in the mid 1980's when the drought was especially severe; during 1983–87, the June to August average dust concentrations exceeded 25 μg m^{-3}; the maximum June–August mean concentration for the entire 27 year record occurred in 1984, 37.8 μg m^{-3}. It is clear from Figure 4 that there is considerable variability in annual dust transport as measured on the short term (i.e., year-to-year) and over the longer term (i.e., decades).

The dust concentrations at Barbados are inversely related to prior year's rainfall in Africa as indicated by Lamb's rainfall index (inset in Figure 4). This index (P. Lamb, pers. comm.) uses as a reference base the rainfall amounts for 20 sub-Saharan stations from 1941 to 1982; index numbers indicate the standard deviation in annual rainfall from the mean with lower numbers corresponding to drier conditions. Mean summer dust concentrations at Barbados increase by about 10 μg m^{-3} for every standard deviation of rainfall deficit in West Africa. The relationship between dust concentrations and rainfall is less evident during the winter when the transport of dust to Barbados is more sporadic. In the winter and early spring, the large-scale wind systems are displaced to the south, the transport paths move toward more southerly latitudes and dust is transported to South America (Prospero et al. 1986; Swap et al. 1992).

There is a tendency to regard dust emissions as being a "natural" process. However, just as in the case of biomass burning, we can distinguish between the "natural" and "anthropogenic" component of the flux on the basis of the role that humans have played in the process. There is an extensive literature on large scale wind erosion (Middleton 1990; Goudie & Middleton 1992; Pye 1987; Golitsyn & Gillette 1993; Glantz 1994). Clearly, arid and semi arid regions are major sources. However, deserts are not necessarily good dust sources; for example, Australia is a very weak source despite the fact that it is largely arid. High rates of dust emission are generally associated with semi-arid regions where marginal lands are used for agriculture and herding (Middleton 1990; Littmann 1991); during periods of drought, the denuded and broken soil surface is easily carried away. The "dust bowl" in the midwestern United States in the 1930's was a good example of this process. The fact that African dust concentrations in the North Atlantic are correlated with rainfall deficits in the Sahel argues against a Saharan source for the dust. This suggests that much of the increased dust transport over the North Atlantic can be attributed to human actions and thus might be considered to be an anthropogenic pollutant.

Past and future scenarios

Records of past fluxes

Ice cores from permanent glaciers in polar regions can provide a detailed history of changes in atmospheric chemistry (Legrand 1995). Annual cycles can be identified in the core layering and these can be accurately dated thereby permitting the chemical composition to be related to emission histories over the past hundred years or so. The record from Greenland shows that nss-$SO_4^=$ (non-sea-salt $SO_4^=$) concentrations have increased by a factor of two to four since the last century (Mayewski et al. 1990, 1994; Laj et al. 1992); the trend is similar to SO_2 emissions in the United States (Mayewski et al. 1990). Similar increases are noted in the Arctic (Barrie & Bottenheim 1991) and central Greenland (Whitlow et al. 1992) with sources attributed to central Europe and Eurasia. Increases are also seen in the Alps (Wagenbach et al. 1988). In contrast, nss-$SO_4^=$ in Antarctic cores show no such increase. The northern hemisphere records conclusively demonstrate that the long range transport of pollutants has increased deposition levels by a substantial amount over a huge area; however, the time record of deposition is complicated by changes in meteorology so that the interpretation in terms of source/receptor is not always obvious (Mayewski et al. 1994).

Furthermore, there are problems in interpreting the NO_3^- deposition record. There is evidence that NO_3^- can be lost from snows through the volatilization of HNO_3. Nonetheless, on the basis of the nss-$SO_4^=$ record in cores, it is probably safe to assume that the long range transport of NO_3^- to the high latitudes parallels that of $SO_4^=$, following the emission scenario for NO_x.

Ammonium concentrations in Greenland do not show an increase over the past century that could be attributed to anthropogenic emissions. This is surprising in the context of the large increase in nss-$SO_4^=$. The absence of a trend in NH_4^+ over time could be due to losses by volatilization from the snow. Sharp peaks in NH_4^+ concentration are observed in cores; these coincide with peaks in organic acids (formate, acetate, oxalate, etc.) and soot; these peaks are attributed the deposition of products from boreal forest fires (Legrand et al. 1992).

Mineral dust is often an important component in ice cores, especially in the northern hemisphere (Mayewski et al. 1994; Laj et al. 1992; Pfirman et al. 1989) and dust concentrations in cores have been used as a paleoclimate indicator (Hansson 1994). In modern times, substantial amounts of Saharan dust have been detected at Dye 3, Greenland, suggesting that African dust could be a significant contributor to the soil components in high-latitude ice cores (Mosher et al. 1993).

Over the longer term (tens of thousands of years), dust is the most variable of core constituents. In Greenland, dust concentrations were 30–50 times higher during glacial times than today; a similar trend is observed in the Antarctic but absolute concentrations are much lower than in Greenland. The increased dust concentration is related to a number of factors including: increased aridity (which is predicted by a number of climate models and supported by other paleoclimate data); increased availability of fine soil material without vegetative cover in glacial outwash areas and on exposed continental shelves; the presence of more vigorous wind systems and more efficient transport to the high latitudes. However, models have been unable to simulate the transport of dust to the polar regions in glacial times.

Projections of future transport

On a global basis, of the 110 Tg N yr^{-1} created by human processes, about 20 Tg N yr^{-1} are transferred to the oceans by rivers and 16 Tg N yr^{-1} by atmospheric deposition (Galloway 1996). The amount of N transferred to the world's oceans will certainly increase in the future, given the projected increases in fertilizer use and combustion. Galloway (1996) estimates that on a global basis the N fixation rate due to combustion and fertilizer use will increase by about 80 Tg N yr^{-1} in the year 2020. However, most of this increase (about 70%, 59 Tg N yr^{-1}) will occur in Asia, with much of the remaining increase in Africa (8%) and Central and South America (9%).

In the United States, NO_x emissions increased from about 3 Tg N yr^{-1} in 1950 to about 6 Tg N yr^{-1} in 1970 and then remained relatively constant. Ammonia emissions were in the range of 1.8 to 2.2 Tg N yr^{-1} over the same period. Projections to the year 2020 do not show any substantial increase in either NO_x or NH_3 emissions (Galloway et al. 1994, 1995, 1996).

In Europe, nitrogen sources increased after 1960 and reached a maximum in the 1980's. Between 1980 and 1988 (the year of the maximum) the emissions of NO_x from Europe increased by 4%. Since 1988 emissions have declined, partly as a consequence of the NO_x protocol to the Geneva Convention. NH_x emissions during the same period have been essentially unchanged (Tuovinen et al. 1994).

Therefore, we anticipate that over the next few decades, the atmospheric transport and deposition of nutrient-N to the NAO will remain relatively constant. Although some increases in emissions are projected for Africa and Central and South America, these should not contribute substantially to the N cycle of the North Atlantic Ocean.

It is not possible to project the future trend in mineral dust emissions. As stated above, soil erosion is very sensitive to climatic factors coupled with land use practices. Because of the great social and economic changes taking place in North Africa, it is not possible to forecast land-use trends. Climate variability in arid regions is inherently great. These complex factors preclude any assessment of dust transport trends other than they will most likely increase over the longer term (Duce 1995).

Conclusions

Model results suggest that large amounts of aerosol material are being transported by wind to the NAO from the bordering continents. About 13% of the estimated global emissions (natural and anthropogenic) of NO_y are deposited in the NAO and about 6% of the global continental emissions of NH_x. We would expect that most of the deposited material is derived from sources in North America and Europe which jointly emit about half of the global total.

Our deposition estimates are based on model estimates that are subject to considerable uncertainty which is difficult to quantify. While agreement amongst the models was quite good, the comparison between the mean model results and the measurements made at various island stations yielded inconsistent results, in some cases excellent and in others poor. There are problems in attempting to "validate" deposition models by comparing results with island measurements. The rainfall amounts on the island can differ greatly from those over the open ocean because of enhanced convection due to island heating effects. Because deposition rates of chemical species are highly correlated with rainfall amounts, the increased rainfall over the island will result in increased deposition as is evident in the rainfall data from Bermuda. It is not clear how such "validation" comparisons can be easily improved. Nonetheless, at worst the model results are within a factor of two of measured amounts; in general, the results are probably better. Also, the estimated deposition of NO_3^- to the NAO is a "reasonable" fraction of the estimated emissions from the surrounding land masses, for NO_y about a quarter.

There were two classes of materials for which it was particularly difficult to quantify deposition: organic species and mineral dust. Evidence suggests that the deposition rate of organic nitrogen species is quite large, comparable to that of inorganic nitrogen. In order to improve estimates for DON and PON, we will need much more data on their concentration in precipitation over the oceans; this will be difficult to do in a quantitative manner because of the large number of species that are most likely involved. Mineral dust presents a different set of problems. First is the large temporal and spatial variability of dust transport and deposition due to seasonal changes in meteorology. Second, there is a long-term variability that is linked to climate coupled with land-use factors. Third, because of the relatively large size distribution of mineral dust particles, dry deposition (i.e., particle sedimentation) could be an important removal mechanism, especially in ocean regions close to the dust sources; it is much more difficult to measure dry deposition rates than wet deposition.

Because of the huge quantities of mineral dust that are being transported out of North Africa, the NAO is heavily impacted by dust deposition (probably more so than any other region except the Arabian Sea). As shown by Michaels et al. (this volume), the Fe associated with the dust could have a substantial impact on nitrogen cycling in the NAO. Because of the great temporal and spatial variability of dust transport and deposition, the effect on ocean processes should also be highly variable. There is considerable evidence that dust mobilization has been greatly increased because of human activities. If this is true, then the emitted dust must be regarded as a "pollutant" and any resulting effects that this source might have should be thought of as an anthropogenic impact.

References

Andreae MO (1995) Climatic effects of changing atmospheric aerosol levels. In: Henderson-Sellers A (Ed) World Survey of Climatology, Vol. 16, Future Climates of the World (pp 341–392). Elsevier, Amsterdam

Arimoto R, Duce RA & Ray BJ (1989) Concentrations, sources, and air/sea exchange of trace elements in the atmosphere over the Pacific Ocean. In: Riley JP, Chester R & Duce RA (Eds) Chemical Oceanography, Vol. 10, SEAREX. (pp. 107–149). Academic Press, London

Arimoto R, Ray BJ, Duce RA, Hewitt AD, Boldi R & Hudson A (1990) Concentrations, sources, and fluxes of trace elements in the remote marine atmosphere of New Zealand. J. Geophys. Res. 95: 22,389–22,405

Arimoto R, Duce RA, Savoie DL & Prospero JM (1992) Trace elements in aerosol particles from Bermuda and Barbados: Concentrations, sources, and relationships to aerosol sulfate. J. Atmos. Chem. 14: 439–457

Barrie LA & Bottenheim JW (1991) Sulphur and nitrogen pollution in the Arctic atmosphere. In: WT Sturges (Ed) Pollution of the Arctic Atmosphere (pp 2505–2512). Elsevier, London and New York

Benkovitz CM, Dignon J, Scholtz T, Pacyna J, Tarrason L, Dignon J, Voldner EC, Spiro PA, Logan JA & Graedel TE (1995) Global gridded inventories of anthropogenic emissions of sulfur and nitrogen. J. Geophys. Res. (submitted)

Bergametti G, Dutot A-L, Buat-Menard P, Losno R & Remoudaki E (1989a) Seasonal variability of the elemental composition of atmospheric aerosol particles over the northwestern Mediterranean. Tellus 41(B): 353–361

Bergametti G, Gomes L, Coude-Gaussen G, Rognon P & Le Coustumer M (1989b) African dust observed over Canary Islands: Source-regions identification and transport pattern for some summer situations. J. Geophys. Res. 94: 14,855–14,864

Bergametti G, Remoudaki E, Losno R, Steiner E & Chatenet B (1992) Source, transport and deposition of atmospheric phosphorus over the northwestern Mediterranean. J. Atmos. Chem. 14: 501–513

Betzer PR, Carder KL, Duce RA, Merrill JT, Tindale NW, Uematsu M, Castello DK, Young RW, Feeley RA, Breland JA, Bernstein RE & Greco AM (1988) A pulse of Asian dust to the central North Pacific: Long range transport of giant mineral aerosol particles. Nature 336: 568–570

Buijsman E, Jonker PJ, Asman WAH & Ridder TB (1991) Chemical composition of precipitation collected on a weathership in the North Atlantic. Atmos. Environ. 25A(5/6): 873–883

Cornell S, Rendell A & Jickells T (1995) Atmospheric inputs of organic nitrogen to the oceans. Nature 376: 243–246

d'Almeida GA (1986) A model for Saharan dust transport. J. Climate Appl. Meteorol. 25: 903–916

d'Almeida GA & Schütz L (1983) Number, mass and volume distributions of mineral aerosols and soils of the Sahara. J. Clim. Appl. Meteorol. 22: 233–243

Delaney AC & Zenchelsky ST (1976) The organic component of wind-erosion-generated soil-derived aerosol. Soil Sci. 121: 146–155

DeMore WB, Sander SP, Golden DM, Hampson RF, Kurylo MJ, Howard CJ, Ravishankara AR, Kolb CE & Molina MJ (1994) Chemical kinetics and photochemical data for use in stratospheric modeling, JPL/NASA: Evaluation number 11

Dentener FJ & Crutzen PJ (1994) A three dimensional model of the global ammonia cycle. J. Atmos. Chem. 19: 331–369

DiTullio GR, Hutchins DA & Bruland KW (1993) Interaction of iron and major nutrients controls phytoplankton growth and species composition in the tropical North Pacific Ocean. Limnol. Oceanogr. 38: 495–508

Donaghay PL, Liss PS, Duce RA, Kester DR, Hanson AK, Villareal T, Tindale NW & Gifford DJ (1991) The role of episodic atmospheric nutrient inputs in the chemical and biological dynamics of oceanic ecosystems. Oceanography 4: 62–70

Duce RA, Mohnen VA, Zimmerman PR, Grosjean D, Cautreels W, Chatfield R, Jaenicke R, Ogren JA, Pellizzari ED & Wallace GT (1983) Organic material in the global troposphere. Rev. Geophys. & Space Phys. 21: 921–952

Duce RA (1986) The impact of atmospheric nitrogen, phosphorus, and iron species on marine biological productivity. In: Buat-Menard P (Ed) The Role of Air-Sea Exchange in Geochemical Cycling (pp 497–528). Reidel, Dordrecht

Duce RA & Tindale NW (1991) Atmospheric transport of iron and its deposition in the ocean. Limnol. Oceanogr. 36: 1715–1726

Duce RA Liss PS Merrill JT Atlas EL Buat-Menard P Hicks BB Miller JM Prospero, JM Arimoto R, Church T, Ellis M, Galloway JN, Hanson L, Jickells TD, Knap AH, Reinhardt KH, Schneider B, Soudine A, Tokos JJ, Tsunogai S, Wollast R & Zhou M (1991) The atmospheric input of trace species to the world ocean. Global Biogeochem. Cycles 5: 193–259

Duce RA (1995) Sources, distributions, and fluxes of mineral aerosols and their relationship to climate. In: Charlson RJ & Heintzenberg J (Eds) Dahlem Workshop on Aerosol Forcing of Climate (pp 43–72). Berlin

Eppley RW (1980) Estimating phytoplankton growth rates in the central oligotrophic oceans. In: Falkowski PG (Ed) Primary Productivity in the Sea (pp 231–242). Plenum, New York

Fanning KA (1989) Influence of atmospheric pollution on nutrient limitation in the ocean. Nature 339: 460–463

Galloway JN, Knap AH & Church TM (1983) The composition of western Atlantic precipitation using shipboard collectors. J. Geophys. Res. 88: 10,589–10,865

Galloway JN (1985) The deposition of sulfur and nitrogen from the remote atmosphere, background paper. In: Reidel D (Ed) The Biogeochemical Cycling of Sulfur and Nitrogen in the Remote Atmosphere (pp 141–175). Norwell, Mass.

Galloway JN & Whelpdale DM (1987) WATOX–86 overview and western North Atlantic Ocean S and N atmospheric budgets. Global Biogeochem. Cycles 1: 261–281

Galloway JN, Keene WC, Artz RS, Miller JM, Church TM & Knap AH (1989) Processes controlling the concentrations of $SO4^{=}$, NO_3^-, NH_4^+, H^+, HCOOT, and CH_3COOT in precipitation on Bermuda. Tellus 41B: 427–433

Galloway JN, Savoie DL, Keene WC & Prospero JM (1993) The temporal and spatial variability of scavenging ratios for nss sulfate, nitrate, methanesulfonate and sodium in the atmosphere over the North Atlantic Ocean. Atmos. Environ. 27A: 235–250

Galloway JN, Levy H & Kasibhatia PS (1994) Year 2020: Consequences of population growth and development on deposition of oxidized nitrogen. Ambio 23: 120–123

Galloway JN, Schlesinger WH, Levy II H, Michaels A & Schnoor JL (1995) Nitrogen fixation: Anthropogenic enhancement-environmental response. Global Biogeochem. Cycles 9(2): 235–252

Galloway JN (1996). Acid deposition in marine regions. In: DM Whelpdale & Kaiser MS (Eds) Acid Deposition Assessment Report, WMO/UNEP, Geneva (in press)

Galloway JN, Zhao D, Thomson VE & Chang LH (1996) Nitrogen mobilization in the United States of America and the Peoples Republic of China. Atmos. Environ. (in press)

Glantz MH (1994) The West African Sahel. In: Glantz MH (Ed) Drought Follows the Plow (pp 33–43). Cambridge Univ. Press, Cambridge

Golitsyn G & Gillette DA (1993) Introduction: a joint Soviet-American experiment for the study of Asian desert dust and its impact on local meteorological conditions and climate. Atmos. Environ. 27A: 2467–2470

Gorzelska K & Galloway JN (1990) Amine nitrogen in the atmospheric environment over the North Atlantic Ocean. Global Biogeochem. Cycles 4: 309–333

Gorzelska K, Scudlark J & Keene W (1996) Dissolved organic nitrogen in the atmospheric environment. In: Ecological Risk: Science, Policy, Law, and Perception, Society of Environmental Toxicology and Chemistry. Denver (in press)

Goudie AS & Middleton NJ (1992) The changing frequency of dust storms through time. Climatic Change 20: 197–225

Graham WF & Duce RA (1979) Atmospheric pathways of the phosphorus cycle. Geochim. et Cosmochim. Acta 43: 1195–1208

Graham WF & Duce RA (1982) The atmospheric transport of phosphorus to the western North Atlantic. Atmos. Environ. 16: 1089–1097

Hameed S & Dignon J (1988) Changes in the geographical distributions of global emissions of NO_x and SO_x from fossil fuel combustion between 1966 and 1980. Atmos. Environ. 22: 441–449

Hameed S & Dignon J (1992) Global emissions of nitrogen and sulfur oxides in fossil fuel combustion 1970–1986. J. Air Waste Management Assoc. 42: 159–163

Hanseen JE, Pederson U, Schaug J, Dovland H, Pacnya JM, Semb A & Skjalmsen JE (1990) Summary report Chemical Coordinating Centre for the Fourth Phase of European Monitoring and Evaluation Program (EMEP), EMEP Cooperative programme for monitoring and evaluation of the long range transmission of air pollutants in Europe. Meteorol. Syn. Cent.-West. Norw. Meteorol. Inst., Oslo

Hansson ME (1994) The Renland ice core. A northern hemisphere record of aerosol composition over 120,000 years. Tellus 46B: 390–418

Hardy JT & Crecilius EA (1981) Is atmospheric particulate matter inhibiting marine primary productivity? Environ. Sci. Technol. 15: 1103–1105

Hasse L (1983) Introductory meteorology and fluid dynamics. In: Liss PS & Slinn WGN (Eds) Air-Sea Exchange of Gases and Particles (pp 1–52). Reidel, Dordrecht

Howarth RW, Billen G, Swaney D, Townsend A, Jaworski N, Lajtha K, Downing JA, Elmgren R, Caraco N, Jordan T, Berendse F, Freney J, Kudeyarov V, Murdoch P & Zhu Z-L (1996) Regional nitrogen budgets and riverine N & P fluxes for the drainages to the North Altantic Ocean: natural and human influences (this volume)

Ignatov AM, Stowe LL, Sakerin SM & Korotaev GK (1995) Validation of the NOAA/NESDIS satellite aerosol product over the North Atlantic in 1989. J. Geophys. Res. 100: 5123–5132

Jahnke RA (1992) The Phosphorous Cycle. In: Global Biogeochemical Cycles (pp 301–315). Academic Press

Jickells TD, Knap AH, Sheriff-Dow R & Galloway J (1990) No ecosystem shift. Nature 347: 25–26

Jickells T (1995) Atmospheric inputs of metals and nutrients to the oceans: their magnitude and effects. Mar. Chem. 48: 199–214

Jordan TE, Correll DL, Weller DE & Goff NM (1995) Temporal variation in precipitation chemistry on the shore of Chesapeake Bay. Water, Air, Soil Poll. 83: 263–284

Joussaume S (1990) Three-dimensional simulations of the atmospheric cycle of desert dust particles using a general circulation model. J. Geophys. Res. 95: 1909–1941

Knap A, Jickells T, Pszenny A & Galloway J (1986) Significance of atmospheric-derived fixed nitrogen on productivity of the Sargasso Sea. Nature 320: 158–159

Knap AH (1990) The Long-Range Atmospheric Transport of Natural and Contaminant Substances, Kluwer, Dordrecht

Kasibhatla PS, Levy II H & Moxim WJ (1993) Global NO_x, HNO3, PAN, and NO_y distributions from fossil fuel combustion emissions: A model study. J. Geophys. Res. 98: 7165–7180

Lacaux JP, Delmas R, Kouadio G, Cros B & Andreae MO (1992) Precipitation chemistry in the Mayombe Forest of Equatorial Africa. J. Geophys. Res. 97: 6195–6206

Laj P, Palais J & Sigurdsson H (1992) Changing sources of impurities to the Greenland ice sheet over the last 250 years. Atmos. Environ. 26A: 2627–2640

Lawrence MG, Chameides WL, Kasibhatla PS, Levy II H & Moxim WJ (1995) Lightning and atmospheric chemistry: The rate of atmospheric NO production In: Volland H (Ed) Handbook of Atmospheric Electrodynamics, Vol 1 (pp 189–202). CRC Press, Boca Raton, Fl

Legrand M, De Angelis M, Staffelback T, Neftel A & Stauffer B (1992) Large perturbations of ammonium and organic acids content in the Summit-Greenland ice core: Fingerprint form forest fires?. Geophys. Res. Lett. 19: 473–475

Legrand M (1995) Atmospheric chemistry changes versus past climate inferred from polar ice cores. In: Charlson RJ & Heintzenberg J (Eds) Dahlem Workshop on Aerosol Forcing of Climate (pp 123–152). Berlin

Lepple FK (1975) Eolian dust over the North Atlantic Ocean. Ph.D. thesis, University of Delaware
Lepple FK & Brine CJ (1976) Organic constitutents in eolian dust and surface sediments from Northwest Africa. J. Geophys. Res. 81: 1141–1147
Levitus S, Conkright ME, Reid JL, Najjar RG & Mantyla A (1993) Distribution of nitrate, phosphate and silicate in the world oceans. Prog. Oceanogr. 31: 245–273
Levy II H, Mahlman JD & Moxim WJ (1982) Tropospheric N_2O variability. J. Geophys. Res. 87: 3061–3080
Levy II H (1989) Simulated Global Deposition of Reactive Nitrogen Emitted by Fossil Fuel Combustion, In: Proceedings of IAHS Third Scientific Assembly, Symposium I, Atmospheric Deposition (pp 3–9). Baltimore, MD
Levy H & Moxim WJ (1989) Simulated global distribution and deposition of reactive nitrogen emitted by fossil fuel consumption. Tellus 41B: 256–271
Levy II H, Moxim WJ, Kasibhatla PS & Logan JA (1991) The global impact of biomass burning on tropospheric reactive nitrogen. In: Levine JS (Ed) Global Biomass Burning: Atmospheric, Climatic, and Biospheric Implications (pp 363–359). MIT Press, Cambridge, MA
Levy II H, Moxim WJ & Kasibhatla PS (1995) A global 3-dimensional time-dependent lightning source of tropospheric NO_x. J. Geophys. Res. (submitted)
Littmann T (1991) Dust storm frequency in Asia: climatic control and variability. Int. J. of Climatology 11: 393–412
Logan JA (1983) Nitrogen oxides in the troposphere: global and regional budgets. J. Geophys. Res. 88: 10,785–10,807
Mahlman JD & Moxim WJ (1978) Tracer simulation using a global general circulation model: Results from a midlatitude instantaneous source experiment. J. Atmos. Sci. 35: 1340–1374
Martin JH & Fitzwater SE (1988) Iron deficiency limits phytoplankton growth in the north-east Pacific subarctic. Nature 331: 341–343
Mayewski PA, Lyons WB, Spencer MJ, Twickler MS, Buck CF & Whitlow S (1990) An ice-core record of atmospheric response to anthropogenic sulphate and nitrate. Nature 346: 554–556
McCarthy JJ & Carpenter EJ (1983) Nitrogen cycling in near-surface waters of the open ocean. In: Carpenter EJ & Capone DG (Eds) Nitrogen in the Marine Environment (pp 487–512). Academic Press, New York
Merrill JT (1986) Atmospheric pathways to the oceans. In: Buat-Menard P (Ed) The Role of Air-Sea Exchange in Geochemical Cycling (pp 35–63). Reidel D, Norwood, MA
Merrill JT (1989) Atmospheric long-range transport to the Pacific Ocean. In: Riley JP Chester R & Duce RA (Eds) Chemical Oceanography, Vol. 10, SEAREX (pp 15–50). Academic Press, London
Michaels AF, Siegel DA, Johnson RJ, Knap AH & Galloway JN (1993) Episodic inputs of atmospheric nitrogen to the Sargasso Sea: Contributions to new production and phytoplankton blooms. Global Biogeochem. Cycles 7: 339–351
Middleton NJ (1990) Wind erosion and dust-storm contro. In: Goudie AS (Ed) Techniques for Desert Reclamation (pp 87–108). John Wiley & Sons, New York
Miller CB, Frost BW, Booth B, Wheeler PA, Landry MR & Welschmeyer N (1991) Ecological processes in the subarctic Pacific: Iron limitation cannot be the whole story. Oceanography 4: 71–78
Moody JL & Galloway JN (1988) Quantifying the influence of atmospheric transport on the composition of precipitation on Bermuda. Tellus 40: 463–479
Moore RM, Milley JE & Chatt A (1984) The potential for biological mobilization of trace elements from aeolian dust in the ocean and its importance in the case of iron. Oceanol. Acta. 7: 221–228
Mopper K & Zika RG (1987) Free amino acids in marine rain: Evidence for oxidation and potential role in nitrogen cycling. Nature 325: 246–249

Morel FMM, Reuter JG & Price NM (1991) Iron nutrition of phytoplankton and its possible importance in the ecology of ocean regions with high nutrient and low biomass. Oceanography 4: 56–61

Mosher BW, Winkler P & Jaffrezo J-L (1993) Seadonal aerosol chemistry at Dye 3, Greenland. Atmos. Environ. 27A: 2761–2772

NAS (1978) The Tropospheric Transport of Pollutants and Other Substances to the Oceans (p 243). National Academy of Sciences Press, Washington, DC

Nixon SW, Ammerman J, Atkinson L, Berounsky V, Billen G, Boicourt W, Boynton W, Church T, DiToro D, Elmgren R, Garber J, Giblin A, Jahnke R, Owens N, Pilson MEQ & Seitzinger S (1996) The fate of nitrogen and phosphorus at the land-sea margin of the North Atlantic Ocean. (this volume)

OECD (1977) The OECD Programme on Long Range Transport of Air Pollutants: Measurements and Findings, Paris

Oort AH (1983) Global atmospheric circulation statistics, 1958–1973, NOAA professional paper, 14, US Government printing office, Washington DC

Owens NJP, Galloway JN & Duce RA (1992) Episodic atmospheric nitrogen deposition to the oligotrophic oceans. Nature 357: 397–399

Pacyna JM, Larssen S & Semb A (1989) European Survey for NO_x Emissions, 1985. Norsk Institutt for Luftforskning NILU OR 26/89 0–8668. Lillestrom, Norway.

Paerl HW (1985) Enhancement of marine primary productivity by nitrogen enriched acid rain. Nature 315: 747–749

Pfirman S, Wollenburg, I Theide J & Lange MA (1989) Lithogenic sediment on Arctic pack ice: potential aeolian flux and contribution to deep sea sediment. In: Leinen M & Sarnthein M (Eds) Paleoclimatology and Paleometeorology: Modern and Past Patterns of Global Atmospheric Transport (pp 463–493). Kluwer Academic Publ., Boston

Prospero JM (1981) Aeolian transport to the world ocean. In: Emiliani C (Ed) The Sea, Vol. 7 The Oceanic Lithosphere (pp 801–874). Wiley Interscience, New York

Prospero JM & Nees RT (1986) Impact of the North African drought and El Niño on mineral dust in the Barbados trade wind. Nature 320: 735–738

Prospero JM, Glaccum RA & Nees RT (1986) Atmospheric transport of soil dust from Africa to South America. Nature 289: 570–572

Prospero JM, Nees RT & Uematsu M (1987) Deposition rate of particulate and dissolved aluminum derived from Saharan dust in precipitation at Miami, Florida. J. Geophys. Res. 92: 14,723–14,731

Prospero JM, Uematsu M & Savoie DL (1989) Mineral aerosol transport to the Pacific Ocean. In: Riley JP, Chester R & Duce RA (Eds) Chemical Oceanography, 10 (pp 187–218). Academic Press, London

Prospero JM (1990) Mineral aerosol transport to the North Atlantic and North Pacific: the impact of African and Asian sources. In: Knap AH (Ed) The Long-Range Atmospheric Transport of Natural and Contaminant Substances (pp 59–86). Kluwer, Dordrecht

Prospero JM (1996) The atmospheric transport of particles to the Ocean. In: Ittekkott V, Honjo S, Schäfer P & Depetris PJ (Eds) Particle Flux in the Ocean (pp 19–52). 1996 SCOPE Report 57. John Wiley & Sons Ltd

Prospero JM, Barrett K, Dentener F, Galloway JN, Levy II H & Moody J (1995) Models Used to Estimate Atmospheric Transport and Deposition to the North Atlantic Basin. Unpublished workshop manuscript

Pye K (1987) Aeolian Dust and Dust Deposits (p 334). Academic Press, London

Quinn PK, Charlson RJ & Bates TS (1988) Simultaneous observations of ammonia in the atmosphere and ocean. Nature 335: 336–338

Quinn PK, Bates TS, Johnson JE, Covert DS & Charlson RJ (1990) Interactions between the sulfur and reduced nitrogen cycles over the central Pacific Ocean. J. Geophys. Res. 95: 16,405–16,416

Quinn PK, Asher WE & Charlson RJ (1992) Equilibria of the marine multiphase ammonia system. J. Atm. Chem. 14: 11–30

Quinn PK, Barrett KJ, Dentener FJ, Lipschultz F & Kurtz KD (1996) Estimation of the air/sea exchange of ammonia for the North Atlantic Basin. (this volume)
Rao CRN, Stowe LL & McClain EP (1989) Remote-Sensing of Aerosols over the Oceans Using AVHRR Data: Theory, Practice, and Applications. Int. J. Remote Sens. 10: 743–749
Rendell AR, Ottley CJ, Jickells TD & Harrison RM (1993) The atmospheric input of nitrogen species to the North Sea. Tellus 45B: 53–63
Riley JP, Chester R & Duce RA (1989) Chemical Oceanography, SEAREX, 10: 404. Academic Press, London
Risbo T, Claussen HB & Rasmussen KL (1981) Supernovae and nitrate in the Greenland ice sheet. Nature 294: 637–639
Rodhe H (1985) The transport of sulfur and nitrogen through the remote atmosphere. In: Galloway JN, Charlson RJ, Andreae MO & Rodhe H (Eds) The Biogeochemical Cycling of Sulfur and Nitrogen in the Remote Atmosphere (pp 105–124). Reidel, Dordrecht
Sandnes H (1993) Calculated budgets for airborne acidifying components in Europe, 1985, 1987, 1988, 1989, 1990, 1991 and 1992. EMEP/MSC-W Report 1/93. Det Norske Meteorologiske Institutt Technical Report 109
Savoie DL, Prospero JM, Oltmans SJ, Graustein WC, Turekian KK Merrill JT & Levy II H. (1992) Sources of nitrate and ozone in the marine boundary layer of the tropical North Atlantic. J. Geophys. Res. 97: 11,575–11,589
Schlesinger WH & Hartley AE (1992) A global budget for atmospheric NH_3. Biogeochem. 15: 191–211
Schütz L, Jaenicke R & Pietrek H (1981) Saharan dust transport over the North Atlantic Ocean. Geol. Soc. Am. Special Paper 186: 87–100
Schütz LW, Prospero JM, Buat-Menard P, Carvalho RAC, Cruzado A, Harriss R, Heidam NZ & Jaenicke R (1990) The long-range transport of mineral aerosols: Group report In: Knap AH (Ed) The Large-Scale Atmospheric Transport of Natural and Contaminant Substances, Chapter 10 (pp 197–229). Kluwer Academic, Dordrecht, Holland
Spokes LJ, Jickells TD & Lim B (1994) Solubilisation of aerosol metals by cloud processing: A laboratory study. Geochim. et Cosmochim. Acta 58(15): 3281–3287
Stedman DH & Shetter R (1983) The global budget of atmopsheric nitrogen species. In: Schwartz SS (Ed) Trace Atmospheric Constituents: Properties, Transformations and Fates. John Wiley & Sons, New York
Swap R, Garstang M, Graco S, Talbot R & Kallberg P (1992) Sahara dust in the Amazon basin. Tellus 44B: 133–149
Talbot RW, Harriss RC, Browell V, Gregory GL, Sebacher DI & Beck SM (1986) Distribution and geochemistry of aerosols in the tropical North Atlantic troposphere: Relationship to Saharan dust. J. Geophys. Res. 91: 5173–5182
Taylor SR & McLennan SM (1985) The Continental Crust: Its Composition and Evolution (pp 312). Blackwells, Oxford, England
Tegen I & Fung I (1994) Modeling of mineral dust in the atmosphere: Sources, transport, and optical thickness. J. Geophys. Res. 99: 22,897–22,914
Tegen I & Fung I (1995) Contribution to the atmospheric mineral aerosol load from land surface modification. J. Geophys. Res. 100: 18,707–18,726
Tuovinen J-P, Barrett K & Styve H (1994) Transboundary Acid Pollution in Europe: calculated fields and budgets 1985–1993. EMEP/MSC-W Report 1/94. Det Norske Meteorologiske Institutt Technical Report. Oslo
Uematsu M, Duce RA & Prospero JM (1985) Deposition of atmospheric mineral particles to the North Pacific. J. Atmos. Chem. 3: 123–138
Wagenbach D, Munnich KO, Schotterer U & Oeschger H (1988) The anthropogenic impact on snow chemistry at Colle Gnifetti, Swiss Alps. Annal. Glaciol. 10: 183–187
Warneck P (1988) Chemistry of the Natural Atmosphere (pp 757). Academic Press, San Diego
Whelpdale DM & Moody JL (1990) Large-scale meteorological regimes and transport process In: Knap AH & Marston MS (Eds) The Large-Scale Atmospheric Transport of Natural and Contaminant Substances (pp 3–36). D. Reidel, Dordrecht, Holland

Whelpdale DM & Galloway JN (1994) Sulfur and reactive nitrogen oxide fluxes in the North Atlantic atmosphere. Global Biogeochem. Cycles 8: 481–493

Whelpdale DM & Kaiser MS (1996) Acid Deposition Assessment Report. WMO/UNEP, Geneva

Whitlow S, Mayewski PA & Dibb JE (1992) A comparison of major chemical species input timing and accumulation at South Pole and Summit, Greenland. Atmos. Environ. 26: 2045–2054

Willey JD & Paerl HW (1993) Enhancement of chlorophyll a production in Gulf Stream surface seawater by synthetic versus natural rain. Marine Biology 116: 329–334

Williams PM (1967) Sea surface chemistry: Organic carbon and organic and inorganic nitrogen and phosphorus in surface films and subsurface waters. Deep Sea Res. 14: 791–800

Yienger JJ & Levy II H (1995) Global inventory of soil-biogenic NO_x emissions, J. Geophys. Res. 100: 11,447–11,464

Young RW, Carder KL, Betzer PR, Costello DK, Duce RA, DiTullio GR, Tindale NW, Laws EA, Uematsu M, Merrill JT & Feely RA (1991) Atmospheric iron inputs and primary productivity: Phytoplankton responses in the North Pacific. Global Biogeochem. Cycles 5: 119–134

Zhu XR, Prospero JM, Millero FJ, Savoie DL & Brass GW (1992) The solubility of ferric ion in marine mineral aerosol solutions at ambient relative humidities. Mar. Chem., 38: 91–107

Zhu X, Prospero JM, Savoie DL, Millero FJ, Zika RG & Saltzman ES (1993) Photoreduction of Fe(III) in marine mineral aerosol solutions. J. Geophys. Res. 98: 9039–9046

Zhuang GS, Duce RA & Kester DR (1990) The solubility of atmospheric iron in surface seawater of the open ocean. J. Geophys. Res. 95: 16,207–16,216

Zhuang G, Yi Z, Duce RA & Brown PR (1992) Chemistry of iron in marine aerosols. Global Biogeochem. Cycles 6: 161–173

Zimmermann PH (1988) Moguntia: A handy global tracer model. In: van Dop H (Ed) Air Pollution Modeling and its Applications VI. Plenum, New York

Biogeochemistry **35:** 75–139, 1996.

Regional nitrogen budgets and riverine N & P fluxes for the drainages to the North Atlantic Ocean: Natural and human influences

R.W. HOWARTH[1], G. BILLEN[2], D. SWANEY[1], A. TOWNSEND[1], N. JAWORSKI[3], K. LAJTHA[4], J.A. DOWNING[5], R. ELMGREN[6], N. CARACO[7], T. JORDAN[8], F. BERENDSE[9], J. FRENEY[10], V. KUDEYAROV[11], P. MURDOCH[12] & ZHU ZHAO-LIANG[13]

[1]*Ecology & Systematics, Cornell University, Ithaca, NY 14853 USA;* [2]*Groupe de Microbiol. des Milieux Aquatiques, Universite Libre de Bruxelles, B-1050 Bruxelles, BELGIUM;* [3]*U.S. EPA Lab, 27 Tarzwell Drive, Narragansett, RI 02882–1197 USA;* [4]*Department of Botany and Plant Pathology, Oregon State University, Corvallis OR 97331-2902 USA;* [5]*Animal Ecology, Iowa State University, Ames, Iowa 50011-3221 USA;* [6]*Department of Systems Ecology and Centre for Marine Research, Stockholm University, Stockholm, Sweden;* [7]*Institute of Ecosystem Studies, Millbrook, NY 12545-0129 USA;* [8] *Smithsonian Environmental Research Center, Edgewater, MD 21037-0028 USA;* [9]*CABO DLO, Wageningen, The Netherlands;* [10]*CSIRO, Division of Plant Industry, Canberra ACT 2601, Australia;* [11]*Institute of Soil Science and Photosynthesis, 142292 Pushchino on Oka, Moscow Region, Russia;* [12]*Water Resources Division, U.S. Geological Survey, Albany, NY 12201 USA;* [13]*Lab of Material Cycling in Pedosphere, Institute of Soil Science, Nanjing, People's Republic of China*

Received 10 August 1995; accepted 9 November 1995

Key words: anthropogenic, atmospheric deposition, eutrophication, fertilizer, nitrogen, nitrogen budget, nitrogen fixation, N:P ratio, phosphorus, pristine, rivers, temperate, tropical

Abstract. We present estimates of total nitrogen and total phosphorus fluxes in rivers to the North Atlantic Ocean from 14 regions in North America, South America, Europe, and Africa which collectively comprise the drainage basins to the North Atlantic. The Amazon basin dominates the overall phosphorus flux and has the highest phosphorus flux per area. The total nitrogen flux from the Amazon is also large, contributing 3.3 Tg yr^{-1} out of a total for the entire North Atlantic region of 13.1 Tg yr^{-1}. On a per area basis, however, the largest nitrogen fluxes are found in the highly disturbed watersheds around the North Sea, in northwestern Europe, and in the northeastern U.S., all of which have riverine nitrogen fluxes greater than 1,000 kg N km^{-2} yr^{-1}.

Non-point sources of nitrogen dominate riverine fluxes to the coast in all regions. River fluxes of total nitrogen from the temperate regions of the North Atlantic basin are correlated with population density, as has been observed previously for fluxes of nitrate in the world's major rivers. However, more striking is a strong linear correlation between river fluxes of total nitrogen and the sum of anthropogenically-derived nitrogen inputs to the temperate regions (fertilizer application, human-induced increases in atmospheric deposition of oxidized forms of nitrogen, fixation by leguminous crops, and the import/export of nitrogen in agricultural products). On average, regional nitrogen fluxes in rivers are only 25% of these anthropogenically derived nitrogen inputs. Denitrification in wetlands and aquatic ecosystems is probably the dominant sink, with storage in forests perhaps also of importance. Storage of nitrogen in groundwater, although of importance in some localities, is a very small sink for nitrogen inputs in all regions. Agricultural sources of nitrogen dominate inputs in many regions, particularly the Mississippi basin and the North Sea drainages. Deposition of oxidized nitrogen, primarily of industrial origin, is the major control over river nitrogen export in some regions such as the northeastern U.S.

Using data from relatively pristine areas as an index of change, we estimate that riverine nitrogen fluxes in many of the temperate regions have increased from pre-industrial times by 2 to 20 fold, although some regions such as northern Canada are relatively unchanged. Fluxes from the most disturbed region, the North Sea drainages, have increased by 6 to 20 fold. Fluxes from the Amazon basin are also at least 2 to 5 fold greater than estimated fluxes from undisturbed temperate-zone regions, despite low population density and low inputs of anthropogenic nitrogen to the region. This suggests that natural riverine nitrogen fluxes in the tropics may be significantly greater than in the temperate zone. However, deforestation may be contributing to the tropical fluxes. In either case, projected increases in fertilizer use and atmospheric deposition in the coming decades are likely to cause dramatic increases in nitrogen loading to many tropical river systems.

Introduction

Human activity has greatly altered the nitrogen cycle on land, in aquatic systems, and in the atmosphere (Berner & Berner 1987; Galloway et al. 1995). Currently, global fixation of atmospheric N_2 for fertilizer, in combustion of fossil fuels, and by leguminous crops exceeds that by all natural sources, and changes in land use cause large additional amounts of nitrogen to be released from long-term reservoirs in both vegetation and soil (Vitousek 1994). These disturbances have been linked to a number of environmental concerns, including coastal eutrophication (Howarth 1988; Nixon 1992, 1995; National Research Council 1993; Gabric & Bell 1993; Paerl 1993; Justic et al. 1995; Rabalais et al., in press), acidification of freshwater lakes and streams (Driscoll et al. 1987; Henriksen & Brakke 1988; Kelly et al. 1990; Murdoch & Stoddard 1992), forest decline (Schulze 1989), climate change (Keller et al. 1986; Khalil & Rasmussen, 1992; Vitousek & Matson 1993), and shifts in community structure (Tilman 1984; Crabtree & Bazzaz 1993; Bowman et al. 1995) and in ecosystem function (McNulty et al. 1991; Aber et al. 1993; Burton et al. 1993; Neff et al. 1994). Accelerated nitrogen cycling has also been suggested to increase the sequestration of carbon in forests, slowing the rise of atmospheric carbon dioxide (Schindler & Bayley 1993). However, gaps in our understanding of nitrogen dynamics over large spatial scales often limit our ability to clearly identify the areas of greatest concern, and to predict the response of systems to future change.

The foundations of ecosystem ecology and biogeochemistry are based in part on studies of small-scale watersheds. Much less attention has been given to evaluating material fluxes through the landscape at larger scales, and we are only beginning to understand the rules that govern regional exchanges of nitrogen between atmospheric, terrestrial and aquatic systems. These exchanges may be especially crucial to the functioning of freshwater and coastal ecosystems, as even relatively small changes in the processing and retention of nitrogen applied to the terrestrial landscape could have a large impact on the downstream aquatic environment. In fact, several studies have noted much greater fluxes of nitrate in rivers over the past few decades. Nitrate concentrations in the Mississippi River have more than doubled since 1965 (Turner & Rabalais 1991; Justic et al. 1995; Rabalais et al., in press), and Pacés (1982) estimated that nitrate in many European rivers has probably increased 5- to 10-fold since the turn of the century. Such increases are presumably in response to increased use of nitrogen fertilizer, cultivation of nitrogen-fixing crops, greater population density, and increased atmospheric deposition of nitrogen. Total nitrogen fluxes have also probably increased, but historical data on total nitrogen in rivers are extremely rare.

Our overall objective in this study was to estimate total nitrogen export (TN) in all of the major rivers draining into the North Atlantic Ocean, to assess how much that export has changed since pre-industrial times, and to identify the principal agents of that change. The work presented here is part of a larger collaborative effort to assemble a nitrogen budget for all components of the North Atlantic region: atmosphere, watersheds, coastal systems, and the open ocean. Although the focus of the paper is on nitrogen, estimates of phosphorus (TP) are also developed because of the potentially valuable information provided by considering N:P ratios. In some cases, we have had to extrapolate estimates based on fluxes of organic carbon or other indirect approaches, but most of our estimates are derived from direct data on water fluxes and concentration of TN and TP (including dissolved and particulate inorganic and organic forms). These data are summarized by dividing the total study area into 14 major watershed regions. In addition to the estimates of TN and TP export by rivers, we present data on population density, sewage and wastewater inputs, fertilizer application, atmospheric deposition, fixation of nitrogen by leguminous crops, and the net import or export of nitrogen in agricultural products.

We chose the North Atlantic region for several reasons. One, when all components of the region are considered, it is probably the best studied area on the planet, and we therefore had the best chance to derive a reasonably accurate nitrogen budget for an entire ocean basin. In discussing transport of C, N, and P in the world's rivers, Meybeck (1982) notes that "it is very difficult to get information on more than half the river water discharging to the ocean" as a whole, and fewer data are available for TN than for C and P. However, for the North Atlantic Basin, we have direct data for TN fluxes in the majority of watersheds representing approximately 80% of the total freshwater discharge. Two, much of the region has been and continues to be significantly affected by human activity, and we wished to assess the impacts of such activity on regional-scale biogeochemistry. Finally, the region is comprised of a great diversity of environments, thereby allowing comparisons of many of the world's major biome types within the confines of our study boundaries.

Regional boundaries and data sources

We divide the watersheds which flow into the North Atlantic Ocean into 14 regions (Figure 1; Table 1). Seven of these regions are in North America, 2 are in Central and South America, 4 are in Europe, and one is in Africa. Regions are selected to coincide with discrete portions of the coastal ocean. The definition of the size and boundaries of regions is dictated in part by the availability of data and by previous studies. All estimates reported here are at the scale of these large regions, although estimates for many of the regions are calculated by summing fluxes for watersheds and smaller areas within the regions. Most nutrient flux estimates are for average periods during the 1980's, although estimates for a few regions are based on late 1970's data. We have explicitly not considered inputs from land into the Mediterranean Sea; net nitrogen fluxes between the Mediterranean and the Atlantic Ocean proper are addressed by the oceanic working group (Michaels et al., this volume).

North Canadian rivers

The "north Canadian rivers" region is defined to be all drainages into Hudson Bay plus the drainages which flow into the Atlantic from Labrador and Newfoundland; this is a large area which includes 40% of Canada (Laycock 1987). Data for area, population, and freshwater discharges are all from Laycock (1987). Annual fluxes of total nitrogen (TN) and total phosphorus (TP) from this region are estimated as the product of the average TN and TP concentrations in 6 rivers in the region (17 μM N and 0.44 μM P; Paré & Goulet 1980; Schetagne 1981; Schetagne & Roy 1985) and annual water discharge to Hudson Bay and coastal discharges of streams and rivers in Labrador and Newfoundland (Laycock 1987).

St. Lawrence

The "St. Lawrence" region includes the drainage basins of the Great Lakes (van der Leeden et al. 1990) plus areas of Canada labeled by Laycock (1987) as "St. John-St. Croix," "maritime coastal," and "North Shore-Gaspé." The area of this region is calculated from information in Laycock (1987) and van der Leeden (1975). Freshwater discharge is the sum of discharge for the St. Lawrence River (Pocklington & Tan 1987) and regional discharge from the

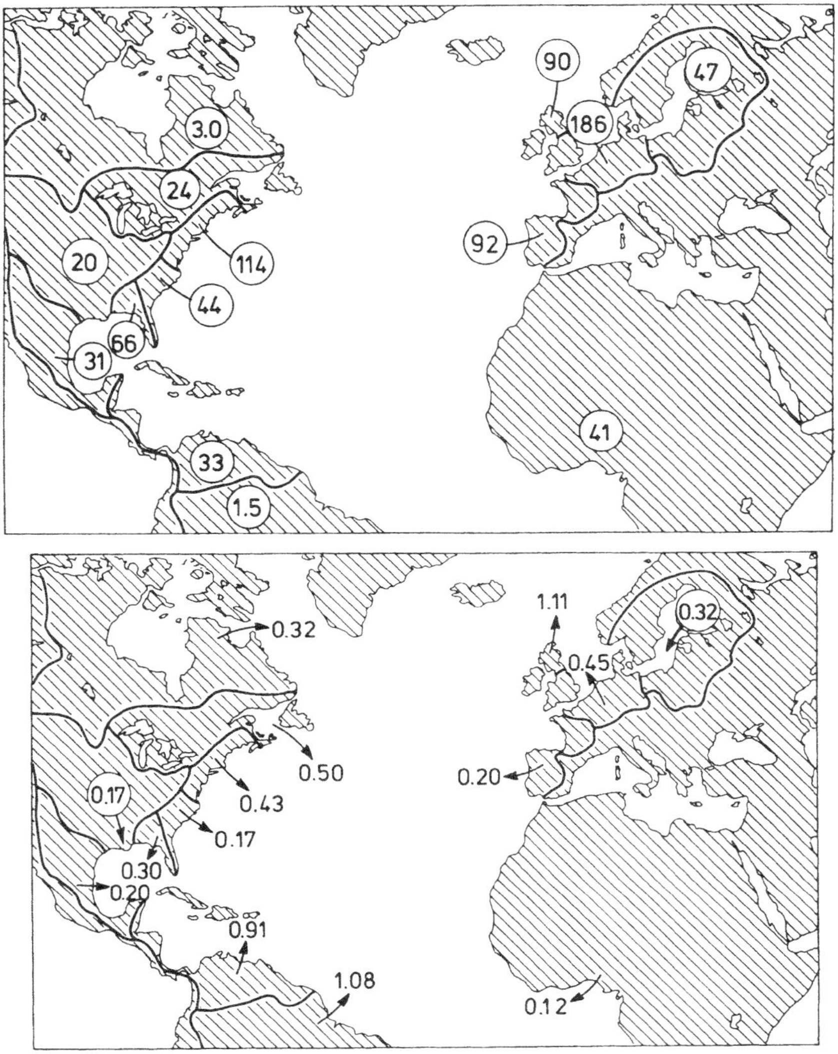

Figure 1. Population density of each of the 14 regions in individuals per km^2 (top, a), and water discharge per area in $m^3\ m^{-2}\ yr^{-1}$ (bottom, b).

St. John-St. Croix, maritime coastal, and North Shore-Gaspé areas (Laycock 1987). Population for the St. Lawrence region is estimated by summing the population of the St. John-St. Croix, maritime coastal, and North Shore-Gaspé areas (Laycock 1987) and the St. Lawrence and Great Lakes basins; population in the St. Lawrence and Great Lakes basins is the product of land

Table 1. Summary of regional data on drainage area, population, and water discharge.

	Drainage area (10^6 km^2)	Population (10^6 ind.)	Water discharge (10^9 m^3 yr^{-1})
North Canadian rivers	3.98	12	1,260
St. Lawrence basin	1.60	38	801
NE coast of US	0.48	55	208
SE coast of US	0.35	16	59
Eastern Gulf of Mexico	0.35	23	106
Mississippi basin	3.23	64	546
Western Gulf of Mexico	1.42	44	285
North America – total	11.4	252	3,260
Caribbean Islands and Central America (incl. Orinoco)	2.28	75	2,070
Amazon & Tocantins basin	6.49	10	7,000
Central & South America – total	8.77	85	9,070
Baltic Sea	1.50	70	475
North Sea	0.84	156	380
NW European coast	0.34	30	378
SW European coast	0.55	50	110
Western Europe – total	3.23	306	1,340
NW Africa	3.53	146	416
Total – all basins	26.9	790	14,100

Note: Watersheds flowing into Mediterranean Sea are not included.

area and mean population density in the states and provinces of this region (22 individuals km^{-2}; Espenshade 1990). Annual fluxes of TN from the region are taken as those of the St. Lawrence River at Quebec City (Comité d'étude sur le fleuve Saint-Laurent 1978) plus contributions from the North Shore-Gaspé, St. John-St. Croix, and Maritime Coastal regions calculated as the product of regional water discharges (Laycock 1987) and average nutrient concentrations. Annual fluxes of TP are estimates from a phosphorus budget for the upper estuary of the St. Lawrence (Lucotte 1989) plus contributions from the three coastal regions. Nutrient concentrations for these regions are assumed to be the same as in the North Canadian Rivers region.

NE coast of US

The NE coast of the US corresponds to a water resource region defined by the U.S. Geological Survey (Crippen 1969). Area and freshwater discharge are from Crippen (1969). Population data and estimates of TN and TP fluxes are from Jaworski and Howarth (in press, and ms. in prep.). Jaworski and Howarth (in press, and ms. in prep.) use many data sources but rely most heavily on data from characterization studies for individual estuaries in the region as part of the Natural Estuary Program of the U.S. Environmental Protection Agency and state environmental agencies.

SE coast of US

The SE Coast of the U.S. also corresponds to a U.S.G.S. water resource region (Crippen 1969). Area and freshwater discharge data are from Crippen (1969). The population is estimated from state population data (Espenshade 1990) using a delineation by Crippen (1969) for assignment of what percent of each state lives within the region. Annual flux of TN from the region is estimated as the sum of independent estimates for nitrate fluxes per area, dissolved organic nitrogen per area, and particulate organic nitrogen per area multiplied by the area of the drainage. The nitrate flux per area is from Smith et al. (1993). Dissolved and particulate organic nitrogen fluxes are estimated from dissolved and particulate organic carbon fluxes in the region (Degens et al. 1984, as cited in Meybeck et al. 1989) by assuming a C:N molar ratio of 23.3 for dissolved organic matter (Meybeck 1982) and of 9.7 for particulate organic matter (Ittekot & Zhang 1989). The flux of TP is estimated from Smith et al. (1993).

Eastern Gulf of Mexico

Boundaries are those defined by the U.S.G.S. water region (Crippen 1969). Area, freshwater discharge, population, and annual fluxes of TN and TP from the region are estimated as for the "SE coast of U.S." region using the same data sources. These two regions have much in common; we consider them separately since one drains into the Gulf of Mexico while the other drains into the North Atlantic Ocean proper.

Mississippi basin

This region is delineated by the natural drainage basin of the Mississippi River and its tributaries. It is a large region, making up 41% of the area of the lower 48 states of the United States (Wright 1993). Data on population, area, and freshwater discharge are from Meybeck et al. (1989). Annual fluxes of TN and TP are from Turner & Rabalais (1991).

Western Gulf of Mexico

The "western Gulf of Mexico" region includes another U.S.G.S. water resource region (Crippen 1969) but also watersheds within Mexico from the Rio Grande to the tip of the Yucatan which flow into the Gulf of Mexico (del Rio & Wilson 1969). This includes most of Mexico. Population in this region is given by the population of Texas (Espenshade 1990) plus the product of the drainage area within Mexico (del Rio & Wilson 1969) and the average population density of Mexico (Espenshade 1990). Regional discharge is calculated by summing the flows within the U.S. (Crippen 1969) and those within Mexican drainages to the Gulf (del Rio & Wilson 1969). Fluxes of TN and TP are estimated as for the "SE Coast of U.S.," using the same data sources. For TP and the nitrate portion of TN, values of Smith et al. (1993) for the portion of this region within the U.S. are extrapolated to the portion of the region within Mexico as well.

Caribbean Islands and Central America

This region extends from the Yucatan Peninsula to the French Guiana/Brazil border and includes all islands in the Caribbean Sea. Area in the region is the sum of the area of all of the major Caribbean islands (Espenshade 1990) and the area on the mainland, consisting of a Central American portion and a South American portion. The Central American area is estimated from the fractional area of the Central American isthmus which drains into the Caribbean (0.692; Ahlgren et al. 1969) multiplied by the area of Guatemala, Honduras, Nicaragua, Costa Rica, and Panama (Espenshade 1990). The amount of area in this region in South America is obtained by summing the principal drainage areas, including the Orinoco, the Magdalena, and the principal rivers of Surinam, Guyana, and French Guiana (van der Leeden 1975; Lewis & Saunders 1989). Population is the sum of the population of the Caribbean Islands (Espenshade 1990), the population of countries in the South American portion of this region (Colombia, Venezuela, Guyana, Surinam, French Guiana; Espenshade 1990), and population of the Central American countries in the region (Espenshade 1990) multiplied by the percentage area of these countries which flow to the Atlantic (Ahlgren et al. 1969).

Discharge data for the Central American area are from Ahlgren et al. (1969) and for the South American area are from van der Leeden (1975) and Lewis & Saunders (1989). Fluxes of TN and TP from the region are based on a discharge-based extrapolation of estimates for the Orinoco River (Lewis

& Saunders 1989) plus estimates for sewage and wastewater inputs. The Orinoco is the largest river in the region and contributes 55% of total freshwater discharge for the region. Sewage and wastewater inputs are added to this extrapolation of the Orinoco data since the Orinoco watershed is relatively unpopulated compared to much of the region (particularly the Caribbean Islands). Wastewater inputs of nitrogen are estimated as the product of population, the fraction of the population which is sewered (WRI/UNEP 1988), and per capita nitrogen production in wastewater (3.3 kg N yr^{-1} $person^{-1}$; Meybeck et al. 1989). Wastewater inputs of phosphorus were handled similarly, assuming 0.4 kg P yr^{-1} per person (Meybeck et al. 1989), except that the estimate of Ramirez (1991) was used for Venezuela. The percent of the population which is sewered is taken as 70%, 55%, and 50% respectively in the Central American component, the South American component, and the Caribbean islands component of the region (WRI/UNEP 1988).

Amazon and Tocantins

This region includes all of the Amazon River basin plus the basin of the Tocantins River, a river immediately to the south of the Amazon. The region is very large and comprises 36% of South America (Wright 1993). Estimates for area, population, and freshwater discharge are from Meybeck et al. (1989). The Amazon dominates discharge in the region (96% of total discharge) and fluxes of TN and TP. This river is well studied, and very good nutrient flux estimates are available. We used the estimates of Salati et al. (1982) for TN and Devol et al. (1991) for TP. The Amazon data are extrapolated to the Tocantins basin on the basis of discharge. Note that while the Amazon discharges right at the equator, oceanic currents carry its discharge north into the North Atlantic Ocean, so we include it in the North Atlantic basin.

Baltic Sea

The Baltic Sea region includes all drainages flowing into the Baltic (Larsson et al. 1985). Data for area, population, and discharge are from Wulff et al. (1990). Nutrient flux estimates are taken as the total contributions of municipal, industrial and riverine inputs into the sub-basins of the Baltic Sea, excluding nitrogen fixation in the Baltic Sea itself (Larsson et al. 1985).

North Sea

The North Sea region follows the boundaries defined for purposes of nutrient loading by the 1990 North Sea Conference. It includes all Norwegian drainages south of latitude 62° N, all watersheds in the United Kingdom draining into the North Sea and English Channel, and the following watersheds and areas on the continent: Elbe, Weser, Ems, Rhine, Scheldt, Yser, Nord Pas de Calais, Somme, Seine, and Normandy/Brittany. For our overall estimates for the North Sea, we sum separate estimates for population, area, freshwater discharge, and nutrient fluxes for the Norwegian drainages, the U.K. drainages, and the continental drainages. All of our estimates for the Norwegian drainages are from Holtan et al. (1992a). For the U.K. drainages, area is from the Institute of Hydrology/British Geological Survey (1993); population is estimated from this area and the average population density of the U.K. (Espenshade 1990); freshwater discharge is estimated from this area and the average discharge per area of 20 major rivers in the U.K. (Ward 1981). Estimates of regional fluxes of TN and TP for the U.K. portion of this region include river fluxes plus direct sewage and wastewater discharges (HMSO 1992, 1994). For the portion of the North Sea region in continental Europe, population is estimated from national average population densities (Espenshade 1990). Freshwater discharges, and fluxes of TN and TP are taken as the sum of estimates for individual drainages and areas: Elbe, Wester, Ems, Rhine, and Yser Rivers (Rijkswaterstaat 1992); Scheldt estuary (Billen et al. 1985, and unpublished data); Nord Pas de Calais and Somme (Chaussepied et al. 1989; extrapolation from Nord Pas de Calais to Somme based on watershed area); Seine River (Allardi & Billen 1993); Normandy and Brittany (extrapolation for entire area based on watershed areas and data from the Morlaix River; Wafar et al. 1989). Ibrekk et al. (1991) provide independent estimates for TN and TP fluxes to the North Sea region (not including fluxes from France) which are 7% less than ours for TN and 8% greater than ours for TP.

NW European coast

The NW European coast region includes 3 components: Ireland, Norwegian drainages north of Latitude 62° N (including those to the Norwegian and Barents Seas), and the westward drainages of the United Kingdom (including those to the Atlantic, the Irish Sea, and the Celtic Sea). Data and estimates for each of these components are summed to give overall regional estimates. All data and estimates for the Norwegian portion of the region are from Holtan et al. (1992a). Estimates for the U.K. portion of this region are derived as for the U.K. portion in the North Sea, using the same sources. For Ireland, drainage areas and population are national totals (Espenshade 1990), freshwater discharge data are from WRI/UNEP (1988), and fluxes of TN and TP are from the Research Unit of the Department of the Environment of Ireland (ERU 1991).

SW European coast

This region goes from the western boundary of the North Sea through the Guadalquivir drainage of Spain, including watersheds of the following rivers: Loire, Sévre, Charente, Garonne/Dordogne, Adour, Douro, Tagus, Guadiana, and Guadalquivir. Population is estimated by summing data for the individual watersheds using appropriate national average population densities (Espenshade 1990); for drainages that overlap national boundaries, we use an approximate area-weighted average population density. Freshwater discharge and regional fluxes of TN are estimated by summing estimates for the individual river basins: Loire and Sévre (Agence de l'eau Loire-Bretagne 1989); Charente (Zhong Cai (1991); Garonne (Probst 1985); Adour (extrapolated from Garonne based on watershed areas); Tagus and Guadiana (Kempe et al 1991); Douro, Mino, and Guadalquivir (extrapolated from the Tagus and Guadiana using watershed areas and mean areal fluxes). The regional flux for TP is also estimated by summing values for each river basin: Loire and Sévre (Agence de l'eau Loire-Bretagne 1989); Charente, Garonne, and Adour (extrapolated from the Loire and Sévre using watershed areas and mean areal fluxes); Tagus (Kempe et al. 1991); Guadiana, Douro, and Guadalquivir (extrapolated from the Tagus using watershed areas).

NW Africa

The northwestern Africa region includes the drainages from Morocco south to the Zaire River basin, but not including the Zaire (which flows into the South Atlantic). We follow Robertson & Rosswall (1986) and ignore the area of the region in the western Sahara since there is virtually no riverine export there. Drainage area, population, and freshwater discharge are estimated by summing data for the following rivers: Bandama, Cassamance, Cavally, Cestos, Cross, Gambia, Kouilou, Mono, Niger, Ntem, Nyong, Ogooue, St. Paul, Sanaga, Sassandra, Senegal, and Volta (UNESCO 1992). Regional flux of TN is taken from Robertson & Rosswall (1986). The TP flux is estimated from the TN flux by assuming a molar N:P ratio of 38:1 as reported for the Gambia River (Lesack et al. 1984).

Methodological concerns and limitations with river flux estimates

The problems associated with assembling data to construct large-scale estimates of nutrient fluxes have been recognized by previous investigators (Meybeck 1988). We briefly mention a few here to alert the reader to the limitations of our analysis. First, we estimate only annual mean fluxes and ignore inter-annual variability. The nature of such variability for nitrogen fluxes are discussed by Alexander et al. (in press). Second, we have mixed two general approaches for estimating regional freshwater discharge: estimates based on summing riverine flows, and estimates based on regional water balances. The approach of summing individual river flows will tend to underestimate regional discharge by the amount of flow in streams and small rivers which are too small to be reported. Mixing the two estimates of regional flow as we have done is inconsistent, but unavoidable where large data gaps exist. Third, there are obvious errors associated with extrapolation of nutrient fluxes to areas where primary data are not available. We have used a variety of different approaches for extrapolation, such as sometimes using systems which have similar geology, climate, and land use, and other times extrapolating TN fluxes from inorganic nitrogen and organic carbon fluxes. We have used our best judgment in each case, but obviously our use of different approaches introduces some additional inconsistencies. Fourth, where direct estimates of riverine flux of TN and TP are available, they are often based on calculations using monthly data (or sometimes even mean annual values). These values should be interpreted with caution since the concentration of suspended materials is generally correlated with discharge (Muller & Forstner 1968; Lewis & Saunders 1989), leading to underestimates of fluxes calculated with mean values of discharge and concentration. The solution to avoiding biased estimates of nutrient fluxes is to use flow-weighted "average" concentrations, but these are often not available for the systems of interest.

Estimation of anthropogenic nitrogen inputs to regions

We consider four major anthropogenic fluxes into and out of regions: the application of nitrogenous fertilizers, fixation by leguminous crops, the net import or export of nitrogen in agricultural products to or from each region, and atmospheric deposition of nitrogen from pollution sources other than those related to agriculture. Sewage and animal wastes are not considered as inputs to regions as they do not represent newly fixed or imported nitrogen, but rather a redistribution or recycling of nitrogen within a region.

Estimates of fertilizer inputs to basins in Europe, Africa and Latin America are derived from Matthews (1994). The inputs are estimated by taking the reported values per country multiplied by the fraction of that country in each watershed. In some instances, the $1° \times 1°$ map of fertilizer application shown in Matthews (1994) allowed us to determine if all or none of the fertilizer applied to the country occurred within the watershed boundaries, but we did not attempt any finer scaling of per country values. Fertilizer estimates for North American regions are taken from Jordan and Weller (in press), as it is not possible to partition the single U.S. value given in Matthews (1994) among the watershed regions.

We estimate crop nitrogen fixation only for the regions within North America and Europe, partly due to data constraints and partly because these regions are by far the most altered by human activity. Although all types of agricultural land – including pastures – exhibit some amount of nitrogen fixation, we restrict our estimates to five main classes of leguminous crops: i) soybeans, ii) other beans (including broad and dry beans), iii) peas and lentils, iv) peanuts, and v) alfalfa, vetch and lupine. Fixation rates for each of these crop classes were estimated from reviews by Keeney (1979), Stevenson (1982), and Messer & Brezonik (1983). Total fixation inputs for each region are derived by multiplying the rate of fixation for each crop type by the crop area within the watershed. Crop areas for Europe and Canada are taken from 1985 FAO statistics; those for regions within the U.S. come from the 1993 Bureau of Census. As in the fertilizer estimates, the values for Europe are reported on a per country basis, and therefore are scaled for each country by the fraction of country area contained within the watershed boundaries. We do not include nitrogen fixation by non-leguminous crops or pastures largely because of highly variable data on fixation rates (particularly in pastures and rangeland), but also because the range in fixation rates reported for many of these systems (Woodmansee 1978; Messer & Brezonik 1983; Barry et al. 1993) is comparable to that for natural temperate ecosystems (Bormann et al. 1977; Henderson et al. 1978; Boring et al. 1988; Schlesinger 1991). Thus, fixation in such areas is unlikely to be a large source of fixed nitrogen above background natural rates.

Estimates for the net import or export of nitrogen in agricultural products (food and feed) for U.S. portions of the regions are taken from Jordan and Weller (in press). Those for Canada and Europe are estimated by using 1985 FAO statistics on per country imports and exports of the following agricultural categories: i) cereals, ii) leguminous crops, iii) meat, and iv) milk. Average nitrogen contents for each category are taken from Blanck (1955), and per country values are then scaled by watershed areas as above. As for the crop nitrogen fixation estimates, values for net food and feed nitrogen fluxes are derived for only the North American and European watersheds.

The values for deposition of atmospheric nitrogen to each region are described in detail by the atmospheric working group (Prospero et al., this volume). Briefly, Prospero et al. present modeled estimates of wet and dry deposition of oxidized (NO_y) and reduced (NH_x) forms of nitrogen. Those for oxidized nitrogen are based on the GCTM model (Levy & Moxim 1989; Galloway et al. 1994; Levy et al., ms. in prep.), which produces estimates of deposition at a 2.5° × 3.25° latitude-longitude resolution. The NH_x portion is based on a model described by Dentener and Crutzen (in press), which estimates emissions, transport and deposition of ammonia species at 10° × 10° resolution.

Units for presentation

In most of this manuscript, nutrient fluxes are presented in weight units, as is typically done for terrestrial studies. We also provide summaries of overall nutrient inputs to the North Atlantic for each region in molar units to better interface with the oceanographic and coastal working groups of the Block Island workshop. Fluxes per area are given per km^2; for readers more accustomed to hectares, 100 kg km^{-2} = 1 kg $hectare^{-1}$.

Characterization of the regions

Characteristics of the regions bordering the North Atlantic Ocean such as drainage area, population, and freshwater discharge are summarized in Table 1 and Figure 1. Overall, these regions make up 18% of the surface area of all land on earth and have 15% of the world's population. Population density in the regions averages 29 individuals km^{-2}, or 20% less than the world average of 36.5 individuals km^{-2} (Wright 1993). Freshwater discharge from these North Atlantic regions contributes 30 to 33% of the world's total freshwater discharge into the oceans (42 to 46 × 10^{12} m^3 yr^{-1}; Schlesinger 1991; Maidment 1993), with half of this coming from the Amazon (Table 1).

The North Atlantic regions vary greatly in discharge and population density. The Amazon makes up 24% of the total drainage area feeding into the North Atlantic and 50% of the freshwater discharge but has only 1% of the population (Table 1). At the other extreme, the drainages flowing into the North Sea comprise only 3% of the total drainage area and contribute 3% of the freshwater discharge but have 20% of the population. The regions vary in population density from 1.5 person km^{-2} in the Amazon to 186 persons km^{-2} in the North Sea region (Figure 1a). Freshwater discharge per area of watershed varies from 0.12 m^3 m^{-2} yr^{-1} in northwestern Africa to 1.1 m^3 m^{-2} yr^{-1} for northwestern European coast (Figure 1b).

Estimated fluxes of N & P

We estimate nitrogen and phosphorus inputs to the North Atlantic Ocean from terrestrial sources to be approximately 13 Tg N yr^{-1} (935 Gmoles N yr^{-1}) and 2.3 Tg P yr^{-1} (73 Gmoles P yr^{-1}; Table 2). Our estimate does not include inputs into the Mediterranean Sea, which must be quite large: net export of nitrogen from the Mediterranean to the North Atlantic Ocean proper is estimated to be approximately 2.8 Tg N yr^{-1} (200 Gmoles yr^{-1}; Michaels et al., this volume). For nitrogen, our estimate of 13 Tg yr^{-1} is 6.5% of the estimate of the report by the Joint Group of Experts on the Scientific Aspects of Marine Pollution (GESAMP 1987) for riverine input of nitrogen to all the world's oceans, is 32% of total world riverine flux estimate of Galloway et al. (1995), but is 33 to 130% of the total world riverine flux estimated by Smil (1990). The Smil (1990) estimate for the world must clearly be wrong, if one accepts our estimates for the North Atlantic. Our analysis may also suggest that the GESAMP (1987) estimate for the world is high, since the North Atlantic regions have 18% of the world's land surface area, 15% of the world's population, and 30% of the world's freshwater discharge and yet contribute only 6.5% of the GESAMP (1987) estimate for world nitrogen flows into oceans. Nitrogen flux on land (defined as nitrogen use in fertilizer plus livestock and human waste per area) in the North Atlantic regions is typical of that the world as a whole (Nixon 1995). These considerations lead us to favor the estimate of Galloway et al. (1995) as the best currently available for global riverine nitrogen flux to the oceans. Our estimate of P inputs to the North Atlantic from terrestrial sources is 10.5% of the estimated total riverine input of phosphorus to the world's oceans, which is perhaps logical since that global estimate is strongly dominated by the high erosional input of material to the oceans from Asia and Oceania (Howarth et al. 1995).

The Amazon basin dominates the riverine phosphorus inputs to the North Atlantic, making up 68% (Table 2). The Amazon is also the single largest source of nitrogen to the North Atlantic, comprising 25% of the total riverine input (Table 2), although most of the other regions also contribute significantly to nitrogen inputs. In addition to the Amazon, the Mississippi basin, Caribbean and Central American region, the North Sea, and northwestern Africa all contribute more than 1 Tg N yr^{-1} to the North Atlantic (Table 2).

When expressed per area of watershed, nutrient fluxes from the various regions vary markedly (Figure 2a, 2b). The highest fluxes per area for nitrogen are in the watersheds surrounding the North Sea (1,450 kg N km^{-2} yr^{-1}) with high fluxes also along the northwest coast of Europe (1,300 kg N km^{-2} yr^{-1}) and the northeast coast of the United States (1,070 kg N km^{-2} yr^{-1}). The lowest fluxes are from the northern Canadian rivers (76 kg N km^{-2} yr^{-1}). The average nitrogen flux per area for all of the regions flowing into the North Atlantic Ocean is 486 kg N km^{-2} yr^{-1}. Surprisingly, the highest phosphorus flux per area is in the Amazon basin (236 kg P km^{-2} yr^{-1}), followed by the northeast coast of the U.S. (139 kg P km^{-2} yr^{-1}), the watersheds of the North Sea (117 kg P km^{-2} yr^{-1}), and the southwestern European coast (101 kg P

Table 2. Summary of regional data on inputs of N and P to the North Atlantic from land.

	TN		TP		N:P
	Tg yr^{-1}	(Gmol yr^{-1})	Tg yr^{-1}	(Gmol yr^{-1})	(molar)
North Canadian rivers	0.30	(22)	0.018	(0.57)	37
St. Lawrence basin	0.66	(47)	0.02	(0.63)	73
NE coast of US	0.51	(37)	0.067	(2.1)	17
SE coast of US	0.24	(17)	0.011	(0.35)	47
Eastern Gulf of Mexico	0.23	(17)	0.011	(0.35)	46
Mississippi basin	1.82	(130)	0.107	(3.4)	38
Western Gulf of Mexico	0.85	(61)	0.007	(0.22)	271
North America – total	4.61	(331)	0.241	(7.65)	42
Caribbean Islands and Central America (incl. Orinoco)	1.09	(78)	0.141	(4.5)	17
Amazon & Tocantins basin	3.28	(234)	1.53	(49.)	4.8
Central & South America – total	4.37	(312)	1.67	(53.)	5.8
Baltic Sea	0.74	(53)	0.072	(2.3)	23
North Sea	1.22	(87)	0.099	(3.1)	28
NW European coast	0.44	(32)	0.028	(0.89)	36
SW European coast	0.20	(14)	0.054	(1.7)	8.3
Western Europe – Total*	2.60	(186)	0.253	(8.0)	23
NW Africa	1.48	(106)	0.087	(2.8)	38
Total*	13.1	(935)	2.25	(71.)	13

* not including Mediterranean Sea.

km^{-2} yr^{-1}). The lowest phosphorus fluxes per area are from the northern Canadian rivers (4.5 kg P km^{-2} yr^{-1}) and the western Gulf of Mexico region (5 kg P km^{-2} yr^{-1}; Figure 2b). The high phosphorus flux from the Amazon basin is mostly particulate, and is likely a consequence of high rates of erosion and weathering in the geologically young Andean headwaters (Stallard 1985; Devol et al. 1991). Much of this phosphorus flux from the Amazon may consist of apatite, which is insoluble in seawater and therefore considered to be biologically unavailable upon reaching the ocean (Howarth et al. 1995).

N:P ratio of nutrient inputs

The riverine nutrient inputs vary tremendously in their N:P molar ratio among the regions of the North Atlantic basin (Table 2). The highest ratio is found in

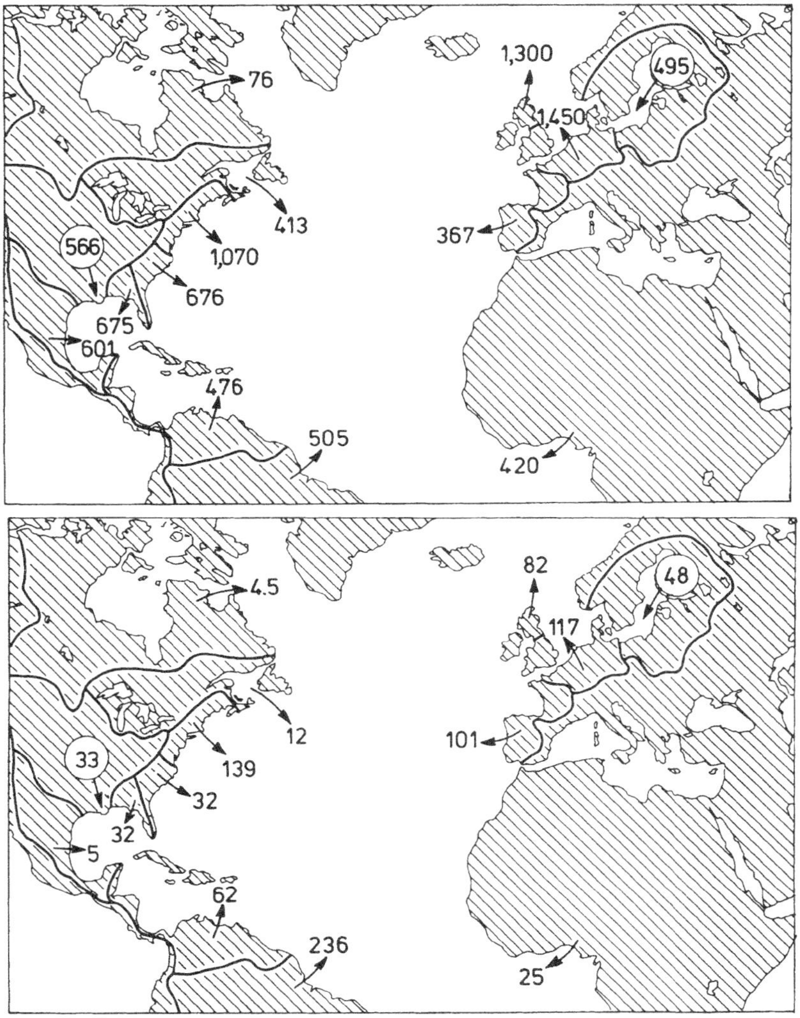

Figure 2. Total nitrogen (TN) export in rivers in kg N km^{-2} yr^{-1} (top, a) and total phosphorus (TP) export in rivers in kg P km^{-2} yr^{-1} (bottom, b). See text for explanation of estimates.

the western Gulf of Mexico (271:1), while the lowest value is in the Amazon basin (4.8:1). The average for the North Atlantic Ocean as a whole (excluding inputs from the Mediterranean Sea) is 13:1 (Table 2). Billen et al. (1991) noted that the N:P ratio of nutrient exports from watersheds is influenced by land use and disturbance, with the ratio for watersheds tending to decrease as total nutrient load increases in response to disturbance. In their analysis, nutrient exports from forested watersheds typically had N:P ratios well above 200:1, exports from agricultural watersheds had ratios between 30:1 and 300:1, and urbanized watersheds had ratios between 5:1 and 40:1. Given the fairly low N:P ratios in Table 2, this analysis might suggest that virtually all of the regions draining into the North Atlantic Ocean are quite disturbed. We find this unlikely, however; note the low population density in the north Canadian rivers region and Amazon basin (Figure 1a), two regions with low to extremely low N:P ratios. We conclude that at the scale of large regions, the N:P ratio of riverine nutrient fluxes is not a good indicator of disturbance. In part, the high N:P ratio noted by Billen et al. (1991) for forested watersheds may reflect high atmospheric deposition onto these forests: two of the three forests they considered are in areas of high atmospheric deposition of nitrogen. As is discussed further below, the nitrogen export (and therefore N:P ratio) from truly undisturbed forests is likely to be much less than observed in forests receiving high rates of nitrogen deposition.

The N:P ratio of nutrient inputs is one important factor influencing whether primary production in estuaries and coastal waters is more likely to be limited by nitrogen or phosphorus (Howarth 1988; Billen et al. 1991; Howarth et al. 1995; Justic et al. 1995). Phytoplankton require nitrogen and phosphorus in the approximate molar ratio of 16:1, the "Redfield" ratio. Consequently, unless there are biogeochemical processes within an ecosystem which alter the N:P ratio (such as nitrogen fixation, denitrification, and phosphorus sorption by sediments), net primary production in ecosystems receiving nutrient loads with an N:P ratio for available nutrients above 16:1 will be phosphorus limited while those receiving loads with an N:P ratio below 16:1 will be nitrogen limited. The N:P ratio for land-based nutrient inputs from most of the regions surrounding the North Atlantic Ocean are within a factor of 2 or 3 of the Redfield ratio. This suggests that estuaries in most of these regions could well be either nitrogen or phosphorus limited, depending upon how biogeochemical processes within the estuary alter the N:P ratio. An important qualification of this argument must be made: not all nutrient fluxes are biologically available, and the N:P ratio for TN and TP fluxes (as reported in this paper) can vary significantly from the N:P ratio for available nitrogen and phosphorus (which is more determining of nutrient limitation). For nitrogen, very little is known about the fate of organic nitrogen inputs to estuaries and coastal oceans; a critical question is to determine how much is biologically available, and how much is simply buried. For phosphorus, Howarth et al. (1995) estimated that only 25% to 70% of the phosphorus inputs to the world's ocean are biogeochemically active, and the rest is biologically unavailable. When human activity increases phosphorus fluxes, the percentage availability of the phosphorus probably also increases (Meybeck 1982; Howarth et al. 1995).

Nutrient fluxes on the continental shelves off most of the North Atlantic regions are dominated by cross-shelf transport from the deep ocean (Michaels et al., this volume; Nixon et al., this volume). Thus, the N:P ratio of riverine nutrient fluxes probably has little influence on whether primary production on much of the shelf area is nitrogen or phosphorus limited. However, riverine inputs can be important in some regions. For example, a large increase in the N:P ratio in discharges from the Mississippi River over the past few decades may have had substantial effects on nutrient limitation on the shelf in the Gulf of Mexico (Justic et al. 1995; Rabalais et al., in press).

Biogeochemical processes within estuaries alter the N:P ratio, with denitrification lowering the ratio and accumulation of phosphorus in sediments raising the ratio. As discussed by the coastal working group (Nixon et al., this volume), rates of denitrification tend to exceed rates of nitrogen fixation in most estuaries; on average, about half of the nitrogen inputs from rivers is consumed within estuaries. Residence time of water within the estuary appears to be the major factor regulating the extent of nitrogen loss through denitrification (Nixon et al., this volume), although there may be a slight trend towards more denitrification in more eutrophic systems. The percentage of phosphorus entering estuaries from rivers that is "permanently" buried in sediments is quite variable, ranging from 10% or less up to more than 100% (in the case of estuaries which are trapping large amounts of phosphorus from offshore; Howarth et al. 1995). Often, the percentage accumulation of phosphorus in sediment decreases as an estuary becomes more eutrophic. Since the percent phosphorus storage is often sensitive to the trophic status of the estuary (with increasing phosphorus availability in more eutrophic estuaries) while nitrogen consumed in an estuary is relatively insensitive to trophic status, biogeochemical processes may be more likely to increase the N:P ratio in pristine oligotrophic estuaries and lower the N:P ratio in more eutrophic estuaries (Howarth et al. 1995). We might expect nitrogen limitation to predominate in many of the North Atlantic regions having high total nutrient fluxes, since estuaries in these regions would tend to be mesotrophic and eutrophic. On the other hand, phosphorus limitation might be more prevalent in the regions having lower total nutrient fluxes, where estuaries are likely to have lower average rates of primary production.

Nutrient limitation in the North Atlantic Ocean as a whole is determined by the total suite of factors influencing the availability of nitrogen and phosphorus. While the riverine input of nutrients has a fairly low N:P ratio (13:1; Table 2), nutrient inputs from the atmosphere tend to have a high N:P ratio, averaging 70:1 for all of the world's oceans (Duce et al. 1991). The importance of nitrogen fixation and losses of nitrogen and phosphorus in the deep ocean may well be the most important factors regulating nitrogen and phosphorus availability (Michaels et al., this volume), with the riverine inputs being quite insignificant to the North Atlantic as a whole.

Direct sources of nitrogen to rivers – sewage and human waste

Human sewage and wastewaters are obvious sources of nitrogen to rivers. In an analysis of 42 major world rivers, Cole et al. (1993) concluded that sewage inputs alone are sufficient to account for the increased flux of nitrate observed in rivers whose watersheds have a higher population density. Although they acknowledged that deforestation, atmospheric deposition, and fertilizer application can all contribute significantly to nitrogen export from rivers, they stated that "watersheds with moderate to high human population will likely be dominated by sewage" rather than other inputs. On the other hand, agricultural sources of nitrogen and nitrogen deposition from the atmosphere are frequently cited as the major causes of increased nitrogen loading to rivers and estuaries (National Research Council 1993). Turner & Rabalais (1991) proposed that the observed increase in nitrogen flux in the Mississippi River over the past few decades is due to increased use of fertilizer in the watershed, and Fisher & Oppenheimer (1991) concluded that atmospheric deposition onto the watershed may be the single largest input to Chesapeake Bay, accounting for 37% of all inputs (Fisher & Oppenheimer 1991).

We estimate the contribution of sewage and wastewater to the exports of nitrogen from each of the regions bordering the North Atlantic Ocean from data on what percentage of the population in each region is sewered. Where available, we use data from OECD (1991). Elsewhere, we use data from WRI/UNEP (1988) on access to waste treatment facilities in urban areas. Actual nitrogen fluxes due to sewage wastewater are calculated from sewered population estimates by assuming a per capita nitrogen load in wastewaters of 3.3 kg N yr^{-1} per person (Meybeck et al. 1989).

The percent contribution of sewage and wastewaters to total nitrogen inputs varies among the regions, from a high of 34% in the North Sea region to virtually none in the Amazon basin (Table 3). The percentage contribution of sewage and wastewaters to total nitrogen flux from a region is strongly correlated with population density ($p < 0.05$; compare Table 3 and Figure 1a). Nonetheless, this analysis suggests that nitrogen sources other than sewage and wastewater inputs dominate fluxes in all regions, and quite strongly in most regions. Overall, nitrogen from sewage and wastewater inputs is only 11% of the total nitrogen inputs to the North Atlantic Ocean from land (Table 3).

Table 3. Sewage and wastewater inputs to the North Atlantic and its river basins.

	Sewage N input ($Tg\ yr^{-1}$)	% of Total N input
North Canadian rivers	0.022	7%
St. Lawrence basin	0.072	11%
NE coast of US	0.13	26%
SE coast of US	0.034	14%
Eastern Gulf of Mexico	0.049	21%
Mississippi basin	0.16	9%
Western Gulf of Mexico	0.082	10%
North America – total	0.55	12%
Caribbean Islands and Central America (incl. Orinoco)	0.140	13%
Amazon & Tocantins basin	0.0003	0.01%
Central & South America – total	0.14	3%
Baltic Sea	0.10	14%
North Sea	0.42	34%
NW European coast	0.073	16%
SW European coast	0.047	23%
Western Europe – total*	0.64	25%
NW Africa	0.096	6%
Total – all basins*	1.43	11%

* Mediterranean Sea not included.

There may be some error in our assessment of sewered populations, and human waste which does not enter a sewer system can also contribute to nitrogen fluxes from regions to the North Atlantic. To obtain a maximum estimate of human waste contribution to nitrogen inputs to the North Atlantic, we use total population values (Table 1) and an assumption that humans excrete 4.4 kg N yr^{-1} per person (Vollenweider 1968; note that this is slightly more nitrogen than the per capita load in sewage, referenced above, because of nitrogen losses in sewage treatment). Given the total population in the North Atlantic basin of 790×10^6 (Table 1), the maximum estimate for nitrogen excretion by humans in the regions surrounding the North Atlantic Ocean is 3.5 Tg N yr^{-1}. This is a little over twice our estimate for release of nitrogen in sewage and wastewaters, which seems reasonable since sewage treatment removes some nitrogen and since not all human waste is treated in wastewater treatment plants. Even if all human waste reached the North Atlantic, it cou account for at most 27% of total nitrogen inputs (Tables 2 and 3). For tl North Sea region, the most populous region surrounding the North Atlantic, we similarly estimate that total nitrogen production in human waste migh be 0.69 Tg N yr^{-1}, or roughly half of the total regional flux of nitrogen to the ocean (Tables 2 and 3). Clearly non-sewage sources of nitrogen dominate fluxes from all regions. These non-point sources include agricultural runoff, atmospheric deposition onto watersheds, and runoff from animal feedlots.

Peierls et al. (1991) noted that on a log-log plot, nitrate fluxes in the major rivers of the world are correlated with human population density (Figure 3). Total nitrogen fluxes per area for the 14 regions of the North Atlantic basin are also correlated with human population density on a log-log scale ($r^2 = 0.45$; $p = 0.01$; Figure 3). However, the slope of the relationship for total nitrogen fluxes (our data for North Atlantic regions) is less than that for nitrate fluxes in major world rivers (Peierls et al. 1991) by almost a factor of 2 (Figure 3). This suggests that disturbance associated with human population density preferentially mobilizes nitrate over other nitrogen forms. These correlations between population density and either nitrate or total nitrogen fluxes could reflect the importance of human waste as one nitrogen input to watersheds or could reflect correlations between human population density and accelerated nitrogen cycling through fertilizer use, food movements, atmospheric pollution, and land disturbance. This latter explanation is perhaps more likely since non-point sources of nitrogen dominate fluxes of nitrogen in the rivers of the North Atlantic basin.

Net nitrogen inputs to the regions from human activity

We estimate total net anthropogenic nitrogen inputs to the watersheds of the North Atlantic basin to be approximately 30 Tg N yr^{-1} (calculated from input date in Table 4 and areas in Table 1); this value includes leguminous crop fixation and net fluxes of food and feed for temperate but not for tropical regions. This represents an extremely large flux of nitrogen into these watersheds; it is 2.5 fold greater than the total river flux to the North Atlantic (Table 2), and roughly one-third of global nitrogen fixation from all natural sources (Soderlund & Rosswall 1982). Fertilizer and deposition dominate total inputs, accounting for 21.8 and 8.2 Tg N yr^{-1}, respectively. Fixation by leguminous crops (4.6 Tg N yr^{-1}) and net food and feed movement (-3.8 Tg N yr^{-1}) nearly balance at this scale.

For atmospheric inputs of fixed nitrogen to each region from human activity, we consider only deposition of oxidized nitrogen compounds (NO_y, including both wet and dry deposition), and only the increase in this deposition

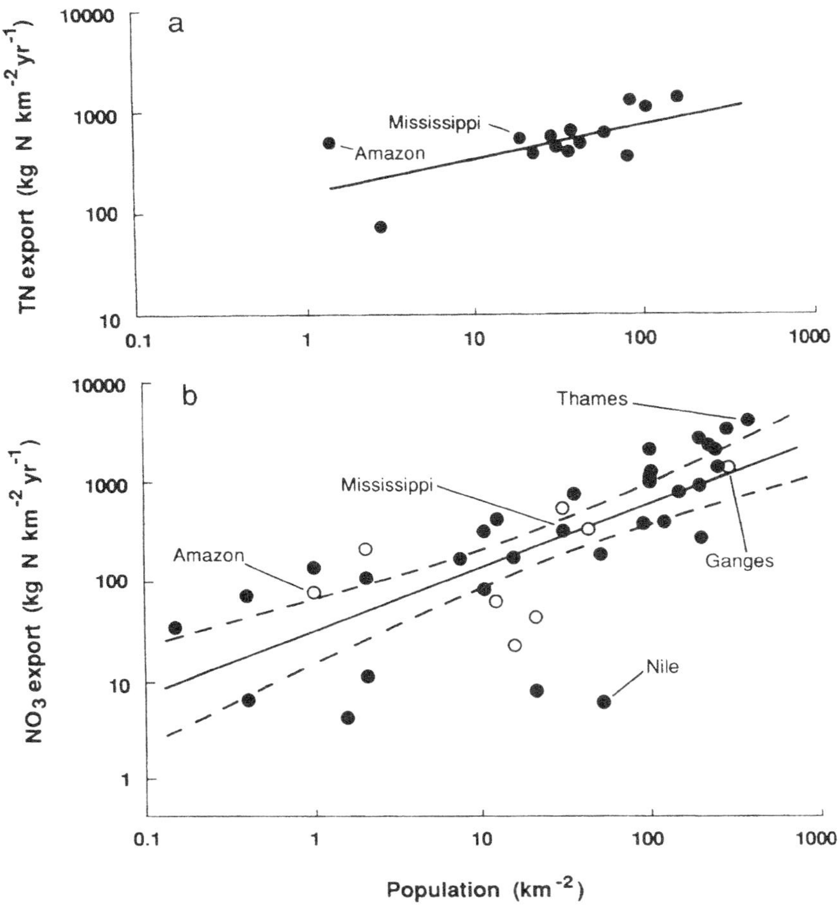

Figure 3. Relationship between log of population density and log of total nitrogen export from regions of the North Atlantic basin (top, a) and of nitrate in major world rivers (bottom, b; redrawn and reprinted by permission from Peierls et al. 1991). Both relationships are significant, but the relationship for nitrate fluxes in world's rivers (bottom) is more significant, has less scatter, and has a steeper slope. For TN fluxes in the North Atlantic basin (top), log TN = 2.2 + 0.35 log (population density); $r^2 = 0.45$; $p = 0.01$. For nitrate fluxes in world rivers (bottom), log NO_3^- = 1.15 + 0.62 log (population density); $r^2 = 0.53$; $p = 0.00001$.

due to human activity; we estimated this by subtracting modeled estimates of pre-industrial deposition from those for current levels estimated by Prospero et al. (this volume). Estimates for modern and pre-industrial deposition of both NO_y and NH_x are shown in Table 5. NH_x deposition is excluded from the estimate of atmospheric nitrogen input to a region because ammonia and ammonium do not travel far in the atmosphere before being deposited back to the ground, and because the principal sources of high levels of ammonia in the atmosphere are volatilization from fertilizer and animal waste products (Schlesinger & Hartley 1991). Both of these sources are accounted for as inputs to a region in our estimates of fertilizer, nitrogen fixation by crops, and net movement of nitrogen in feeds. Adding NH_x deposition would be a form of double accounting and would cause an over-estimate of total nitrogen inputs at the scale of regions since most of the NH_x deposition is driven by NH_x volatilization in the same region.

In contrast to NH_x, the principal sources of NO_y are probably from combustion of fossil fuels and are not otherwise accounted for as inputs of nitrogen to a region. Some NO_y may come from soil processes, such as denitrification, but we suspect that this is small compared to the fossil-fuel source. That NO_y deposition is largely independent of our other source estimates and is an additional input to each region, while NH_x deposition is related to other inputs, can be seen by regressing both NO_y and NH_x deposition against our estimates for fertilizer inputs to a region: NO_y deposition shows no relationship, whereas at the scale of these large regions, NH_x deposition and fertilizer inputs are significantly correlated ($r^2 = 0.53$; $p < 0.05$; data in Table 5 and Figure 2a). It is important to note, however, that while at the scale of an entire region NH_x deposition is not an additional new input of nitrogen, volatilization and subsequent deposition of agricultural nitrogen can cause significant redistribution of nitrogen within a region, much of which may fall on forests and other natural ecosystems and be transported into rivers. This is discussed below in the section on "Fates of Anthropogenic N in Non-Agricultural Lands."

As with the river nitrogen fluxes, there is substantial variation among the regions in the intensity of nitrogen inputs. Human alteration of the nitrogen cycle is clearly much greater in the temperate portions of the North Atlantic basin, with nearly 90% of total anthropogenic nitrogen inputs occurring in the temperate-zone North American and European watersheds (Table 4). The enormous range among the watersheds in the intensity of human disturbance becomes especially clear when inputs are expressed on a per area basis: in the Amazon basin and the Central America/Caribbean region, total net anthropogenic nitrogen inputs average 174 and 482 kg N km^{-2} yr^{-1}, while in North America they average 1,389 kg N km^{-2} yr^{-1} (Table 4). From this perspective, the European regions are the most altered; their average total net anthropogenic input of 3,913 kg N km^{-2} yr^{-1} is more than twice that of the North American watersheds and an order of magnitude greater than the tropical regions (Table 4). There is a tremendous range in total inputs among individual watershed regions even within the temperate zone, from a low in Northern Canada of 216 kg N km^{-2} yr^{-1} to a high of 7,044 kg N km^{-2} yr^{-1} in the North Sea (Table 4).

Table 4. Major net anthropogenic nitrogen inputs to the North Atlantic watersheds by region. All values are in kg N km^{-2} yr^{-1}.

	Anthropogenic NO_y deposition	N fertilizer	Leguminous crop fixation	Import (+) or Export (−) of food and feeds	Total inputs
North Canadian rivers	72.	161.	32.9	−50.2	216.
St. Lawrence basin	612.	331.	256.	−31.3	1,168.
NE coast of US	1,204.	600.	748.	998.	3,550.
SE coast of US	1,023.	1,170.	369.	454.	3,016.
Eastern Gulf of Mexico	763.	1,260.	248.	576.	2,847.
Mississippi basin	620.	1,840.	1,060.	−1,300	2,220.
Western Gulf of Mexico	318.	1,254.			1,572.
North America – total	431.	878.	397.	−317.	1,389.
Caribbean Islands and Central America	140.	342.			482.
Amazon and Tocantins basin	111.	63.			174.
Central and South America – total		118.	136.		254.
Baltic Sea	479.	1,730.	26.7	20.7	2,256.
North Sea	1,085.	5,960.	4.7	−4.2	7,044.
NW European Coast	1,087.	2,870.	55.0	−324.	3,688.
SW European Coast	462.	3,370.	14.7	−64.0	3,783.
Western Europe – total	698.	3,230.	21.9	−36.5	3,913.

Note: Total N inputs for the Western Gulf region and all of Latin America do not include estimates of crop fixation or of net food and feed movement; thus, the continental totals are based on somewhat different components for each continent.

Table 5. Atmospheric deposition of oxidized (NO_y) and reduced (NH_x) forms of nitrogen to the watershed regions. Modern values are total current deposition as reported in Prospero et al. (this volume), pre-industrial values are modeled estimates described in text, and anthropogenic values are the difference between modern and pre-industrial. All values are in kgN km^{-2} yr^{-1}.

	Modern NO_y	Pre-industrial NO_y	Anthropogenic NO_y	Modern NH_x	Pre-industrial NH_x	Anthropogenic NH_x
North Canadian rivers	84.0	11.6	72.4	70.0	28.7	41.3
St. Lawrence basin	630.	18.3	612.	266.	43.3	223.
NE coast of US	1,232.	28.0	1,204.	238.	43.7	194.
SE coast of US	1,078.	54.6	1,023.	350.	55.5	295.
Eastern Gulf of Mexico	812.	49.0	763.	280.	58.5	222.
Mississippi basin	658.	37.8	620.	266.	45.8	220.
Western Gulf of Mexico	364.	46.2	318.	224.	28.7	195.
Caribbean Islands and Central America	210.	70.0	140.	182.	83.2	98.8
Amazon and Tocantins basin	196.	85.4	111.	140.	86.7	53.3
Baltic Sea	490.	11.3	479.	560.	47.5	513.
North Sea	1,092.	7.3	1,085.	742.	57.5	684.
NW European coast	1,092.	4.8	1,087.	742.	60.1	682.
SW European coast	476.	14.0	462.	322.	53.9	268.

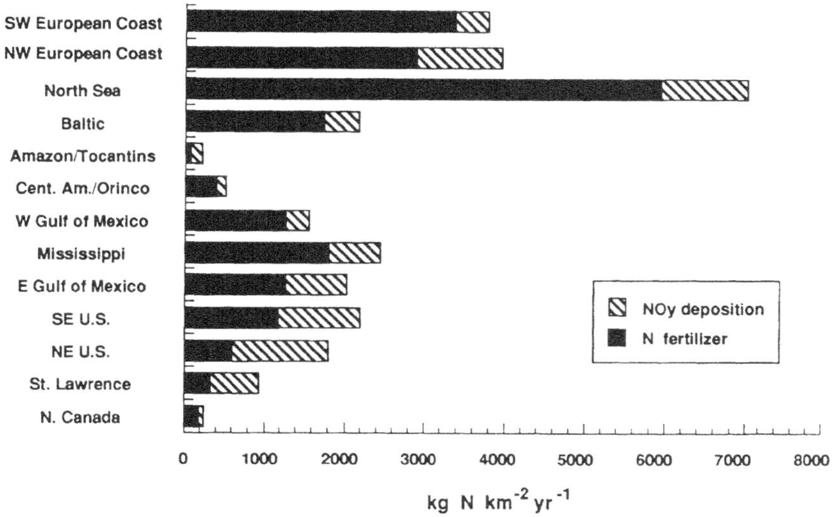

Figure 4. Application of nitrogen in fertilizers and by atmospheric deposition of NO_y to 13 of the watershed regions. Values are in kg N km^{-2} yr^{-1}. Fertilizer inputs dominate net anthropogenic nitrogen inputs in most, but not all, regions of the North Atlantic basin.

The relative importance of different anthropogenic nitrogen sources also varies substantially among the individual watersheds. Nitrogen fertilizer is the major input to most regions, although inputs through atmospheric deposition of NO_y are greater in the St. Lawrence basin and the northeast and southeast coastal regions of the United States (Table 4). The net movement of nitrogen in agricultural food stocks is important only in the northeastern United States, where it is a major input, and the Mississippi basin, where it is a major export. While never the major input, nitrogen fixation by leguminous crops is often a significant input in the North American regions but is generally quite minor in the European regions (Table 4; Isermann 1993). That agricultural sources are dominant inputs to the Mississippi basin is expected since this region drains the heartland of American agriculture. More surprising is the fertilizer dominance in Europe, despite notoriously high rates of atmospheric deposition in some areas; fertilizer inputs dominate in all of the European regions, accounting for 83% of total net anthropogenic inputs to Europe as a whole. Much of the atmospheric deposition in Europe is deposition of NH_x which originates from fertilizer and animal wastes (and we consider to be recycled nitrogen within the region). Per area rates of fertilizer application are far higher in Europe than elsewhere (Figure 4; Table 4).

Comparison of riverine fluxes with anthropogenic inputs – temperate regions

This section focuses on a comparison of net anthropogenic inputs of nitrogen and river fluxes for regions in the temperate regions only. The tropical regions within the North Atlantic basin, in our opinion, are currently much less disturbed and are likely to respond to changing inputs much differently than temperate ecosystems. Further, natural fluxes appear to be greater in the tropics. Fluxes in the tropics are discussed separately below in the section on "Nitrogen Balance in the Humid Tropics."

Fluxes of total nitrogen per area from the temperate regions of the North Atlantic basin are strongly correlated with net anthropogenic inputs of nitrogen per area to these regions ($r^2 = 0.73$; $p = 0.002$; Figure 5a). Note that Figure 5a shows the relationship between linear river fluxes and linear inputs, rather than the weaker log-log relationship between river fluxes of nitrogen and human population density shown in Figure 3. To analyze which nitrogen inputs are most related to riverine fluxes, we can separate the total anthropogenic fluxes into two categories: those related to the combustion of fossil fuels (represented by NO_y deposition) and agricultural inputs (consisting of fertilizer inputs, nitrogen fixation by crops, and the net movement of nitrogen in feeds). The resultant two variable linear model (see Figure 5b) shows that both the sum of the agricultural inputs and NO_y deposition are significant predictors of river nitrogen export ($R^2 = 0.89$, $p = 0.0005$). Given that NO_y deposition is a much smaller input than agricultural sources in most regions, and that nitrogen limitation of production is prevalent in the natural systems throughout most of these regions, we find it especially intriguing that deposition is so well correlated with river fluxes. This point is further illustrated in Figures 5c and 5d, which compare simple regressions of river nitrogen export against NO_y deposition ($r^2 = 0.79$; $p = 0.0006$) and fertilizer ($r^2 = 0.39$; $p = 0.05$), respectively. While neither leguminous fixation nor the net movement of nitrogen in feeds was a significant individual variable in any of the analyses, it is clear from the regional data that they may be important in some watersheds, such as the Mississippi (see Table 4).

The strength of these correlations seems somewhat surprising, considering the wide range in magnitude of fluxes and inherent limitations with the quality of data. Given the enormous range in nitrogen inputs being considered, it is even more surprising that the correlations suggest a *linear* relationship between net anthropogenic inputs and riverine fluxes. The capacity for processes within the landscape to provide a "sink" for excess nitrogen before it reaches the coasts seems inherently limited; therefore, one might expect to see river nitrogen fluxes increase exponentially with increasing nitrogen inputs,

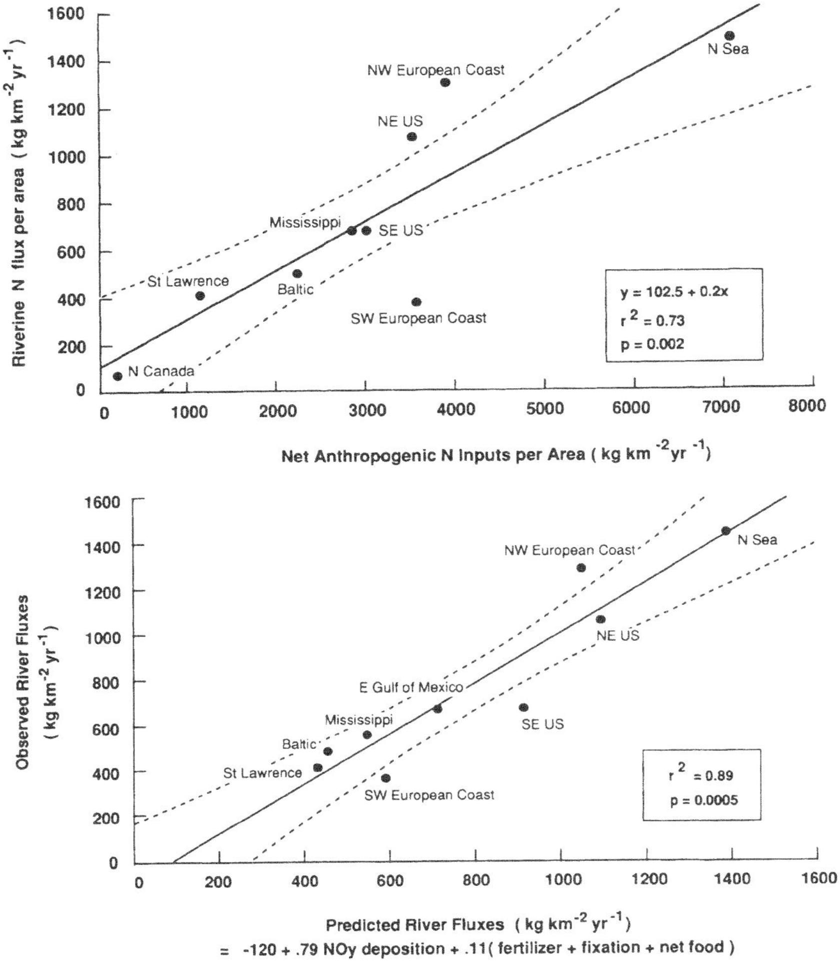

Figure 5. A. Plot of river nitrogen export versus net human-derived nitrogen inputs to each of the temperate regions (all European and North American regions, except the Western Gulf of Mexico); net inputs are equal to the sum of anthropogenic NO_y deposition, fertilizer inputs, nitrogen fixation by crops, and the net import or export of nitrogen in food and feed.

B. Modeled versus observed river nitrogen export for the same set of regions as in 5A. The model is a multiple linear regression of river nitrogen flux against "agricultural" inputs (fertilizer, plus nitrogen fixation by leguminous crops, plus net import or export of nitrogen in food and feed) and inputs from combustion of fossil fuels (anthropogenic NO_y deposition).

yet we find highly significant linear relationships across a 30-fold range in per area inputs.

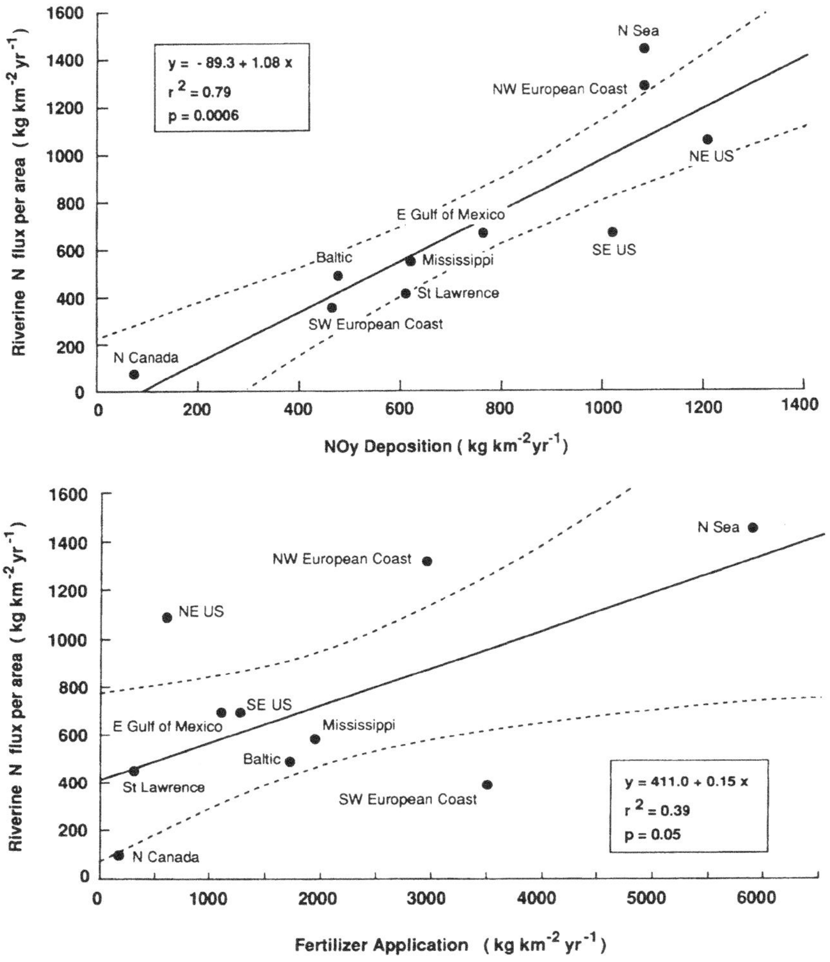

Figure 5. Continued.

C. Simple regression of river nitrogen export per area versus anthropogenic NO_y deposition for the same regions as in 5A.

D. Same as for 5C, but regressor is application of nitrogen fertilizer to each region.

The ratio of riverine nitrogen flux to net anthropogenic nitrogen inputs to the temperate regions of the North Atlantic basin is consistently low, ranging from 0.1 (southwestern European coast) to 0.35 (northwestern European coast, St. Lawrence basin, and northern Canadian rivers region) and averages 0.25 (calculated from Tables 2 and 4). That is, on average 75% of the net anthropogenic nitrogen inputs do not appear in riverine fluxes but rather must be denitrified or stored in the landscape; the total amount of nitrogen removed or stored in the landscape is even greater than this since some nitrogen is also input to each region through natural processes. Careful mass balance studies of partially forested watersheds with substantial agricultural activity have reached similar conclusions, finding some 80% of net anthropogenic nitrogen inputs stored within the watershed or denitrified and only 20% exported in rivers (Jaworski et al. 1992; Lowrance & Leonard 1988).

It is critical to identify the processes behind the apparent pattern of anthropogenic nitrogen loss upstream from the coastal estuaries, as some pathways may be relatively stable across a large range of inputs whereas others may quickly saturate. Denitrification loss as N_2 is the most benign and perhaps the least limited pathway; Neff et al. (1994) present evidence of increasing denitrification in an upland terrestrial system with increases in nitrogen deposition. However, changes in the landscape such as removal of wetland and riparian areas could reduce the total denitrification capacity of a given region (Krug 1993; Downing et al. 1993). Other nitrogen sinks include storage in groundwater and storage in organic matter in terrestrial ecosystems. Both of these might tend to saturate with time and continued high rates of nitrogen input, perhaps eventually resulting in higher rates of riverine nitrogen flux to the North Atlantic.

Fate of fertilizer N in agricultural lands

Fertilizer is the single largest input of nitrogen to the watersheds of the North Atlantic basin, and so it is important to determine the fate of nitrogen applied in fertilizer. In general, somewhat less than 50% of nitrogen fertilizer applied to fields is removed in crop harvest, and the rest is stored in the soil, is volatilized to the atmosphere, or moves into groundwater or surface waters (Bock 1984; Nelson 1985). Obviously, movement of nitrogen into surface waters is one pathway for increasing riverine fluxes, and the magnitude of this pathway is discussed below. However, nitrogen which is harvested in crops can also affect riverine fluxes at the scale of the region, as crops are fed to livestock or to humans with resulting nitrogen flows in wastes. Nitrogen volatilized into the atmosphere from agriculture lands can also affect riverine fluxes as it is deposited back onto terrestrial systems elsewhere in the region. This and the storage of nitrogen in groundwater are discussed in later sections.

A variety of factors can influence the rate of leaching of nitrogen from agricultural systems into surface waters, including water runoff, rate of fertilizer application, type and texture of soil, and the type of land use. Figure 6 summarizes a number of small watershed or lysimeter studies aimed at quantifying nitrogen losses to the hydrosystem from agricultural systems within the North

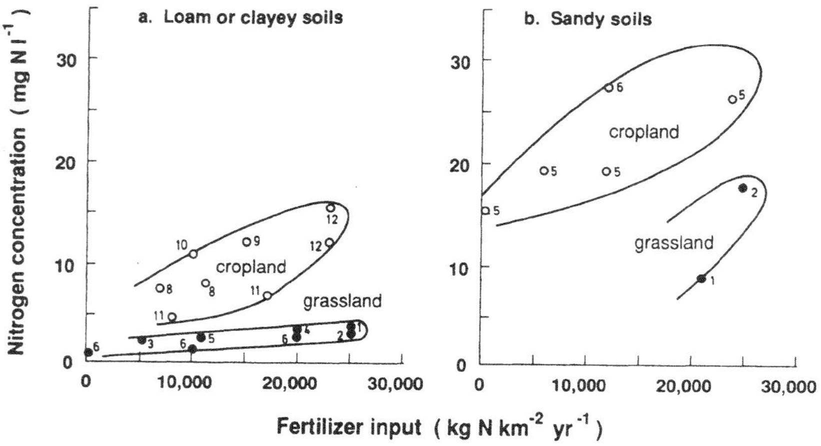

Figure 6. Mean total nitrogen concentration in water leached from agricultural watersheds of different soil type and varying rate of nitrogen fertilizer application. a) Cropland and grassland over loamy or clayey soils. b) Cropland and grassland over sandy soils.

a. Grassland and cropland over loamy and clayey soils: 1. Meadow (NL) (Jurgens-Gschwind & Jung 1979); 2. Meadow (NL) (Kolenbrander 1969); 3. Meadow (B) (De Becker et al. 1984); 4. Grassland (UK) (Webb & Walling 1985); 5. Grassland (UK) (Smith et al. 1982); 6. Meadow (Cz) (Ulehlova 1987); 7. Cropland (NL) (Kolenbrander 1973); 8. Cropland (B) (De Becker et al. 1984); 9. Cereals (F) (Mariotti 1982); 10. Cereals (F) (Belamie 1980); 11. Cereals (Cz) (Bashkin 1987); 12. Corn (England) (Gold et al. 1990); 13. Bluegrass (Ontario) (Baldwin et al., 1977); 14. Corn (Ontario) (Baldwin et al. 1977); 15. Corn (Vermont) (Baldwin et al.,1977); 16. Corn (Ohio) (Baldwin et al. 1977).

b. Grassland and cropland over sandy soil: 1. Meadows (B) (De Becker et al. 1984); 2. Meadows (NL) (Kolenbrander 1973) 3. Arable land (NL) (Sluijsmans et al. 1978); 4. Coastal farmland (S) (Fleisher et al. 1987); 5. Cereals (Sweden) (Bertilsson 1988); 6. Cereals (G) (Strebel et al. 1984); 7. Corn/oats (Iowa) (Baldwin et al. 1977).

Atlantic watershed. Runoff-weighted average concentration of total nitrogen in leachate is plotted against fertilizer inputs. Collectively, these data clearly illustrate the much higher susceptibility to nitrogen leaching of sandy versus loamy or clayey soils, and of arable land compared to grasslands. For runoff rates between 0.2 and 0.7 $m^{-3}\ m^{-2}\ yr^{-1}$, and for loamy or clayey soils, nitrogen export varies between 3 and 10% of fertilizer input for grasslands, and between 10 and 40% for arable land. On sandy soils, corresponding figures are 15 to 50% for grasslands and 25 to 80% for arable land. These percentages are misleading, however, because other sources of nitrogen than added fertilizers (such as atmospheric deposition and nitrogen fixation) contribute to the nitrogen economy of agricultural lands and because leaching losses are not necessarily proportional to total nitrogen input. Relatively high leaching rates (compared to those observed for forests or unfertilized grassland) are often observed for arable land receiving no fertilizer input, perhaps due to lowered evapotranspiration. There exists evidence that this was already the case in agricultural systems even before the advent of inorganic fertilizer use. Thus, a long-term set of observations of the water composition of the Seine River upstream from Paris (Naves et al. 1991; Meybeck 1993) shows that by the beginning of the 20th century (when fertilizer input and atmospheric deposition were negligible), nitrate concentrations were already about 100 μM, or about 5 times higher than the concentration presently found in forested watersheds in the same region (Thibert 1994), probably exposed to higher atmospheric inputs of nitrogen than in 1900.

Using the data in Figure 6, as well as the mean runoff rate, agricultural land use, fertilizer application and soil types of the different North Atlantic watershed regions, we can roughly estimate the potential magnitude of leaching of fertilizer nitrogen from agricultural lands in temperate regions. We are unable to make such a calculation for tropical regions because of a lack of data on soil leaching rates for agricultural systems in these areas. For Europe, we estimate that nitrogen leaching from agricultural soils is approximately 3.6 Tg N yr^{-1} (Table 6), or nearly one third of the total net nitrogen inputs to the region (Table 4) and 138% of our estimated total delivery to the coastal zone in rivers (Table 2), suggesting that some further removal processes occur within the river system (see below). For North America, our estimated leaching rate of agricultural soil is 2.2 Tg N yr^{-1} (Table 6), or only half the total riverine delivery of 4.6 Tg N yr^{-1} (Table 2), stressing the importance of sources of nitrogen other than leaching of fertilizer from agricultural lands. These other sources might include nitrogen applied in fertilizer, but then harvested as crops or volatilized to the atmosphere, as well as nitrogen from fossil-fuel combustion.

Fates of anthropogenic N in non-agricultural lands

Anthropogenic nitrogen reaches non-agricultural lands largely by deposition from the atmosphere, including both NO_y and NH_x (which although not considered here to be a net anthropogenic input to a region can be a significant internal flux of nitrogen to non-agricultural lands; Table 5). Forests and other non-agricultural lands have the potential to store nitrogen in vegetation and soils. Prior to human influence, most of the forest and grassland systems in our study region were probably nitrogen limited (Vitousek & Howarth 1991); thus, some amount of nitrogen inputs from the atmosphere might be tightly retained in vegetation and soils. However, as such systems are continuously exposed to elevated levels of deposition, their capacity for retention of those inputs may "saturate," and losses of nitrogen in both aqueous and gaseous

Table 6. Estimated inputs of nitrogen to surface water by leaching from agricultural soils in the temperate watershed regions of the North Atlantic basin.

	Percent crop land	Percent grass land	Nitrogen concentration in cropland runoff (μM)	Nitrogen concentration in grass land	Runoff $m^3\ m^{-2}\ yr^{-1}$	Nitrogen export Tg N yr^{-1}
St. Lawrence basin	20%	3%	290	70	500	0.66
NE coast of US	11%	4%	430	110	430	0.15
SE coast of US	13%	6%	500	140	170	0.06
Eastern Gulf of Mexico	9%	11%	500	140	300	0.09
Mississippi basin	32%	30%	360	70	170	1.04
Western Gulf of Mexico	8%	44%	360	70	200	0.24
North America	14%	15%			290	2.24
Baltic Sea	9%	1%	1,800	710	320	1.11
North Sea	33%	29%	710	140	450	1.45
NW European coast	6%	10%	1,300	710	1,110	0.76
SW European coast	40%	25%	360	140	200	0.26
Europe	20%	13%			420	3.58

Note: assumed mean concentration in leachate deduced from the data of figure 6, according to the mean fertilizer application rate and the dominant type of soil; Baltic soils are sandy, SW European soils are loamy or clayey, North Sea is one-third sandy and two-thirds loamy or clayey, and NW Europe is two-thirds sandy and one-third clayey or loamy.

forms may increase rapidly (Aber et al. 1989; Gundersen & Bashkin 1994). A number of studies have shown extremely high nitrogen leaching losses from European forests in high deposition regions (van Breemen et al. 1982; Schulze et al. 1989; Durka et al. 1994). A switch from nitrogen retention and accumulation to high nitrogen losses may also be a self-accelerating process: rising concentrations of mineral nitrogen can lead to soil acidification and parallel increases in concentrations of potentially toxic cations such as aluminum, possibly leading to ecosystem decline and even greater nitrogen losses (Skibe et al. 1989; Schulze et al. 1989; Willison et al. 1990). In addition, high nitrogen deposition is often associated with diminishing availability of other elements such as magnesium or calcium. This may limit the ability of vegetation to take up nitrogen and thereby further accelerate nitrogen losses and system decline (Schulze 1989), although there is no evidence for increasing concentrations of base cations in rivers of the northeastern U.S. (Jaworski and Howarth, ms. in prep.).

At the scale of large regions, it is very difficult to predict what fraction of nitrogen deposition onto the natural portion of the terrestrial landscape might reach the aquatic systems. In a review of European data, Hauhs et al. (1989) showed that leaching losses of nitrogen from forests tended to be either very small (<100 kg N km^{-2} yr^{-1}) or relatively large (>500 kg N km^{-2} yr^{-1}). Johnson (1992) summarized data on a number of U.S. forests in which nitrogen losses were shown to vary from a tiny fraction to more than 100% of total inputs, with no apparent relationship to the size of the inputs. However, Hedin et al. (1995) point out that many attempts to use stream export of nitrogen as an index of forest nitrogen retention have not measured organic or particulate nitrogen. In their Chilean sites, organic nitrogen is the major form of export. Thus, the lack of a correlation between inputs and losses in Johnson (1992) may in part arise from a mix of studies in which not all forms of stream nitrogen were measured. In addition, some of the variability may be due to the fact that retention in any given system may depend not only on the amount and prior history of deposition, but also on species composition, soil type, and successional stage (Vitousek & Reiners 1975). Young forests which are accumulating biomass may retain much of the available nitrogen even in areas of high inputs, whereas more mature stands can show losses of nitrogen equal to or greater than inputs even in areas of low deposition (Johnson 1992). However, even young forests can show high nitrogen losses; Lajtha et al. (1995) showed that only 40 to 62% of nitrogen inputs were retained in a young, N-limited forest on Cape Cod, Massachusetts, USA. Losses from neighboring, mature forests were similar, and the authors concluded that the sandy soils of these forests "may have an inherently low ability to retain nitrogen" at any successional stage. The seasonal timing of inputs may also

Table 7. Estimate of maximum possible retention by forests of net anthropogenic nitrogen additions in temperate-zone regions.

	Percent land forested	Increased deposition due to human activity (NO_y and NH_x) (kg N km^{-2} yr^{-1})	Percent contribution of anthropogenic NO_y and NH_x deposition to total net anthropogenic N inputs	Maximum percent of net anthropogenic N inputs to regions retained in forests
North Canadian rivers		114.	53%	
St. Lawrence basin	77%	835.	71%	55%
NE coast of US	85%	1,398.	39%	33%
SE coast of US	81%	1,318.	44%	36%
Eastern Gulfof Mexico	80%	985.	35%	28%
Mississippi basin	38%	840.	38%	14%
Western Gulf of Mexico	48%	513.	33%	16%
Baltic Sea	90%	992.	44%	40%
North Sea	38%	1,769.	25%	10%
NW European coast	84%	1,769.	48%	40%
SW European coast	35%	730.	19%	7%
Average*	70%	730.	37%	26%

* Assuming 90% of northern Canadian region is forested.

influence nitrogen losses; in the Cape Cod sites uptake and retention were high in the summer months, but losses tended to equal or even exceed inputs during the dormant season.

The nitrogen added to a region through human activity can be stored in forests only to the extent that it is deposited on forests. By assuming that all nitrogen deposited onto a forest is retained within a forest, we can estimate the maximum percentage of "anthropogenic nitrogen" which may be thus stored. In Table 7 we show such estimates for each of the regions of the North Atlantic basin that are in the temperate zone. Note that for these calculations, deposition of both NO_y and NH_x above background pristine levels are used; including the NH_x deposition accounts for the volatilization of ammonia from agricultural soils and feedlots within a region with subsequent deposition onto forests. This is recycling within the region, and not a new input of nitrogen to the region, but is critical in determining whether anthropogenically added nitrogen may be stored in forests.

The amount of the net anthropogenic nitrogen added to each region which is deposited from the atmosphere (including the redistribution of NH_x) varies from a low of 19% in southwestern Europe to 71% in the St. Lawrence watershed. By multiplying these values by the percent land area in each region which is forested, and assuming that deposition on forested land occurs at the same rate as on non-forested land, we can estimate the percentage of the net anthropogenic nitrogen added to each region which is deposited onto forests (above background levels of deposition). This is the maximum amount of the net anthropogenic nitrogen which could potentially be stored in forests, assuming that no nitrogen leaves forests. This percentage varies from only 7% in southwestern Europe to 55% in the St. Lawrence watershed (Table 7). From similar calculations for the entire land mass in the temperate zone of the North Atlantic basin, we conclude that a maximum of 26% of net anthropogenic nitrogen might be retained and stored in forests. While not trivial, this is only one third of the amount of net anthropogenic nitrogen which is not exported to the North Atlantic in rivers (see discussion in "Comparison of Riverine Fluxes with Anthropogenic Inputs – Temperate Regions," above). Clearly, the nitrogen added to the North Atlantic basin through human activity is not only stored in forests or exported in rivers; a large amount of nitrogen which never contacts forests must also be stored or denitrified elsewhere in the landscape.

Storage of nitrogen in groundwater

In assessing the fate of increased nitrogen inputs to the terrestrial systems of the North Atlantic drainage basin, the increased storage of nitrogen in groundwater must be considered. A rapid increase in nitrate concentrations

Table 8. Rate of increase of nitrate concentration in groundwaters during the 1980's in different areas in Northwestern Europe and North America. Some data from rivers during periods of baseflow are also included; these are marked by *.

	Aquifer Characteristic	Nitrate increase ($\mu M\ yr^{-1}$)	Reference
Europe			
Jutland (DK)	Sand	6.2	Overgaard 1984
Limburg (NL)	Sand	36.	Strebel et al. 1989
Bayern (G)	Sand/gravel	24.	Resch 1991
Wallongy (B)	Sand	26.	Fond. Roi Baudouin 1992
"	Chalk	24.	"
"	Limestone	10.	"
Champagne	Chalk	34.	Strebel et al. 1989
Seine River basin*		5.7	Naves et al. 1991
North America			
Mississippi River basin*		5.0	Turner and Rabalais 1991
South Wisconsin	Sandstone	7.9	Mason et al. 1990
Nebraska	Alluvial	36.	Schepers et al. 1983

in groundwater has become a major concern in most areas with intensive agriculture in both Europe and North America (Table 8). Schroder (1985) observed that the rate of increase in nitrate contamination of groundwater in Denmark is approximately the same as found in the streams. Although this might not be true in all situations, some data on the rate of increase in river nitrate concentration, measured at base flow, are included in Table 8 for areas where groundwater data were not available. The data presented by Strebel et al. (1989) suggest that in some cases, as for the chalk aquifer of eastern central England, a major increase in nitrate contamination occurred in the early 1970's but is now stabilized. This also seems to be the case in the eastern part of the United States (Smith et al. 1993). The data of Table 8 show a maximum value of nitrate accumulation rate in groundwater of 36 $\mu M\ yr^{-1}$.

The extent of groundwater affected by this increase is difficult to evaluate. A crude estimate of groundwater reservoirs by continents and hydrogeological zones was presented by UNESCO (1978). Taking into account the two uppermost zones, located above sea level and characterized by the most active water exchange, the average groundwater stock is estimated as 47 $\times 10^3\ m^3\ km^{-2}$ for Europe and 78 $\times 10^3\ m^3\ km^{-2}$ for North America. These values may seem low when compared with local estimates for important aquifers: in Europe, the 3 most important aquifers are those associated respectively with Permo-triassic sandstone, Cretassic chalk, and Tertiary sand

and gravels. These have stocks of water which range between 100 and 650 $\times$ 10^3 m^3 km^{-2} (Comm. European Communities 1982). An even higher stock (25×10^6 m^3 km^{-2}) is estimated for the upper Rhine alluvial aquifer. However, these highly productive aquifers cover less than 20% of the area of the continent.

Combining the estimates for the maximum nitrate concentration increase and the groundwater stock yields a maximum estimate of the rate of nitrogen storage in groundwater of intensively cultivated areas of the order of 25 kg N km^{-2} yr^{-1} for Europe and 40 kg N km^{-2} yr^{-1} for North America. This is only 1% of the net anthropogenic inputs of nitrogen to Europe and 3% for North America (Table 4). Groundwater storage may, however, be significant in some localized areas where high rates of increase in nitrate concentrations are observed in some large aquifers, as in the region of intensive agriculture in northwestern Europe. The sand aquifer of central Belgium is an example, with an estimated storage of nitrate of approximately 200 kg N km^{-2} yr^{-1}. The upper Rhine aquifer is another example.

Denitrification and storage of nitrogen in wetlands

Wetlands are important sites of denitrification and in some conditions can greatly reduce nitrogen movements through the landscape. For instance, in a study of an agricultural system in Maryland (USA), Peterjohn and Correll (1984) observed that nitrate discharged from a corn field in surface runoff and groundwater flow into a riparian forest was reduced by 90% before emerging into a stream. Unlike lakes where both accumulation of nitrogen in sediments and denitrification contribute significantly to nitrogen losses from the water (see next section on "Denitrification and Retention of N in Streams, Rivers and Lakes"), loss of nitrogen in wetlands is usually thought to be due largely to denitrification (Jansson et al. 1994; but see Brunet et al. 1994). Unfortunately, it is extremely difficult to evaluate overall nitrogen removal in wetlands at the scale of large regions, since the percentage nitrogen loss in wetlands varies with local physiographical, and hydrological conditions (Arheimer & Wittgren 1994). Also, with increasing area of wetlands in a watershed, nitrogen removal efficiency per area tends to decrease since the wetland area which first "sees" high nitrogen waters has the most impact (Arheimer & Wittgren 1994). Further, at least some wetlands seem to retain little if any nitrogen; in a study of 5 wetlands in Ontario, Devito et al. (1989) reported retentions ranging from −12% to +4% of inputs and not significantly different from zero. These wetlands transformed inorganic nitrogen into organic nitrogen, indicating the importance of measuring all nitrogen species to adequately determine nitrogen retention in wetlands. The complexity of nitrogen transformations within wetlands is further emphasized in the work of McClain et al. (1994) and Triska et al. (1994).

Strips of riparian wetlands between agricultural fields and streams can be particularly effective in removing nitrogen (Peterjohn & Correll 1984; Lowrance et al. 1984; Vought et al. 1994). However, often agricultural runoff reaches streams without passing through such buffers; for instance, even 25 years ago when fertilizer applications were relatively modest, more than half of the nitrate in surface waters in Illinois is estimated to have entered streams directly from agricultural drainage through drainage tiles (Kohl et al. 1971). Much agriculture in the North Atlantic basin occurs on former wetlands, and drainage is necessary to maintain agricultural productivity. This drainage not only facilitates nitrogen movement through drain tiles but has often removed the wetland characteristic of buffer strips between streams and agricultural fields (Vought et al. 1994). Krug (1993) notes that in southern Sweden, the conversion of the last 10–15% of land into agricultural use often has disproportionate effects on nutrient fluxes to aquatic systems since this involves removal of riparian fringes. He attributes up to half of nitrogen fluxes in streams to removal of these fringing wetlands.

Restoration and creation of wetlands offer mechanisms for potentially lowering nitrogen fluxes. Rosenberg et al. (1991) argue that wetland restoration, where feasible, is by far the cheapest alternative to lowering nitrogen fluxes in rivers. Haycock et al. (1993) urge the use of riparian wetland buffers along streams, particularly first and second order streams, as the most effective means of lowering nitrogen fluxes through the landscape.

Denitrification and retention of N in streams, rivers and lakes

Once in aquatic systems, nitrogen can be further lost through denitrification or storage in the sediments of lakes and reservoirs rather than delivered to the ocean. A number of budget studies which provide an estimate of nitrogen retention within the aquatic ecosystem portion of drainage basins are summarized in Table 9. These budgets consider edge-of-field exports of nutrient and point discharges as inputs to the river system. They therefore do not take into account nutrient retention by riparian buffer zones, discussed in the previous sections. The budgets do, however, include retention within lakes, ponds, and impoundments which are part of the aquatic continuum of the river systems considered.

In moderately loaded river systems, in-stream and in-river processes account for losses of approximately 10 to 20% of total nitrogen inputs. Nitrogen retention seems lower in rural or forested streams with total nitrogen inputs less than 1,000 kg N km^{-2} yr^{-1} per area of watershed, while it might amount to nearly 50% in heavily polluted rivers like the Scheldt (Table 9). Processes responsible for the observed nitrogen retention may include permanent deposition of particulate nitrogen in flood plains and impoundments, and denitrification either in the water column or in the sediments. Deposition of particulate nitrogen in the alluvial plain of large rivers during flood events or in reservoirs may be considerable: values as high as 10 to 100 g N m^{-2} yr^{-1} have been reported (Johnston et al. 1984; Billen et al. 1989). The significance of these processes for total nitrogen removal from river systems depends upon the existence and location of permanent deposition sites.

Denitrification in the water column occurs in case of severe oxygen depletion. This was and still is the case in some large, heavily polluted rivers of Western Europe, such as the Thames, the Scheldt, and the Seine Rivers, below major discharges of untreated sewage. Denitrification in such cases can sometimes eliminate all nitrate originating form upstream sources, such as leaching from agricultural soils. Paradoxically, improvement of sewage treatment can lead to increased nitrogen delivery to the ocean by rivers due to improved oxygen status and lowered rates of denitrification (Billen 1990; National Research Council 1993).

In well oxygenated water systems, which comprise the majority of rivers draining into the North Atlantic Ocean, benthic denitrification is the major process removing nitrogen. Some representative values of the observed rates of nitrogen removal by denitrification in lakes and river-bed sediments are shown in Table 10. Areal rates measured in stream beds are often higher than in lake sediments, probably because of a higher supply of organic matter in the streams. The net impact of benthic denitrification in streams versus lakes on the overall nitrogen budget of a river system depends upon a number of factors. Kelly et al. (1987) proposed a model of nitrogen removal through benthic denitrification in lakes with oxic hypolimnia based on simple mass balance calculations. Their model relates nitrogen loss through denitrification to water residence time (t), mean depth (z), and an average mass transfer coefficient (S_n, in m yr^{-1}) representing the height of the water column from which nitrate can be removed per unit time. S_n can be approximated by the ratio between mean areal denitrification rate and mean nitrate concentration in the water column, if first order kinetics are assumed for the rate of denitrification. For lakes, the data in Table 10 suggest S_n is typically in the range of 5 to 20 m yr^{-1}, which allows a prediction of nitrogen retention as shown in Figure 7. This is in very good agreement with the retention data deduced from budget studies in a number of lakes and impoundments, also shown in Figure 7, which range from 20 to 80% of inputs. To apply the model of Kelly et al. (1987) to river and streams, the data in Table 10 suggest S_n values in the range of 20 to 50 m yr^{-1}. The ratio between depth and residence time can be

Table 9. Nitrogen budgets for selected river basins. Rivers are listed in order of nitrogen inputs to the watersheds. Losses with rivers are calculated by the difference between measured inputs and exports; this includes losses within lakes, ponds, and impoundments in the watershed.

	Area (km^2)	Population density (km^{-2})	Inputs (kg N km^{-2} yr^{-1})	In-River Losses (kg N km^{-2} yr^{-1})	% N Loss in river	Reference
Viroin	554	50	650	0	0	De Becker 1986
Aare	8,096	119	735	145	19	Santschi et al. 1977
Seine	64,565	195	1,440	140	10	Billen et al. (unpubl)
Marne	12,730	91	1,710	145	8.5	Billen et al. 1994
Meuse	2,080	159	2,308	770	33	De Becker 1986
Dyle	275	279	2,690	190	7	De Becker 1986
Scheldt	16,200	426	4,490	2,045	45	Billen et al. 1985

Table 10. Rates of denitrification in the sediments of selected lakes and rivers.

	Denitrification (mmol m^{-2} hr^{-1})	Method*	Nitrate conc. (μM)	Sn (M yr^{-1})	Reference
Lakes					
Ontario lakes	0.01 to 0.025	c	7 to 20	5 to 13	Kelly et al. 1987
Lac du Der	0.3 to 0.7	a	100	10 to 25	Garnier et al. 1994
NL Reservoir	0.12 to 0.4	b	285	3.6 to 12	Admiraal et al. 1988
Bryup Langso	0.17 to 0.33	a,b	114	13 to 25	Anderssen 1977
Kvind so	0.10 to 0.25	a,b	71	12 to 30	Anderssen 1977
Rivers					
Dyle	2 to 4	a	500 to 600	30 to 70	De Becker 1986
Orneau	1	a	300	30	De Becker 1986
Meuse	0.9	a	150	52	De Becker 1986
Seine	0.5 to 3	a	150 to 300	30 to 87	Chesterikoff et al. 1992
Oise	1.6	a	210 to 285	50 to 67	Billen et al. 1989
Nottawasaga	0.2 to 0.6	a	150 to 350	5 to 35	Hill 1983
Carnagigue	0.15 to 0.27	d	200 to 650	2 to 12	Chatarpaul et al. 1980
Grachten	0.3 to 4.3	b	350	7.5 to 107	Van Kessel 1977
Swift's Brook	1.3	b			Kaushik and Robinson 1976
Gr.Ouse Trent	0.1 to 3.8	b			Owens et al. 1972
Ivel+Gade	0.3 to 4.3	b			Edwards & Rolley 1965
Potomac	0.21 to 0.24	e	100	18 to 21	Seitzinger 1987
Ontario streams	0.58	b	350	15	Wyer and Hill 1984

* Methods:
a. Net nitrate consumption measured with belljar or incubated cores;
b. Net nitrate consumption estimated from input-output balance;
c. Gross nitrate consumption measured by acetyelene blockage technique;
d. Gross nitrate consumption measured by ^{15}N technique;
e. Gross nitrate consumption measured by N_2 flux.

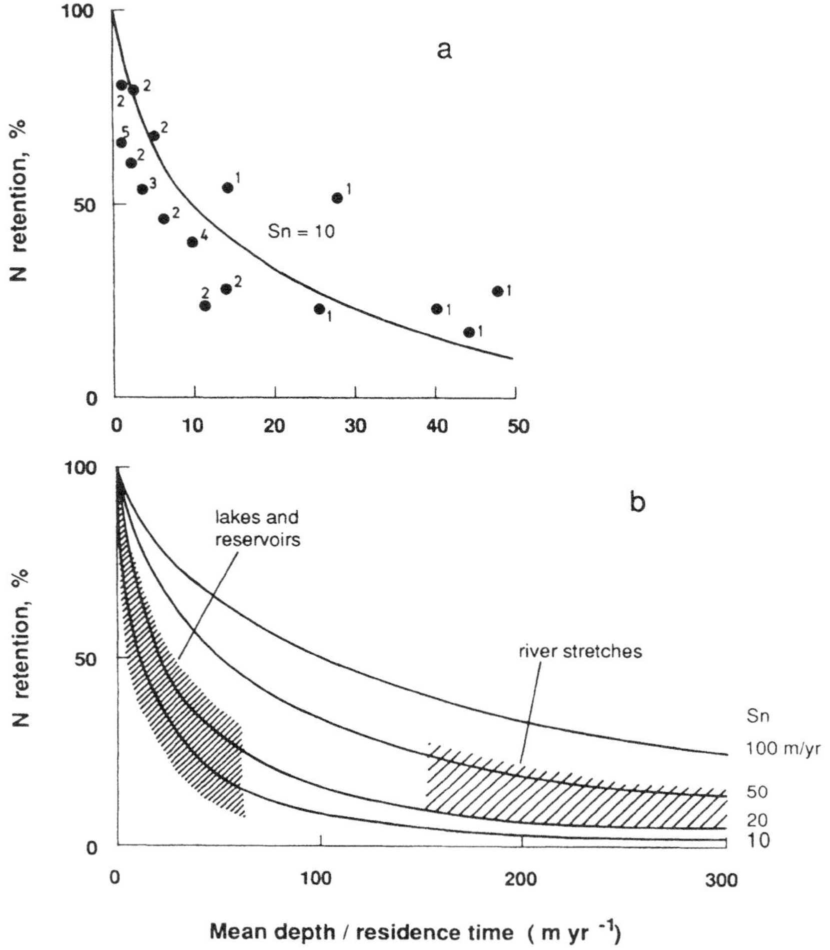

Figure 7. Kelly et al.'s (1987) model of nitrogen retention in unstratified water bodies; S_n is a mass transfer coefficient depending on benthic denitrifying activity and water column nitrate concentration; see text for model explanation.

a. Nitrogen retention expected for S_n = 10 m yr^{-1}, compared with observed data for a number of lakes; 1. Anderssen (1977); 2. Dillon & Molot (1990); 3. Garnier et al. (1994); 4. Calderoni et al. (1978); 5. Schelske (1985); 6. Galicka & Penczak (1989).

b. Nitrogen retention expected for a range of S_n values for the conditions representative of lakes and reservoirs and for streams and rivers in a typical river system of the North Atlantic basin.

calculated by stream order for a typical Atlantic river system. As depth and residence time both increase with increasing stream order, this ratio varies little and is typically in the range of 150 to 300 m yr^{-1}. We therefore predict that nitrogen retention in river channels is only 5 to 20% of inputs (Figure 7), in spite of higher denitrification rates, because of the much lower residence time characterizing river systems in comparison to lakes.

Reservoirs and lakes slow the movement of water and may, therefore, be sites of substantial nitrogen removal. For instance, reservoirs on the Nile and Colorado Rivers have average water residence times of 3.5 and 1.8 years respectively (van Der Leeden et al. 1990). However, most of the river basins feeding into the North Atlantic Ocean have few major reservoirs, and average water residence times are seldom increased by more than 0.33 years (van Der Leeden et al. 1990). Assuming that large reservoirs have average depths of at least 10–20 meters, it is highly unlikely that more than 20% of nitrogen inputs are removed in reservoirs in the regions draining into the North Atlantic (Figure 7). This is probably an overestimate, since most reservoirs in the regions draining to the North Atlantic Ocean are located on smaller order rivers relatively upstream in the watersheds. Areas very rich in lakes, such as the formerly glaciated Baltic and Northern Canada regions (Meybeck 1994, 1995), may show higher retention. Much of the nitrogen loading to rivers (particularly from agricultural and domestic sources) probably occurs downstream of most lakes and reservoirs.

Riverine fluxes in the temperate zone: Comparison to pristine conditions

What were nitrogen fluxes to the North Atlantic like prior to the widespread practice of agriculture and the industrial revolution? How can such fluxes be estimated? One approach is to assume that pristine watersheds, or the most pristine that can be found, represent the pre-agricultural and pre-industrial condition. Meybeck (1982) used this approach to estimate total riverine nitrogen flux to the world's oceans in the absence of human influence; he determined an average nitrogen export from relatively pristine watersheds of 355 kg km^{-2} yr^{-1}, corresponding to an average concentration (dissolved and particulate organic and inorganic nitrogen) of 67 μM (Table 11). Lewis (1986) compiled another set of data on "minimally disturbed" forested watersheds; he found N exports ranging from 84 to 998 kg km^{-2} yr^{-1}, corresponding to average total N concentrations of 5.9 to 82 μM (Table 11). Meybeck's (1982) reference level for pristine conditions, and at least two of the "minimally disturbed" sites compiled by Lewis (1986) – one temperate and one tropical – exhibit nitrogen exports per watershed area that are comparable to or greater than several of the regions in the North Atlantic basin (Figure 2a and Table 11).

Table 11. Concentration of total nitrogen (dissolved and particulate) and nitrogen export for "minimally disturbed" forested watersheds (after Lewis 1986, Lewis 1981, and Hedin et al., 1995).

	Total N (μM)	N export (kg km^{-2} yr^{-1})	N deposition (kg km^{-2} yr^{-1})	Old growth forest?
Venezuela	29.	998	745	Yes
West Africa	82.	126	1,910	??
Northwest Ontario (Canada)	24.	94	635	No
Oregon, USA	5.9	118	150	Yes
New Hampshire, USA	34.	401	880	No
Colorado, USA	14.	84	480	??
Chile, high mountains	11.*		<100	Yes
Chile, coastal	17.*		<100	Yes
Global average (Meybeck 1982)	67.**	355**		No

* dissolved forms only for Chilean study.
** calculated from Tables 2 & 6 of Meybeck (1982).

Several factors affect nitrogen export from forested watersheds, including the maturity of the forest (Vitousek & Reiners 1975; Emmett et al. 1993; Reynolds et al. 1994), the amount of nitrogen fixation (Binkley et al. 1992), and the amount of nitrogen deposition (Johnson 1992). Many forests which are typically considered pristine in fact have relatively high rates of nitrogen deposition from human activity, and therefore might be expected to export more nitrogen than would have been the case prior to the industrial revolution. Hedin et al. (1995) have demonstrated remarkable differences in nitrate concentrations in streams draining old-growth forests, and these differences seem related to differences in atmospheric deposition (but see Johnson 1992). Where deposition is low, as in some forests in Chile, nitrate concentrations are 3 or more orders of magnitude less than in forests in the northeastern U.S. receiving high levels of atmospheric deposition. We suggest that the best reference sites for export of nitrogen from pristine conditions would be old growth forests receiving very low rates of atmospheric deposition of nitrogen (Hedin et al. 1995). This approach is still problematic, as natural variations in the amount of nitrogen fixation and denitrification could cause nitrogen export from pristine old-growth forests to vary. Also, aggrading forests (which are more likely to retain nitrogen) occur naturally due to periodic forest fires. Further, in some parts of the North Atlantic basin, grasslands are a dominant feature of the undisturbed landscape, and may have high rates of both nitrogen fixation and denitrification.

We have several potential ways to estimate pristine conditions. The Andrews Experimental Forest in Oregon, USA, has a relatively low atmospheric deposition of nitrogen (150 kg km^{-2} yr^{-1}) and an export in stream waters estimated as 118 kg km^{-2} yr^{-1} (Table 11; Lewis 1986; Sollins et al. 1980). The northern Canadian region of our study is another relatively undisturbed region, and reflects a compilation of data that is consistent in scale and method with our other analyses. The value here is 76 kg N km^{-2} yr^{-1} (Figure 2a), and an independent study of several Canadian watersheds by Clair et al. (1994) arrived at an average export value of 79 kg N km^{-2} yr^{-1}. Estimates for fluxes in the Mackenzie and Lena rivers, two large relatively undisturbed rivers, are somewhat higher, averaging 230 kg N km^{-2} yr^{-1} (Meybeck, pers. comm.). Historical data on nitrogen export in the Connecticut River suggest a late 1800's value of 110 kg N km^{-2} yr^{-1} (Jaworski, unpublished). Finally, though not statistically significant, the y-intercept of our regression of nitrogen export in rivers against total net anthropogenic nitrogen inputs (Figure 5a) is consistent with the rest of these values at 103 kg N km^{-2} yr^{-1}.

This 3-fold variation in estimates of pristine inputs is probably very real, and in addition to spatial variation, pristine fluxes may have fluctuated both higher and lower in time. The detailed nitrogen budget for the Andrews Experimental Forest shows some of the pitfalls of assuming a uniformity among pristine sites in nitrogen inputs or outputs. For example, nitrogen fixation by epiphytic lichens at Andrews was estimated as 280 kg N km^{-2} yr^{-1} (Sollins et al. 1980), perhaps a low estimate of total nitrogen fixation as asymbiotic soil fixation on fallen logs was not measured. Thus, total inputs are similar to moderately impacted sites in low-elevation forests of the southeastern U.S. and upper New England, particularly if we add a small unmeasured dry deposition input. However, outputs of nitrogen from Andrews were still estimated to be significantly lower than total inputs (Sollins et al. 1980). Even in this old-growth site where carbon and nitrogen storage in vegetation is thought to be decreasingly slightly over time (- 280 kg N km^{-2} yr^{-1}), accumulation of nitrogen in fallen logs and in soil organic matter is probably substantial (780 kg N km^{-2} yr^{-1}), and the ecosystem is thought to be still accumulating nitrogen (Sollins et al. 1980).

Rates of nitrogen loss from eastern Canadian stands may be lower than rates from the Andrews Forest for several reasons: rates of nitrogen fixation may be lower, the stands may be aggrading more rapidly from more recent fire activity, and denitrification in lakes and rivers may be removing more nitrogen than in the Andrews system. Also, precipitation patterns certainly differ. Andrews has a very high annual precipitation and receives most of its rain during the cooler winter months when plant uptake may be low; retention of atmospherically-deposited nitrogen has been shown to be lower during the dormant season (Lajtha et al. 1995). It is reasonable to assume that at least some patches of forest in large old-growth stands have a positive net ecosystem production, either in recovering vegetation or in soil or woody detritus, but to quite varying degrees; successional history alone could cause several-fold differences in storage, and thus output rates.

Although the north Canadian region value is at the low end of the range for nitrogen export, it is interesting to note that this region has some fertilizer inputs and atmospheric deposition values that are well above pristine. On the other hand, the northern Canadian rivers region contains many lakes and has a relatively long surface water residence time, which would tend to lead to nitrogen losses through denitrification and sediment accretion in lakes; the surface water residence time on the Canadian shield is an order of magnitude larger than in most of the other regions of the North Atlantic basin (Meybeck 1994, 1995).

We therefore chose to use 76 to 230 kg N km^{-2} yr^{-1} as the range for our estimates of pristine river nitrogen flux from the temperate regions of the North Atlantic basin. This range suggests that human activity has increased nitrogen fluxes from temperate regions to the North Atlantic Ocean several fold (Table 12). In some regions, such as the northeast coast of the US and the North Sea, we estimate fluxes have increased 5 to 20 fold, while most of the other regions show 2 to 9-fold increases. In comparison, Larsson et al. (1985) using an independent approach of comparing northern Swedish river fluxes (assumed to be pristine) with other riverine inputs to the Baltic have estimated a 4-fold increase due to human influence, which is quite consistent with our analysis. We do not believe that the above range for pristine export should be used to estimate change in the tropical watersheds, for reasons discussed below in the section on "Nitrogen Balance in The Humid Tropics."

If instead of choosing this range of values as our pristine reference, we had instead used the compilation of Meybeck (1982), the natural pristine flux would be 355 kg N km^{-2} yr^{-1}, or 1.5 to 4 times larger than our pristine assumption. This would imply that many of the regions surrounding the North Atlantic have experienced relatively little change, and that in the case of Northern Canada, the rivers exported much more nitrogen to the ocean in their pristine state than they currently do. This is possible, if nitrogen sinks have increased in the watersheds or rivers, but we find it unlikely. We believe our analysis is the preferred one. We note that among rivers assumed to be relatively undisturbed by human activity, Meybeck (1982) included many which probably receive high atmospheric deposition of nitrogen. In addition, the Amazon is one of the rivers considered by Meybeck (1982) to be pristine, and as discussed below, pristine nitrogen fluxes from a tropical systems are likely to be larger, perhaps much larger, than those typical of the temperate zone.

Table 12. Estimated changes in river N export from temperate regions expressed as a ratio of current fluxes to a range of estimates of pristine fluxes. Our estimates of pristine fluxes (all in values of kg N km^{-2} yr^{-1}), ranged from 76 (the current flux from the Northern Canadian watersheds) to 230 (average value for Mackenzie and Lena Rivers). The values in the table reflect those two extremes. Other pristine estimates included an independent study of several Canadian watersheds (=79), the measured value at the Andrews Experimental Forest (=118), historical data from the Connecticut River (=110), and the y-intercept of the relationship between river N export and total anthropogenic N inputs (=103; Figure 4A). See text for full references and details.

Region	*Current river N export* "Pristine" river N export
North Canadian rivers	0.33 to 1
St. Lawrence basin	1.8 to 5.4
NE coast of US	4.7 to 14.1
SE coast of US	2.9 to 8.9
Eastern Gulf of Mexico	2.9 to 8.9
Mississippi basin	2.5 to 7.4
Western Gulf of Mexico	2.6 to 7.9
Baltic Sea	2.2 to 6.5
North Sea	6.3 to 19.1
NW European coast	5.6 to 17.1
SW European coast	1.6 to 4.8
North America	1.7 to 5.3
Europe	3.5 to 10.6

Nitrogen balance in the humid tropics

In comparison to many of the temperate North Atlantic watersheds, the Amazon and Orinoco basins appear minimally disturbed. Nitrogen deposition and fertilizer application are both relatively low in these regions (Tables 4 and 5; Figure 4), as is human population density (Figure 1a). As well, in sharp contrast to the temperate regions, these tropical regions show river nitrogen fluxes being roughly equal to the sum of anthropogenic NO_y deposition and fertilizer input (compare Figure 2a with Table 4). This might suggest a relatively pristine, steady-state nitrogen cycle, and in fact, Meybeck (1982) used the Amazon River as a baseline for estimating anthropogenic perturbations to riverine fluxes of carbon, nitrogen, and phosphorus in other watersheds. However, several lines of evidence indicate that the nitrogen cycle of the neotropics may be changing rapidly, and even the undisturbed nitrogen dynamics of many tropical systems may make them a misleading standard for measuring change in the rest of the world.

First, consider the unique aspects of the nitrogen cycle in many tropical forests. While nitrogen limitation of plant growth is widespread in temperate and boreal ecosystems (Vitousek & Howarth 1991), nitrogen is in relative abundance throughout much of the humid tropics. For example, N:P ratios in litterfall are around 11:1 in temperate and boreal forests; in the tropics they average approximately 23:1 (Vitousek 1984). This excess of available nitrogen leads to lower retention in plants and microbes, and therefore to a "leakier" terrestrial nitrogen cycle, where losses to both the atmosphere and aquatic systems may be high (Matson & Vitousek 1990). Tropical forests produce the highest reported fluxes of nitrous oxide from undisturbed systems (Keller et al. 1993). Total nitrogen fluxes within the forests of the Amazon are some 5 to 6-fold greater than those at the Andrews Forest (Bentley et al. 1982; Sollins et al. 1980; Jordan et al. 1982), an old-growth forest in the temperate zone with relatively low nitrogen deposition. Thus, even in a completely pristine world, nitrogen inputs from terrestrial ecosystems to rivers would probably be much higher in the humid tropics than in any other region, and therefore the use of the Amazon as an index of anthropogenic perturbations to temperate rivers may seriously underestimate the amount of change in the temperate region and globally.

In addition, the clearing and burning of forests in tropical regions may have a disproportionately large effect on regional nitrogen cycling. Combustion losses of nitrogen during conversion are typically around 50,000 kg km^{-2} (Ewel et al. 1981; Keller & Reiners 1994), and another 50,000 to 250,000 kg km^{-2} is lost over the following year or so (Berish 1983). Some of this "lost" nitrogen may be retained via anion exchange in deep soils (Matson et al. 1987), and some will be deposited onto nearby systems (Andreae et al. 1990), but land conversion clearly mobilizes large amounts of nitrogen. The 1980's rate of deforestation in the Amazon has been estimated at 1.5×10^4 km^2 yr^{-1} (Skole & Tucker 1993); if nitrogen losses from land conversion average 200,000 kg km^{-2} yr^{-1}, then deforestation is mobilizing approximately 3 Tg of nitrogen per year. This value is roughly equal to the annual river flux (Table 2), even though it occurs over only 0.25% of basin area. Thus, even though only about 6% of the basin has been cleared, the entire region may experience the effects.

On the other hand, the fraction of nitrogen mobilized by land conversion which eventually reaches the rivers is not clear, as both total nitrogen loss and the partitioning of those losses among various pathways can vary with climate, soil type and the methods of clearing. In one Costa Rican site, the dominant pathways for nitrogen loss appeared to be gaseous (Keller & Reiners

1994). In a neighboring region, Matson et al. (1987) showed that leaching dominated losses in the months following burning, but also showed that much of the nitrogen leached from surface soils was retained via anion exchange in deeper horizons. Such nitrate adsorption is not an important process in most temperate soils, but it can be substantial in the variable charge clays that are common to the humid tropics (Uehara & Gilman 1981), and therefore nitrogen which might appear to be lost from vegetation and surface soils might not actually enter hydrologic systems (Matson et al. 1987).

Land conversion alone is simply a mechanism for redistributing nitrogen within a region. Though it can substantially increase turnover times for otherwise recalcitrant pools of organic nitrogen, it does not provide new inputs, and therefore is unlikely to produce changes in aquatic nitrogen loading that are comparable to those brought on by the high nitrogen inputs to temperate systems. However, the subsequent urbanization or agricultural use of the converted land may lead to much higher nitrogen inputs to the region. Nitrogen deposition and fertilizer use are both increasing rapidly in equatorial latitudes (Galloway et al. 1995). Thus, the coming decades may see changes in tropical river and coastal systems similar to those already observed throughout North America and Europe.

Summary and conclusions

With currently available data, we are unable to fully account for the fate of anthropogenic nitrogen added to the temperate regions of the North Atlantic Basin. Nonetheless, some general conclusions are possible. Fertilizer and atmospheric deposition dominate total anthropogenic nitrogen inputs to the temperate regions as a whole, with fertilizer accounting for roughly two-thirds of the total. Rates of nitrogen export in rivers vary dramatically among the major regions, and are highly correlated with the amount of human-derived nitrogen applied to the landscape. Although export of N in rivers has clearly increased in many areas, all of the regions show that only a relatively small fraction of total human-derived nitrogen inputs can be accounted for in the rivers (roughly 25%). Most of the nitrogen added to regions through human activity is stored within the region or denitrified. Thus, it is critical to understand the other major controls over loss and/or storage of this anthropogenic nitrogen.

The overall nitrogen fluxes within and between the terrestrial, groundwater and river systems of the North Atlantic watershed are summarized graphically in Figure 8. Assuming that terrestrial primary producers take up some 5,000 to 15,000 kg N km^{-2} yr^{-1} in the North Atlantic basin as a whole (Duvigneaud & Denaeyer-DeSmet 1971), net external anthropogenic inputs of nitrogen represent as much as 7% to 22% of internal cycling. Fertilizer application dominates these inputs, but anthropogenic atmospheric deposition and nitrogen fixation associated with crop vegetation also contribute significantly. Nitrogen is transferred from terrestrial systems to the hydrosystem through soil leaching in both agricultural systems and forests and through direct wastewater discharge. Overall for the North Atlantic watershed, sewage and wastewater discharge only represents about 10% of total riverine delivery. Leaching from animal feedlots, which we have not quantified, may also be important. Storage of nitrogen in groundwater, while difficult to quantify, is probably at most a few percent of the rate of input of anthropogenically derived nitrogen. On the other hand, storage of nitrogen in forests may be significant, perhaps accounting for up to 26% of net anthropogenic nitrogen inputs to the temperate portions of the North Atlantic basin; the redistribution of ammonia from agricultural systems to forests through atmospheric transport may be important in the ability of forests to store this amount of nitrogen. Nonetheless, much of the net anthropogenic nitrogen input to the North Atlantic basin is not stored in forests or groundwater and does not flow to the oceans in rivers; by difference with values for the temperate areas, we calculate that on average at least 340 kg N km^{-2} yr^{-1} is probably denitrified or stored in wetlands and aquatic systems, or one third of net anthropogenic inputs to the temperate region. To the extent we have overestimated storage in forests, denitrification will be even more important. Denitrification in both wetlands and aquatic ecosystems is probably of importance.

This overall view of nitrogen cycling at the scale of the whole North Atlantic watershed masks differences among the watershed regions considered in this paper. Three extreme situations have been represented in Figures 9a, 9b, and 9c. The Amazon & Tocantins region (Figure 9a) is characterized by the lowest anthropogenic inputs of nitrogen. The relatively high nitrogen riverine delivery, with organic nitrogen being the dominant form, probably corresponds to the normal functioning of the tropical rain forest, characterized by high rates of natural nitrogen fixation, and phosphorus limitation of primary production; in a degree difficult to assess, deforestation may be impacting the Amazon riverine transport of nitrogen. Nitrogen fluxes from this region to the ocean actually exceed anthropogenic inputs of nitrogen.

The North Canadian region (Figure 9b) probably provides a better idea of what could have been the pristine nitrogen cycle in temperate regions. External anthropogenic nitrogen inputs and riverine delivery of nitrogen to the ocean both represent less than one fifth of the mean value for the overall North Atlantic when expressed on a per area basis. Export of nitrogen in rivers to the sea from this region represents only one third of the net anthropogenic nitrogen inputs, a value typical for many temperate regions in the North

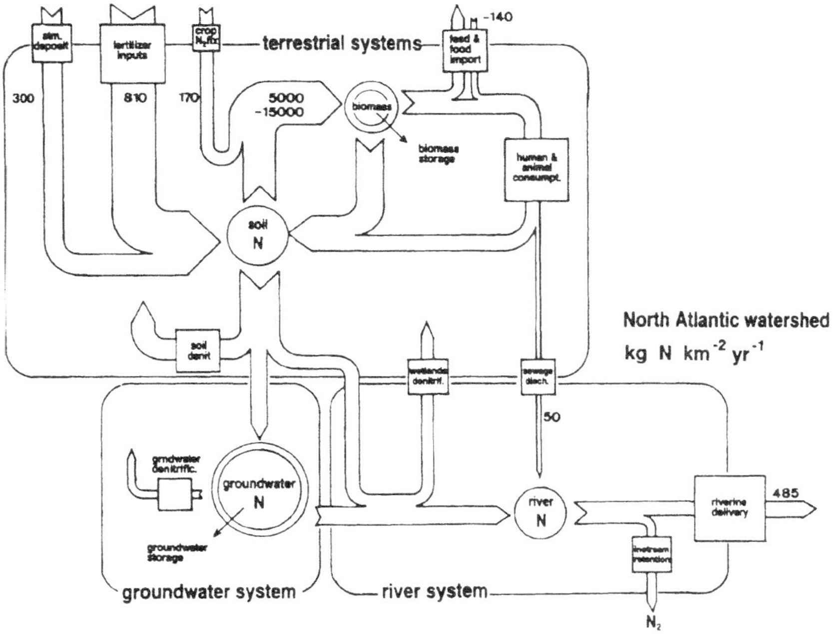

Figure 8. Schematic representation of nitrogen circulation within and between the terrestrial system, the groundwater system, and the river system of the North Atlantic basin as a whole. Values in kg N km^{-2} yr^{-1}. Width of arrows, although not strictly proportional, suggests the relative magnitude of the corresponding fluxes.

Atlantic basin. However, retention of nitrogen in aquatic ecosystems may be higher in this region than in many others because of the high residence time of surface waters (Meybeck 1994, 1995).

The North Sea watershed region (Figure 9c) shows the most perturbed situation, with a largely open nitrogen cycle. Fertilizer inputs (6,000 kg N km^{-2} yr^{-1}) completely dominate external inputs and represent more than one third of nitrogen uptake by the vegetation, resulting in an intense nitrogen soil leaching. The high population density causes an important discharge of sewage and wastewaters into surface water, roughly one third of the total riverine nitrogen flux. Although the rate of nitrogen export in rivers per area of watershed is about 20 times that of the North Canada region, it represents only 20% of total anthropogenic nitrogen inputs. The majority of nitrogen added through human activity is probably eliminated through denitrification, with a small amount stored in forests (Table 7).

The Mississippi basin and the northeastern U.S. coast also deliver respectively 7 and 14 times as much nitrogen per unit area as the North Canadian

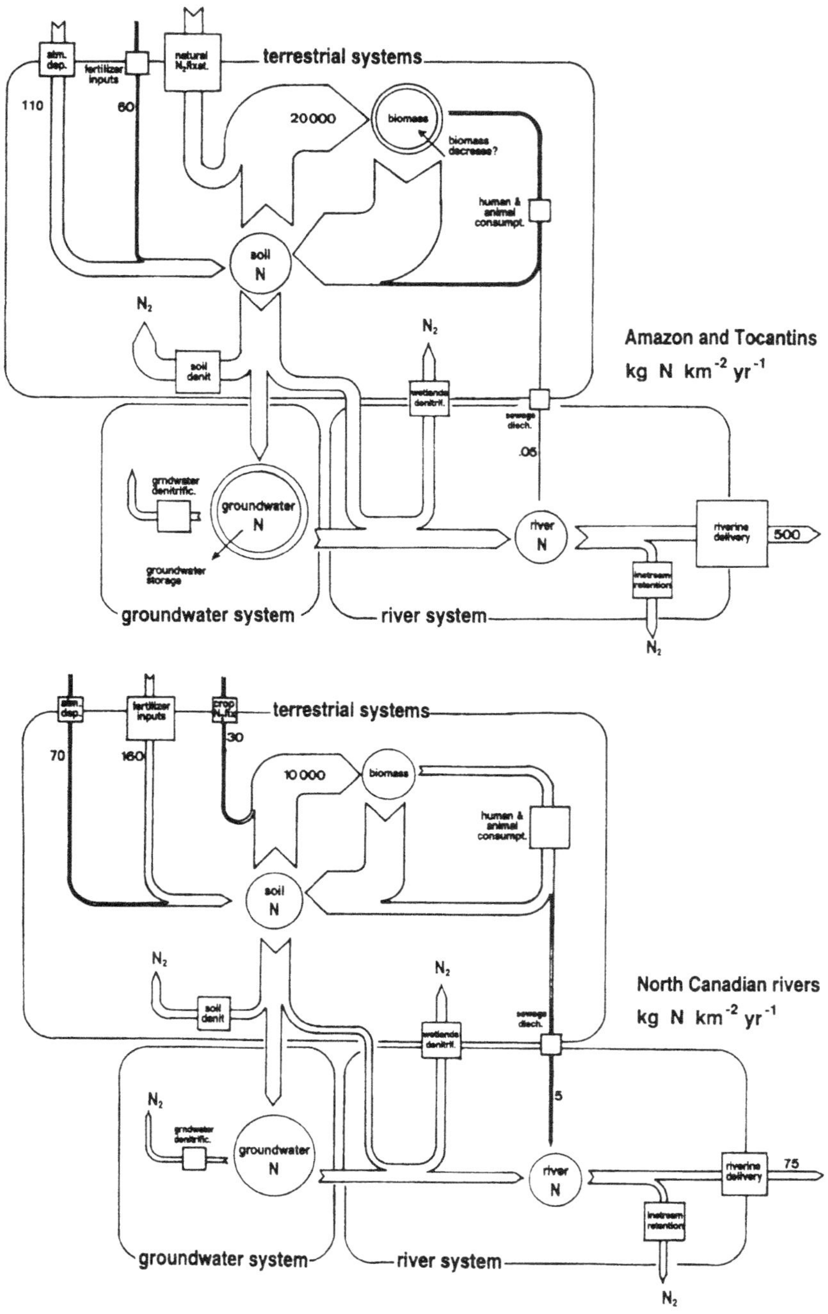

Figure 9. Schematic representation of nitrogen circulation in some watershed regions of the North Atlantic basin: a.) Amazon basin; b.) Northern Canadian rivers; c.) North Sea watershed.

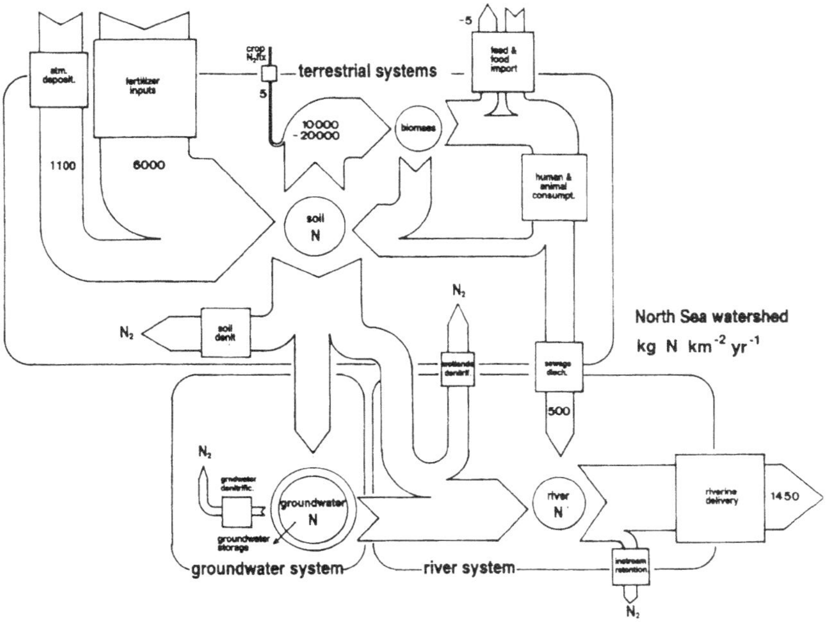

Figure 9. Continued.

region. They differ however from the North Sea basin region in the nature of the external anthropogenic inputs of nitrogen. The Mississippi basin is an intensive agricultural region, with moderate population density. Fertilizer inputs (1,840 kg N km^{-2} yr^{-1}) and nitrogen fixation by crop vegetation (1,055 kg N km^{-2} yr^{-1}) dominate the inputs, but as much as one third of these (1,300 kg N km^{-2} yr^{-1}) is exported as food and feed to other regions. The riverine nitrogen delivery to the coastal zone (565 kg N km^{-2} yr^{-1}) represents 25% of net anthropogenic inputs to the system. By contrast, the northeastern US coast region is characterized by high population density and limited agriculture, and imports of feed and food (1,000 kg N km^{-2} yr^{-1}) and anthropogenic nitrogen atmospheric deposition (1,200 kg N km^{-2} yr^{-1}) represent the two major external nitrogen sources. Inputs of nitrogen as fertilizer and crop N_2 fixation are also important, and the nitrogen delivery to the coastal sea (1,070 kg N km^{-2} yr^{-1}) is nearly twice that of the Mississippi basin.

Acknowledgments

The manuscript benefited greatly from critical reviews by C. Driscoll, L. Hedin, R. Marino, M. Meybeck, N. Rabalais, and W. Schlesinger. We thank A. Widell, F. Clarke, D. Murison, and N. Owens for help in finding sources of data on riverine loads. K. May assisted with preparation of the manuscript and tables. Preparation of parts of this manuscript by R. Howarth, D. Swaney, and A. Townsend was supported by grants from the Mellon Foundation, the Hudson River Foundation, and NOAA. The contribution by G. Billen (Research associate FNRS, Brussels) is part of the Belgian Impulse "Global Change" Programme (Science Policy Office).

References

Aber JD, Magill A, Boone R, Melillo JM, Steudler P & Bowden R (1993) Plant and soil responses to chronic nitrogen additions at the Harvard Forest, Massachusetts. Ecol. Appl. 3: 156–166

Aber JD, Nadelhoffer KJ, Steudler P & Melillo JM (1989) Nitrogen saturation in northern forest ecosystems. Biosci. 39: 378–386

Admiraal W & Van Der Vlugt JC (1988) High rates of denitrification in a storage reservoir fed with water of the river Rhine. Arch. Hydrobiol. 113: 593–605

Agence de l'Eau Loire-Bretagne (1989) Annuaire de la qualite des eaux de surface

Ahlgren L, Basso E & Jovel R (1969) Preliminary evaluation of the water balance in the Central American Isthmus. In: Laycock AH, Francisco M & Fisher T (Ed) Water Balance in North America (pp 288–298). Amer. Water Res. Assn. Urbana, IL

Alexander RB, Murdoch PS & Smith R (1996) Streamflow-induced variations in nitrate flux in tributaries to the Atlantic coastal zone. Biogeochem. In press

Allardi J & Billen G (1993) Rapport de synthese PIREN-Seine Vol 1 Fontionnnement de l'ecosysteme. Seine CNRS, Paris

Anderssen JM (1977) Rates of denitrification of undisturbed sediment for six lakes as a funciton of nitrate concentration, oxygen, and temperature. Arch. Hydrobiol. 80: 147–159

Andreae MO, Talbot RW, Berresheim H & Beecher KM (1990) Precipitation chemistry in central Amazonia. J. Geophys. Res. 95: 16987–16999

Arheimer B & Wittgren HB (1994) Modelling the effects of wetlands on regional nitrogen transport. Ambio 23: 378–386

Baldwin LB, Frere MH, Hjelmfelt AT, McGregor GS, Petri LR, Wirth TL, Storch W & Kelman S (1977) Quality aspects of agricultural runoff and drainage. J. Irrig. Drain Div. 103(IR4): 475–495

Barry DAJ, Goorahoo D & Goss MJ (1993) Estimation of nitrate concentrations in groundwater using a whole farm nitrogen budget. J. Environ. Qual. 22: 767–775

Bashkin VN (1987) Landscape agrogeochemical mass-balance of nitrogen in agricultural regions In Moldan B & Paces T (Ed) GEOMON: International Workshop on Geochemistry and Monitoring in Representative Basins (pp 105–107). Geological Survey, Prague

Belamie R (1980) Influence of cropping practices and hydrological factors on the variation of nitrates in the discharges from a representative basin in the Paris region. Proceedings of the Helsinski Symposium IAMS IAM publ 130

Bentley BJ, Herrera R, Arnason JT, Molina Buck JS, Castilleja R, Garcia LE, Jordan CF, Russell CE, Salati E & Sanhueza E (1982) Report of the working group on Latin American forests. Plant and Soil 67: 415–420

Berish CW (1983) Roots, soil, litter and nutrient changes in simple and diverse tropical successional ecosystems. Dissertation, Univ of Florida, Gainesville

Berner EK & Berner RA (1987) The Global Water Cycle. Prentice-Hall, New Jersey

Bertilsson G (1988) Lysimeter studies of nitrogen leaching and nitrogen balance as affected by agricultural practices. Acta Agric. Scand. 38: 3–11

Billen G (1990) N-budget of the major rivers discharging into the continental coastal zone of the North Sea: The nitrogen paradox. In: Lancelot C, Billen G & Barth H (Eds) Eutrophication and Algal Bloom in North Sea Zones, the Baltic and Adjacent Areas: Prediction and Assessment of Prevention Actions (pp 153–172). Water Pollution Research Reports, Commission of the European Communities

Billen G, Dessery S, Lancelot C & Meybeck M (1989) Seasonal and inter-annual variations of nitrogen diagenesis in the sediments of a recently impounded basin. Biogeochemistry 8(1): 73–100

Billen G, Garnier J & Hanset P (1994) Modelling phytoplankton development in whole drainage networks: The RIVERSTRAHLER model applied to the Seine rive system. Hydrobiol. 289: 119–137

Billen G, Lancelot C & Meybeck M (1991) N, P, and Si retention along the aquatic continuum from land to ocean. In: Matoura RFC, Martin JM & Wollast R (Ed) Ocean Margin Processes in Global Change (pp 19–44). Wiley & Sons, Chichester

Billen G, Somville M, DeBecker E & Servais P (1985) A nitrogen budget of the Scheldt hydrographical basin. Neth J. Sea Res. 19: 223–230

Binkley D, Sollins P, Bell R, Sachs D & Myrold D (1992) Biogeochemistry of adjacent conifer and alder-conifer stands. Ecol. 73: 2022–2033

Blanck FC (1955) Handbook of Food and Agriculture. Reinhold Publishing Co., New York, New York

Bock BR (1984) Efficient use of nitrogen in cropping systems. In: Hauck RD (Ed) Nitrogen in Crop Production (pp 273–294). Amer. Soc. Agron, Madison, WI

Boring LR, Swank WT, Waide JB & Henderson GS (1988) Sources, fates and impacts of nitrogen inputs to terrestrial ecosystems: review and synthesis. Biogeochem. 6: 119–159

Bormann FH, Likens GE & Melillo JM (1977) Nitrogen budget for and aggrading northern hardwood forest ecosystem. Science 196: 981–983

Bowman WD, Theodose TA & Fisk MC (1995) Physiological and production responses of plant growth forms to increases in limiting resources in alpine tundra: implications for differential community response to environmental change. Oecologia 101(2): 217–227

Brunet RC, Pinay G, Gazelle F & Roques L (1994) Role of the floodplain and riparian zone in suspended matter and nitrogen retention in the Adour River, south-west France. Regulated Rivers Research and Management 9: 55–63

Burton AJ, Pregitzer KS & MacDonald NW (1993) Foliar nutrients in sugar maple forests along a regional pollution-climate gradient. Soil Sci. Soc. Amer. 57(1993): 1619–1628

Calderoni A, Mosello R & Tartari G (1978) P, N and Si budget in Lago Mergozzo. Verh. Int. Verein. Limnol. 20: 1033–1037

Chatarpaul L, Robinson JB & Kaushik NK (1980) Effects of tubificid worms on denitrification and nitrification in stream sediment. Can. J. Fish. Aquat. Sci. 37: 656–663

Chaussepied M, Arnal O, Grossel H, Delattre JM, Wartel M (1989) Le littoral de la Region Nord-Pas de Calais, Apports a la Mer IFREMER, rapp Sci Techn No 15

Chesterikoff A, Garban B, Billen G & Poulin M (1992) Inorganic nitrogen dynamics in the River Seine downstream from Paris (France). Biogeochem. 17: 147–164

Clair TA, Pollock TL & Ehrman JM (1994) Exports of carbon and nitrogen from river basins in Canada's Atlantic provinces. Global Biogeochem. Cycles. 8: 441–450

Cole JJ, Peierls BL, Caraco NF & Pace ML (1993) Nitrogen laoding of rivers as a human-driven process. In: McDonnell MJ & Pickett STA (Eds) Humans as Components of Ecosystems: The Ecology of Subtle Human Effects and Populated Areas (pp 141–157). Springer-Verlag, New York, NY

Comite d'etude sur le fleuve Saint-Laurent (1978) Rapport d'etude sur le troncon en alval du Montmagny Vol 4 L'Editeur officiel du Quebec, Quebec
Commission of the European Communities (1982) Groundwater Resources of the European Communities. Schafer, Hanover
Crabtree RC & Bazzaz FA (1993) Seedling response of four birch species to simulated nitrogen deposition: ammonium vs nitrate. Ecol. Appl. 3: 315–321
Crippen JR (1969) Water balance patterns in the Continental United States. In: Laycock AH, Francisco M & Fisher T (Ed) Water Balance in North America (pp 55–61). American Water Resources Assn, Urbana, IL
De Becker E (1986) Apports, transferts, et transformations de l'azote dans les reseaux hydrographiques. Developpement d'une methodologie generale et applications au reseau belge. These Faculte des Sciences, Univ. Libre de Bruxelles
De Becker E, Billen G & Servais P (1984) Evaluation de la contamination des eaux de surface en nutriments (N,P, K) par drainage des sols agricoles en Belgique. Rev. Agric. 37: 117–136
Degens ET, Kempe S & Ittekot V (1984) Monitoring carbon in world rivers. Envir. 26: 29–33
del Rio F & Wilson A (1969) Streamflow distribution patterns in Mexico. In: Laycock AH, Francisco M & Fisher T (Eds) Water Balance in North America (pp 62–73). American Water Resources Assn Urbana, IL
Dentener FJ & Crutzen PJ (1996) A three dimensional model of the global ammonia cycle. J. of Atmos. Chem. In press
Devito KJ, Dillon PJ & Lazerte BD (1989) Phosphorus and nitrogen retention in five Precambrian shield wetlands. Biogeochem. 8: 185–204
Devol AH, Richey JE & Forsberg BR (1991) Phosphorus in the Amazon River mainstream: concentrations, forms, and transport to the ocean. In: Tiessen H, Lopez-Hernandez D & Salcedo IH (Eds) Phosphorus Cycles in Terrestrial and Aquatic Ecosystems Proceedings of Regional Workshop 3: South and Central America (pp 9–23). Saskatchewan Inst of Pedology, University of Sakatchewan, Saskatoon, Ca
Dillon PJ & Molot LA (1990) The role of ammonium and nitrate retention in the acidification of lakes and forested catchment. Biogeochem. 11: 23–44
Driscoll CT, Yatsko CP & Unangst FJ (1987) Longitudinal and temporal trends in the water chemistry of the North Branch of the Moose River. Biogeochem. 3: 37–61
Downing JP, Meybeck M, Orr JC, Twilley PR & Scharpenseel HW (1993) Land and water interface zones. Water, Air & Soil Pollution 70: 123–137
Duce RA, Liss PS , Merrill JT , Atlans EL, Buat-Menard P, Hicks BB, Miller JM, Prospero JM, Atimoto R, Church TM , Ellis W, Galloway JN , Hansen L, Jickells TD, Knap AH, Reinhardt KH, Schneider B, Soudine A, Tokos JJ, Tsunogai S, Wollast R & Ahou M (1991) The atmospheric input of trace species to the world ocean. Global Biogeochem. Cycles 5: 193–259
Durka W, Schulze ED, Gebauer G & Voerkelius S (1994) Effects of forest decline on uptake and leaching of deposited nitrate determined from 15N and 18O measurements. Nature 372: 765–767
Duvigneaud P & Denaeyer-DeSmet S (1971) Cycle des éléments biogenes dans les écosystemes forestiers d'Europe. In: UNESCO: Productivity of forest ecosystem, Actes Coll. Brussels. Ecol. & Cons. 4: 527–542
Edwards RW & Rolley HLJ (1965) Oxygen consumption of river muds. J. Ecol. 53(1): 1–19
Emmett BA, Reynolds B, Stevens PA, Norris DA, Hughes S, Gorres J & Lubrecht I (1993) Nitrate leaching from afforested Welsh catchments – interactions between stand age and nitrogen deposition. Ambio 22: 386–394
ERU (1991) Riverine inputs in 1990. Environmental Research Unit, St Martin's House, Waterloo Road, Dublin 4, Ireland
Espenshade E (1990) Goode's World Atlas 18th edition. Rand McNally, Chicago
Ewel JJ, Erish CB, Brown B, Price N & Raich J (1981) Slash and burn impacts on a Costa Rican wet forest site. Ecol. 62: 816–829

Fisher DC & Oppenheimer MP (1991) Atmospheric nitrogen deposition and the Chesapeake Bay estuary. Ambio 20: 102–108

Fleischer S, Hamrin S, Kindt Rydberg L & Stibe L (1987) Coastal eutrophication in Sweden: reducing nitrogen in land runoff. Ambio 16: 246–251

Fondation Roi Baudouin (1992) Nitrates et qualite des Eaus In Agriculture et Environnement

Gabric AJ & Bell PRF (1993) Review of the effects of non-point nutrient loading on coastal ecosystems. Aust. J. Mar. Freshwater Res. 44: 261–83

Galicka W & Penczak T (1989) Total N and P budgets in the lowland Sulejow Reservoir. Arch. Hydrobiol. 117: 177–190

Galloway JN, Levy II H & Kasibhatla PS (1994) Year 2020: consequences of population growth and development on deposition of oxidized nitrogen. Ambio 23: 120–123

Galloway JN, Likens GE, Keene WC, Gonzales J & Yancz C (1993) The composition of precipitation in a remote Southern Hemisphere location: Torres del Paine National Park, Chile. In: Franklin, Johnson, Behl & Iriarte (Eds) A Patagonia Gem: The Ecology and Natural History of a World Biosphere Reserve Santiago, Chile

Galloway JN, Schlesinger WH, Levy, II V, Michaels A & Schnoor JL (1995) Nitrogen fixation: anthropogenic enhancement–environmental response. Global Biogeochem. Cycles 9(2): 235–252

Garnier J, Billen G, Sanchez N, Leporcq B & Hanset P (1994) Etude du fonctionnement écologique du barrage réservoir de la Marne (Lac de Der). Bilans biogéochimiques et modelisation du fonctionnement du reservoir. Rapport PIREN-Seine, théme "Corridor Fluvial." CNRS, Paris

GESAMP (1987) Land/sea boundary flux of contaminants: contributions from rivers report and studies #32 of the IMCO/FAO/UNESCO/WHOM/WHO/IAEA/UNEP Joint Group of Experts on the Scientific Aspects of Marine Pollution UNESCO, Paris

Gold AJ, DeRagon WR, Sullivan WM & Lemunyon JL (1990) Nitrate nitrogen losses to groundwater from rural and suburban land uses. J. Soil Water Conserv. 45: 305–310

Gundersen P & Bashkin V (1994) Nitrogen cycling. In: Moldan B & Cerny J (Eds) Biogeochemistry of Small Catchments: A Tool for Environmental Research SCOPE 51 (pp 255–283). Wiley and Sons, Chichester

Hauhs M, Rost-Siebert K, Ragen G, Paces T & Vigerust B (1989) Summary of European data. In: Malanchuk JL & Nilsson J (Eds) The Role of Nitrogen in the Acidification of Soils and Surface Waters. Miljorapport 1989: 10 (NORD 1989: 92), Nordic Council of Ministers, Copenhagen

Haycock NE, Pinay G & Walker C (1993) Nitrogen retention in river corridors: european pespective. Ambio 22: 340–346

Hedin LO, Armesto JJ & Johnson AH (1995) Patterns of nutrient loss from unpolluted, old-growth temperate forests: evaluation of biogeochemical theory. Ecol. 76: 493–509

Henderson GS, Swank WT, Waide JB & Grier CC (1978) Nutrient budgets of Appalaichian and Cascade region watersheds: A comparison. Forest Sco. 24: 385–397

Henriksen A & Brakke DF (1988) Increasing contributions of nitrogen to the acidity of surface waters in Norway. Water, Air and Soil Poll. 42(1–2): 183–202

Hill AR (1983) Denitrification: Its importance in a river draining an intensively cropped watershed. Agri., Ecosys. & Envir. 10: 47–62

HMSO (1992) Digest of environmental protection and water statistics 1991 Department of the Environment. Her Majesty's Stationery Office No 15

HMSO (1994) Digest of environmental protection and water statistics 1992 Department of the Environment. Her Majesty's Stationery Office No 16

Holtan G, Berge D, Holtan H & Hopen T (1992a) Paris convention Annual report on direct and riverine inputs to Norwegian coastal waters during the year 1991 A Principles, results and discussion. Norwegian Institute for Water Research Report No 488A/92

Howarth RW (1988) Nutrient limitation of net primary production in marine ecosystems. Ann. Rev. Ecol. & Syst. 19: 89–110

Howarth RW, Jensen H, Marino R & Postma H (1995) Transport to and processing of P in near-shore and oceanic waters. In: Tiessen H (Ed) Phosphorus in the Global Environment Transfers, Cycles and Management SCOPE 54 (pp 323–345). Wiley & Sons, Chichester

Ibrekk HO, Molvaer J & Faafeng B (1991) Nutrient loading to Norwegian coastal waters and its contribution to the North Sea. Wat. Sci. Tech. 24: 239–249

Institute of Hydrology/British Geological Survey (1993) Hydrologic Data United Kingdom Hydrometric Register and Statistics 1986–90. National Environment Research Council/Institute of Hydrology, Wallingford, Oxon

Isermann K (1993) Territorial, continental, and global aspects of C,N, P and S emissions from agricultural ecosystems. In: Wollast R, Mackenzie FT & Chou L (Eds) Interactions of C, N, P, and S Biogeochemical Cycles and Global Change, NATO ASI Series I4 (pp 79–121). Springer Verlag, Berlin

Ittekot V & Zhang S (1989) Pattern of particulate nitrogen transport in world rivers. Global Biogeochem. Cycles 3(4): 383–391

Jansson M, Andersson R, Berggren H & Leonardson L (1994) Wetlands and lakes as nitrogen traps. Ambio. 23: 320–325

Jaworski N & Howarth RW (1996) Preliminary estimates of the pollutant load and fluxes into the northeast shelf LME. In: Sherman K (Ed) The Northeast Shelf Large Marine Ecosystem. Blackwell, Cambridge, MA. In press

Jaworski NA, Groffman PM , Keller AA , Prager JC (1992) A watershed nitrogen and phosphorus balance: The upper Potomac River basin. Estuaries 15: 83–95

Johnson DW (1992) Nitrogen retention in forest soils. J. Envir. Qual. 21: 1–12

Johnston CA, Bubenzer GD, Lee GB, Madison FW, McHenry JR (1984) Nutrient trapping by sediment deposition in a seasonally flooded lakeside wetland. J. Environ. Qual. 13: 283–290

Jordan C, Caskey W, Escalante G, Herrera R, Montagnini F, Todd R & Uhl C (1982) The nitrogen cycle in a "Terra Firme" rainforest on oxisol in the Amazon territory of Venezuela. Plant and Soil 67: 325–332

Jordan TE & Weller DE (1996) Human contributions to terrestrial N flux. BioScience. In press

Justic N, Rabalais NN, Turner RE & Dortch Q (1995) Changes in nutrient structure of river-dominated coastal waters: Stoichiometric nutrient balance and its consequences. Estuarine, Coastal & Shelf Science 40: 339–356

Jürgens-Gshwind S & Jung J (1979) Results of lysimeter trials at the Limburgerhof facility 1927–1977: The most important findings from 50 years of experiments. Soil Sci. 127(3): 146–160

Kaushik NK & Robinson JB (1976) Preliminary observations on nitrogen transport during summer in a small spring-fed Ontario stream. Hydrobiol. 49(1): 59–63

Keeney DR (1979) A mass balance of nitrogen in Wisconsin. Wisconsin Academy of Sciences, Arts and Letters 67: 95–102

Keller M & Reiners WA (1994) Soil-atmosphere exchange of nitrous oxide, nitric oxide, and methane under secondary succession of pasture to forest in the Atlantic lowlands of Costa Rica. Global Biogeochem. Cycles 8(4): 399–410

Keller M, Kaplan WA & Wofsy SC (1986) Emissions of N20, CH4 and CO2 from tropical soils. J. Geophys. Res. 91: 11791–11802

Keller M, Veldkamp E & Weitz AM (1993) Effect of pasture age on soil trace gas emissions from a deforested area of Costa Rica. Nature 365: 244–247

Kelly CA, Rudd JWM & Schindler DW (1990) Acidification by nitric acid: future considerations. Water, Air, and Soil Poll. 50: 49–61

Kelly CA, Rudd JWM, Hesslein RH, Schindler DW, Dillon PJ, Driscoll CT, Gherini SA & Hecky RE (1987) Prediction of biological acid neutralization in acid-sensitive lakes. Biogeochem. 3: 129–141

Kempe S, Pettine M & Cauwet G (1991) Biogeochemistry of European rivers. In: Degens ET, Kempe S & Richey JE (Eds) Biogeochemistry of Major World Rivers SCOPE 42 (pp 169–211). John Wiley and Sons, New York

Khalil MAK & Rasmussen RA (1992) The global sources of nitrous oxide. J. Geophys. Res. 97: 14651–14660

Kohl DH, Shearer GB & Commoner B (1971) Fertilizer nitrogen: contribution to nitrate in surface water in a cornbelt watershed. Science 174: 1331–1334

Kolenbrander GJ (1973) Impact of fertilizers and agricultureal waste products on the quality of waters. OECD Paris, 72 pp

Kolenbrander GJ, (1969) Nitrate content and nitrogen loss in drainwater. Neth. J. Agric. Sci. 17: 246–255

Krug A (1993) Drainage history and land use pattern of a Swedish river system – their importance for understanding nitrogen and phosphorus load. Hydrobio. 251: 285–296

Lajtha K, Seely B & Valiela I (1995) Retention and leaching losses of atmospherically-derived nitrogen in the aggrading coastal watershed of Waquoit Bay, MA. Biogeochem. 28(1): 33–54

Larsson U, Elmgren R & Wulff F (1985) Eutrophication and the Baltic Sea: Causes and consequences. Ambio 14(1): 9–14

Laycock AH (1987) The amount of Canadian water and its distribution. Can. Bull. Fish Aquat. Sci. 215: 13–42

Lesack LFW, Hecky RE & Melack JM (1984) Transport of carbon, nitrogen, phosphorus, and major solutes in the Gambia River, West Africa. Limnol. Oceanogr. 29: 816–830

Levy II H, Kasibhatla PS & Moxim WJ (1996) Global distribution of tropospheric NO_y, past, present, and future. In preparation

Levy II H & Moxim WJ (1989) Simulated global distribution and deposition of reactive nitrogen emitted by fossil fuel combustion, Tellus 41B: 256–271

Lewis WM & Saunders, III JF (1989) Concentration and transport of dissolved and suspended substances in the Orinoco river. Biogeochem. 7: 103–240

Lewis WM (1981) Precipitation chemistry and nutrient loading by precipitation in a tropical watershed. Wat. Res. Res. 17: 169–181

Lewis WM (1986) Nitrogen and phosphorus runoff losses from a nutrient-poor tropical moist forest. Ecol. 67: 1275–1282

Lowrance R & Leonard RA (1988) Stramflow nutrient dynamics on coastal plain watersheds. J. Env. Qual. 17: 734–740

Lowrance R, Todd R, Fail J, Hendrickson O, Leonard R & Asmussen L (1984) Riparian foreests as nutrient filters in agricultural watersheds. BioSci. 34: 374–377

Lucotte M (1989) Phosphorus reservoirs in the St Lawrence upper estuary. Can. J. Fish Aquat. Sci. 46: 59–65

Maidment DR (1993) Handbook of Hydrology. McGraw-Hill, New York

Mariotti A (1982) Apports de la géochimie isotopique à la connaissance du cycle de l'azote. Thesis, Univ Paris VI

Mason JW, Wegner GD, Quinn GI & Lange EI (1990) Nutrient loss via groundwater discharge from small watersheds in southern and south central Wisconsin. J. Soil Water Cons. 1990: 327–331

Matson PA & Vitousek PM (1990) Ecosystem approach to a global nitrous oxide budget. Biosci. 40(9): 667–672

Matson PA, Vitousek PM , Ewel JJ, Mazzarino MJ & Robertson GP (1987) Nitrogen transformations following tropical forest felling and burning on a volcanic soil. Ecol. 68(3): 491–502

Matthews E (1994) Nitrogenous fertilizers: global distribution of consumption and associated emissions of nitrous oxide and ammonia. Global Biogeochem. Cycles 8(4): 411–440

McClain ME, Richey JE & Pimentel TP (1994) Groundwater nitrogen dynamics at the terrestrial-lotic interface of a small catchment in the central Amazon Basin. Biogeochem. 27: 113–127

McNulty SG, Aber JD & Boone RD (1991) Spatialchanges in forest floor and foliar chemistry of spruce-fir forests across New England. Biogeochem. 14: 13–29

Messer J & Brezonik PL (1983) Agricultural nitrogen model: A tool for regional environmental management. Environ. Man. 7: 177–187
Meybeck M (1982) Carbon, nitrogen and phosphorus transport by world rivers. Am. J. Sci. 282: 401–450
Meybeck M (1988) How to establish and use world budgets of riverine materials. In: Lerman A & Meybeck M (Eds) Physical and Chemical Weathering in Geochemical Cycles (pp 247–272). Kluwer Academic Publishers, Dordrecht
Meybeck M (1993) C, N, P, and S in rivers: from sources to global inputs. In: Wollast R, Mackenzie FT & Chou L (Eds) Interaction of C, N, P, and S Biogeochemical Cycles and Global Change (pp 163–193). Springer-Verlag, Berlin
Meybeck M (1994) Les lacs et leur bassin. In: Pourriot R & Meybeck M (Eds) Limnologie Générale (pp 6–59). Masson
Meybeck M (1995) Global distribution of lakes. In: Lerman A, Imboden DM & Gat JR (Eds) Physics and Chemistry of Lakes (pp 1–35). Springer-Verlag, Berlin
Meybeck M, Chapman DV & Helmer R (1989) Global Freshwater Quality: a first assessment. World Health Organization/United Nations Environment Programme Basil Blackwell, Inc., Cambridge, MA
Michaels AF, Olson D, Sarmiento J, Ammerman J, Fanning K, Jahnke R, Knap AH, Lipschultz F & Prospero J (1996) Inputs, losses and transformations of nitrogen and phosphorus in the pelagic North Atlantic Ocean. Biogeochem. 35: 75–139
Muller G & Forstner U (1968) General relationship between suspended sediment concentration and water discharge in the Alpenrhein and some other rivers. Nature 217: 244–245
Murdoch PS & Stoddard JL (1992) The role of nitrate in the acidification of streams in the Catskill Mountains of New York. Wat. Res. Res. 28: 2707–2720
National Research Council (1993) Managing Wastewater in Coastal Urban Areas. Committee on Wastewater Management for Coastal Urban Areas. National Academy Press, Washington, DC
Naves J, Bousquet G, Leroy P, Hubert P & Vilagines R (1991) Evolution de la qualite de l'eau de la Sein a Ivry sur Sein (France) de 1887 a 1986. Comptes Rendus de l'Atelier International UNESCO/AISM/ENIT Application des modeles mathematiques a P evaluation des modifications de la qualite de P eau (pp 35–44). Tunis 1990, ENIT, Tunis Neff JC, Bowman WD, Holland GA, Fisk MC & Schmidt SK (1994) Fluxes of nitrous oxide and methane from nitrogen-amended soils in a Colorado alpine ecosystem. Biogeochem. 27(1): 23–33
Nelson D (1985) Minimizing nitrogen losses in non-irrigated eastern areas. In: Plan Nutrient Use and the Envronment (pp 173–209). Fertilizer Inst, Washington, DC
Nixon SW (1992) Quantifying the relationship between nitrogen input and productivity of marine ecosystems. Adv. Mar. Techn. Conf. 5: 57–83
Nixon SW (1995) Coastal marine eutrophication: A Definition, social causes, and future concerns. OPHELIA 41: 199–219
Nixon SW, Ammerman J, Atkinson L, Berounsky V, Billen G, Boicourt W, Boynton W, Church T, Di Toro, Elmgren R, Garber J, Giblin A, Jahnke R, Owens N, Pilson, MEQ, & Seitzinger S (1996) The fate of nitrogen and phosphorus at the land-sea margin of the North Atlantic Ocean. Biogeochem. 35: 75–139
OECD (1991) OECD Environmental Data, Compendium 1991. Organisation for Economic Co-operation and Development. Paris
Overgaard K (1984) Nitrate pollution of groundwater in Denmark. 20th Nordic Symposium on Water Pollution, NORDFORSK
Owens M, Garland JHN, Hart IC & Wood G (1972) Nutrient budgets in rivers. Symp. Zool. Soc. Land. 29: 21–40
Pacés T (1982) Natural and anthropogenic flux of major elements from central Europe. Ambio. 11(4): 206–208
Paerl HW (1993) Emerging role of atmospheric nitrogen deposition in coastal eutrophication: biogeochemical and trophic perspectives. Can. J. Fish. Aquat. Sci. 50: 2254–2269

Paré C & Goulet M (1980) Evolution physico-chimique des eaux de Radissonie: secteur Nottaway-Broadback-Rupert (1977–1978) Gouvernement du Quebec, Ministere des Richesses Naturelles, Quebec
Peierls B, Caraco N, Pace M & Cole J (1991) Human influence on river nitrogen. Nature 350: 386–387
Peterjohn WT & Correll DL (1984) Nutrient dynamics in an agricultural watershed: Observations on the role of a riparian forest. Ecol. 65(5): 1466–1475
Pocklington R & Tan FC (1987) Seasonal and annual variations in the organic matter contributed by the St Lawrence River to the Gulf of St Lawrence. Geochim Cosmochim Acta 51: 2579–2586
Probst JL (1985) Nitrogen and phosphorus exporation in the Garonne basin (France). J Hydrol. 76: 281–305
Prospero JM, Barrett K, Church T, Dentener F, Duce RA, Galloway JN, Levy II H, Moody J & Quinn P (1996) Atmospheric deposition of nutrients to the North Atlantic basin. Biogeochem. 35: 75–139
Rabalais NN, Turner RE, Justic D, Dortch Q, Wiseman WJ & Sen Gupta BK (in press) Nutrient changes in the Mississippi River and system responses on the adjacent continental shelf. Estuaries
Ramirez AJ (1991) Transport of phosphorus by Venezuelan rivers. In: Tiessen H, Lopez-Hernandez D & Salcedo IH (Eds) Phosphorus Cycles in Terrestrial and Aquatic Ecosystems, Proceedings of Regional Workshop 3: South and Central America (pp 224–33). Saskatchewan Inst of Pedology, University of Sakatchewan, Saskatoon
Resch HN (1991) The balance of nitrogen composition in the FRG. In: Borght P & Tychon B (Eds) Gestion de l'azote agricole et qualite des eaus Vander. Cebedoc, Liege, Belgium
Reynolds B, Ormerod S & Gee A (1994) Spatial patterns in stream nitrate concentrations in upland Wales in relation to catchment forest cover and forest age. Envir. Poll. 84: 27–33
Rijkswaterstaat (1992) Guidance Document for the NSTF Modeling Workshop.
Robertson GP & Rosswall T (1986) Nitrogen in West Africa: The regional cycle. Ecol. Monogr. 56: 43–72
Rosenberg R, Elmgren R , Fleischer S, Jonsson P , Persson G & Dahlin H (1991) Marine Eutrophication Case studies in Sweden. Ambio. 19 : 103–108
Salati E, Sylvester-Bradley R & Victoria RL (1982) Regional gains and losses of nitrogen in the Amazon basin. Plant and Soil 67: 367–376
Santschi PH & Schindler DW (1977) Chemical and geochemical studies of Lake Biel. Schweiz. Zeit. Hydrobiol. 39: 182–200
Schelske CL (1985) Biogeochemical silica mass balances in Lake Michigan and Lake Superior. Biogeochem. 1: 197–218
Schepers JS, Frank KD & Watts DG (1983) Influence of irrigation and nitrogen fertilization on groundwater quality. IAHS Publ 146: 21–29
Schetagne R & Roy D (1985) Reseau de surveillance ecologique du Complexe la Grande 1977–1984 Physico-chimie et pigments chlorophylliens Annexe 1 Region de La Grande 2 Societe d'energie de la Baie James Direction Ingenierie et Environnement, Montreal
Schetagne R (1981) Reseau de surveillance ecologique du Complexe la Grande 1977–1984 Physico-chimie et pigments chlorophylliens Analyses de la Baie James Direction Environnement, Montreal
Schindler DW & Bayley SE (1993) The biosphere as an increasing sink for atmospheric carbon: estimates from increased nitrogen deposition. Global Biogeochem. Cycles 7: 717–733
Schlesinger WH (1991) Biogeochemistry: An Analysis of Global Change. Academic Press, San Diego
Schlesinger WH & Hartley AE (1991) A global budget for atmospheric NH_3. Biogeochem. 15: 191–211
Schroder H (1985) Nitrogen losses from Danish agriculture: trends and consequences. Agric., Ecosys. and Envir. 14: 279–289

Schulze E-D (1989) Air pollution and forest decline in a spurce (*Picea abies*) forest. Science 244: 776–783
Schulze E-D, De Vries W, Hauhs M, Rosén K, Rasmussen L, Tann O-C & Nilsson J (1989) Critical loads for nitrogen deposition in forest ecosystems. Water, Air and Soil Poll. 48: 451–456
Seitzinger SP (1987) Nitrogen biogeochemistry in an unpolluted estuary: The importance of benthic denitrification. Mar. Ecol. Prog. Ser. 37: 65–73
Skibe U, Cresser MS, Derwent RG & Futty DW (1989) Peat acidification in Scotland. Nature 337: 68–69
Skole D & Tucker C (1993) Tropical deforestation and habitat fragmentation in the Amazon: satellite data from 1978 to 1988. Science 260: 1905–1910
Sluijsmans CMF, Van Dijk TA, Kolenbrander GJ, de la Lande Cremer LCN, Smilde KW, Werkoven CME (1978) L'épandage des effluents sur les sols agricoles dans la CEE Informations sur l'agriculture. Vol 47
Smil V (1990) Nitrogen and phosphorus. In: Turner BL, Clark WC, Kates RW, Richards JF, Mathews J & Meyer WB (Eds) The Earth as Transformed by Human Action: Global and Regional Changes in the Biosphere over the Past 300 Years (pp 423–436). Cambridge University Press, Cambridge
Smith RA, Alexander RB & Lanfear KJ (1993) Stream water quality in the coterminous United States – status and trends of selected indicators during the 1980's. In: National Water Summary 1990–91 USGS Water Supply Paper 2400 (pp 111–140). US Govt Printing Office, Washington, DC
Smith RV, Stevens RJ, Foy RH & Gibson CE (1982) Upward trend in nitrate concentration in rivers discharging into Lough Neagh for the period 1969–1979. Water Res. 16: 183–188
Soderlund R & Rosswall T (1982) The nitrogen cycles. In: Huntzinger O (Ed) The Handbook of Environmental Chemistry, vol 1. part B, The Natural Environment and the Biogeochemical Cycles (pp 61–81). Springer Verlag, New York
Sollins P, Grier CC, McCorison FM, Cromack K & Fogel R (1980) The internal element cycles of and old-growth Douglas fir ecosystem in western Oregon. Ecol. Monogr. 50: 261–285
Stallard RF (1985) River chemistry, geology, geomorphology, and soils in the Amazon and Orinoco basins. In: Drever JI (Ed) The Chemistry of Weathering (pp 293–316). Reidel, Dordrecht
Stevenson FJ (1982) Nitrogen in Agricultural Soils. American Society of Agronomy, Inc. Madison, Wisconsin
Strebel O, Duynisveld WHM & Bottcher J (1989) Nitrate pollution of groundwater in Wester Europe. Agric., Ecosys. and Envir. 26: 189–214
Strebel OJ, Bottcher J & Duynisveld WHM (1984) Einfluss von Strandortbedingungen und bodemnutzung auf nitratauswaschung und Nitratkonzentration des Grundwassers Landwirtschaft Forschung Kongressband, Karlsruhe: 33–34
Thibert S (1994) Exportations naturelles et anthropiques des ions majeurs et des elements nutritifs dans le bassin de la Seine. Approches methodologiques. Ph.D. Thesis (p 195) University of Paris VL
Tidal Waters Division of Rijkswaterstaat (1992) Guidance documents for the North Sea Task Force Modelling Workshop, 6–8 May 1992, The Hague
Tilman GD (1984) Plant dominance along an experimental nutrient gradient. Ecol. 65: 1445–1453
Triska FJ, Jackman AP, Duff JH & Avanzino RJ (1994) Ammonium sorption to channel and riparian sediments: a transient storage pool for dissolved inorganic nitrogen. Biogeochem. 26: 67–83
Turner RE & Rabalais NN (1991) Changes in Mississippi River water quality this century. BioSci. 41: 140–147
Uehara G & Gilman G (1981) The minerology, chemistry and physics of tropcial soils with variable charge clays. Westview Press, Boulder, CO

Ulehlova B (1987) Chemical composition of rain and lysimetric waters from grasslands with different rates of fertilization. In: Moldan B & Paces T (Eds) GEOMON: International Workshop on Geochemistry and Monitoring in Representative Basins (pp 105–107). Geological Survey, Prague

UNESCO (1978) World water balance and water resources of the earth. Studies and Reports in Hydrology, UNESCO

UNESCO (1992) Discharge of selected rivers of the world. UNESCO/Gidrometeoizdat, St Petersburg

Van der Leeden F (1975) Water Resources of the world: Selected Statistics. Water Information Center, Port Washington, New York

Van der Leeden F, Troise FL & Todd DK (1990) The Water Encyclopedia Second edition. Lewis Publ, Chelsea, MI

Van Kessel JF (1977) Factors affecting the denitrification rate in two water-sediment systems. Water Research II: 259–267

van Breeman N, Burroughs PA, Velthorst EJ, van Dobben HF, de Wit T, Ridder TB & Reijnders HFR (1982) Soil acidification from atmospheric ammonium sulphate in forest canopy throughfall. Nature 299: 548–550

Vitousek PM (1984) Litterfall, nutrient cycling, and nutrient limitation in tropical forests. Ecol. 65(1): 285–298

Vitousek PM (1994) Beyond global warming: Ecology and global change. Ecol. 75(7): 1861–1876

Vitousek PM & Howarth RW (1991) Nitrogen limitation on land and sea: how can it occur? Biogeochem. 13: 87–115

Vitousek PM & Matson PA (1993) Agriculture, the global nitrogen cycle, and trace gas flux. In: Oremland RS (Ed) The Biogeochemistry of Global Change: Radiative Trace Gases (pp 193–208). Chapman and Hall, New York

Vitousek PM & Reiners WA (1975) Ecosystem succession and nutrient retention: A hypothesis. BioSci. 25: 376–381

Vollenwieder R (1968) Les bases scientifiques de l'eutrophisation des lacs et des eaux courantes sous l'aspect particulier du phosphore et de l'azote comme facteurs d'eutrophisation Rept DAS/CSI/68–27 Paris, OCDE

Vought L, Dahl J, Pedersen C & Lacoursiere J (1994) Nutrient retention in riparian ecotones. Ambio. 23: 342–348

Wafar MVM, Le Corre P & Birrien JL (1989) Transport of carbon, nitrogen and phosphorus in a Brittany river, France. Estuar. Coast. Shelf Sci. 29: 489–500

Ward RC (1981) River systems and river regimes. In: Lewin J (Ed) British Rivers. Allen & Unwin, Ltd, London

Webb BW & Walling DE (1985) Nitrate behaviour in streamflow from a grassland catchment in Devon, UK. Water Res. 19: 1005–1016

Willison TW, Splatt PR & Anderson JM (1990) Nutrient loading of a forest soil: a manipulative approach using rooted experimental chambers. Oecologia 82: 507–512

Woodmansee RG (1978) Additions and losses of nitrogen in grassland ecosystems. Science 28: 448–453

WRI/UNEP (1988) World resources 1988–89. World Resources Institute in collaboration with United Nations Environment Programme. Basic Books Inc., New York

Wright JW (1993) The Universal Almanac. Andrews and McHeel, Kansas City

Wulff F, Stigebrandt A & Rahm L (1990) Nutrient dynamics of the Baltic Sea. Ambio. 19(3): 126–133

Wyer MD & Hill AR (1984) Nitrate transformations in southern Ontario stream sediments. Water Resources Bull. 20: 729–737

Zhong Cai MA (1991) Modelisation du transfert des nitrates: du bassin de recherche au grand bassin (exemples des bassins de Melarchez et de la Charente) Thesis, Universite Louis Pasteur, Strasbourg, France

Biogeochemistry **35:** 141–180, 1996.

The fate of nitrogen and phosphorus at the land-sea margin of the North Atlantic Ocean

S.W. NIXON[1], J.W. AMMERMAN[2], L.P. ATKINSON[3], V.M. BEROUNSKY[1], G. BILLEN[4], W.C. BOICOURT[5], W.R. BOYNTON[5], T.M. CHURCH[6], D.M. DITORO[7], R. ELMGREN[8], J.H. GARBER[9], A.E. GIBLIN[10], R.A. JAHNKE[11], N.J.P. OWENS[12], M.E.Q. PILSON[1] & S.P. SEITZINGER[13]
[1]*University of Rhode Island;* [2]*Texas A&M University;* [3]*Old Dominion University;* [4]*Universite Libre de Bruxelles;* [5]*University of Maryland;* [6]*University of Delaware;* [7]*Manhattan College;* [8]*Stockholm University;* [9]*U.S. Environmental Protection Agency;* [10]*Marine Biological Laboratory;* [11]*Skidaway Instit. of Oceanography;* [12]*University of Newcastle;* [13]*Rutgers University/NOAA CMER Program*

Received 11 August 1995; accepted 13 February 1996

Key words: continential shelf, estuaries, mass balance, nitrogen, North Atlantic, nutrient budget, phosphorus

Abstract. Five large rivers that discharge on the western North Atlantic continental shelf carry about 45% of the nitrogen (N) and 70% of the phosphorus (P) that others estimate to be the total flux of these elements from the entire North Atlantic watershed, including North, Central and South America, Europe, and Northwest Africa. We estimate that $61 \cdot 10^9$ moles y^{-1} of N and $20 \cdot 10^9$ moles y^{-1} of P from the large rivers are buried with sediments in their deltas, and that an equal amount of N and P from the large rivers is lost to the shelf through burial of river sediments that are deposited directly on the continental slope. The effective transport of active N and P from land to the shelf through the very large rivers is thus reduced to $292 \cdot 10^9$ moles y^{-1} of N and $13 \cdot 10^9$ moles y^{-1} of P.

The remaining riverine fluxes from land must pass through estuaries. An analysis of annual total N and total P budgets for various estuaries around the North Atlantic revealed that the net fractional transport of these nutrients through estuaries to the continental shelf is inversely correlated with the log mean residence time of water in the system. This is consistent with numerous observations of nutrient retention and loss in temperate lakes. Denitrification is the major process responsible for removing N in most estuaries, and the fraction of total N input that is denitrified appears to be directly proportional to the log mean water residence time. In general, we estimate that estuarine processes retain and remove 30–65% of the total N and 10–55% of the total P that would otherwise pass into the coastal ocean. The resulting transport through estuaries to the shelf amounts to $172–335 \cdot 10^9$ moles y^{-1} of N and $11–19 \cdot 10^9$ moles y^{-1} of P. These values are similar to the effective contribution from the large rivers that discharge directly on the shelf.

For the North Atlantic shelf as a whole, N fluxes from major rivers and estuaries exceed atmospheric deposition by a factor of 3.5–4.7, but this varies widely among regions of the shelf. For example, on the U.S. Atlantic shelf and on the northwest European shelf, atmospheric deposition of N may exceed estuarine exports. Denitrification in shelf sediments exceeds the combined N input from land and atmosphere by a factor of 1.4–2.2. This deficit must be met by a flux of N from the deeper ocean. Burial of organic matter fixed on the shelf removes only a small fraction of the total N and P input (2–12% of N from land and atmosphere; 1–17% of P), but it may be a significant loss for P in the North Sea and some other regions. The removal of N and P in fisheries landings is very small. The gross exchange of N and P between the shelf and the open ocean is much larger than inputs from land and, for the North Atlantic shelf as a whole, it may be much larger than the N and P removed through denitrification, burial, and fisheries. Overall, the North Atlantic continental shelf appears to remove some 700–950 · 10^9 moles of N each year from the deep ocean and to transport somewhere between 18 and 30 · 10^9 moles of P to the open sea. If the N and P associated with riverine sediments deposited on the continental slope are included in the total balance, the net flux of N to the shelf is reduced by 60 · 10^9 moles y^{-1} and the P flux to the ocean is increased by 20 · 10^9 moles y^{-1}. These conclusions are quite tentative, however, because of large uncertainties in our estimates of some important terms in the shelf mass balance.

Introduction

The word "margin" appears commonly in papers dealing with exchanges between land and sea. It is a conveniently ambiguous term for an interface that is difficult to define. We use it here to refer to the whole complex border of environments between free-flowing fresh water and the edge of the continental shelf, including intertidal wetlands, estuaries, and the shelf. While it would be best to describe the outer boundary of the shelf by a marked change in water depth, as a practical matter we define the shelf as extending seaward to a depth of 100 or 200 m. Areas within each of these depths have been calculated (Pilson & Seitzinger, this volume), and one or the other is frequently used to fix the shelf-slope boundary (Eisma 1988).

If we define the oceans as beginning where fresh and salt water first meet (an admittedly elusive point), then the nitrogen (N) and phosphorus (P) fluxes reported by Howarth et al. (this volume) may be taken as the inputs to the North Atlantic from land. In what may have been the first attempt to develop nutrient budgets for the world ocean, Emery et al. (1955) worked from such a perspective and established a convention for later efforts (for example, the Paris Commission as discussed by Balls 1994). During the last 15–20 years, however, there has been a growing appreciation that intense biogeochemical exchanges in estuaries (including their wetlands) and on the continental shelf may influence markedly the net flux of N and P from land to the open ocean (Edmond et al. 1981; GESAMP 1987; Balls 1994; Howarth et al. 1995; Nixon et al. 1995). In this report we attempt to assess the magnitude of this influence around the land-sea margin of the North Atlantic so that we can estimate the net transport of N and P from land onto the continental shelf and the net exchange between the shelf and the open sea.

Very large rivers

Five major rivers with an average water flow exceeding 3000 m^3 s^{-1} discharge directly onto the continental shelf on the western side of the North Atlantic

Table 1. Estimated flux of total nitrogen and phosphorus in major rivers that discharge directly on the North Atlantic continental shelf.[1]

	Water Discharge[2] $m^3\ s^{-1}$	10^9 moles y^{-1}	
		N	P
Mississippi-Atchafalaya[3]	17,545	130	3.5
Grijalva-Usumacinta[4] (Western Gulf of Mexico)	3,265	13	0.05
Magdelena[4] (Central America)	8,000	9.5	0.6
Orinoco[5] (Central America)	28,000	39	2.3
Amazon[6]	180,000	223	47
Total direct discharge on shelf[7]		414	53

[1] The St. Lawrence is also a very large river (10,400 $m^3\ s^{-1}$), but it does not form a delta and contains an estuarine mixing zone at its mouth (Czaya 1981; Bugden 1981).
[2] Czaya (1981).
[3] Recent values calculated by Howarth et al. (this volume) from data summarized by Turner and Rabalais (1991).
[4] Calculated assuming that the ratio of river nutrient discharge to regional nutrient discharge is equal to the ratio of river water discharge to regional water discharge as reported by Howarth et al. (this volume).
[5] Lewis & Saunders (1989).
[6] N from Solati et al. (1982); P from Devol et al. (1991).
[7] As discussed in text, we estimate that 29% of this N and 75% of this P is buried in deltas and on the continental slope.

(Czaya 1981). Because the flows from these rivers are sufficiently great that fresh and salt water mixing takes place largely on the shelf, rather than in an estuarine embayment, we can consider the N and P they carry as direct additions to the shelf. Unfortunately, the fluxes of total N and total P in two of these rivers (the Grijalva-Usumacinta and the Magdelena) appear not to have been measured, and we have assumed that the discharge of N and P from the regions they drain is partitioned according to regional water discharge as described by Howarth et al. (this volume). The contributions of N and P from the Mississippi, the Amazon, and the Orinoco have been measured directly (Table 1). Together these five river systems appear to carry about 45% of the N and 70% of the P that Howarth et al. (this volume) calculate for the total riverine flux to the North Atlantic.

These fluxes are so large that it is useful to consider them in more detail rather than simply entering them in future tables as inputs to the shelf. This is particularly important for P, because a very large portion of the total P carried by the rivers is associated with suspended sediments. Most of this P will ultimately be buried with the sediments and thus not play an active role in what are geologically speaking short-term biogeochemical processes on the shelf and in the ocean (Froelich 1988).

The fate of the sediment discharged by the very large rivers (Table 2) is not known with a great deal of certainty – some accumulates in the delta formed at the mouth of each of the rivers (Table 2), some may be deposited elsewhere on the shelf away from the delta, and some is carried over the shelf break and buried on the continental slope. For the purpose of this accounting, we assume that 50% of the sediment discharged from each river is buried in its delta as found for the Amazon (Kuehl et al. 1986). We further assume that the remaining sediment is deposited and buried on the slope, thus bypassing the shelf entirely.

Because there have been few reports of the N content of delta sediments, we based our estimate of N burial on the more numerous measurements of organic carbon and an assumed C/N ratio of 10 (molar). Meybeck (1982) reported a C/N (molar) ratio of 9.3–14 for marine sediment and DeMaster et al. (1994) found a range from 7.5 to 10 in rapidly accreting slope sediments off North Carolina. Trefry et al. (1994) reported a C/N (molar) ratio of 9.9 for particulate organic matter in the Mississippi River and Richey & Victoria (1993) found a mean ratio of 10.8 for fine particulates in the Amazon. Our use of a ratio of 10 may underestimate the N burial, however, since Aller & Aller (1986) reported a mean value of 7.4 in 22 samples of sediment (0–45 cm depth) from the Amazon continental shelf.

The organic carbon content of delta sediments has been measured for the Mississippi (1.4%, Trefry et al. 1994), the Orinoco (0.91%, van Andel & Postma 1954), and the Amazon (0.66%, Showers & Angle 1986). Following Berner (1982) we assumed a C value of 0.75% for delta sediments from the remaining two rivers. Based on these assumptions, we estimate that about $61 \cdot 10^9$ moles N y^{-1} are buried in the deltas of the five very large rivers (Table 2), and that an equal amount of N from the rivers is buried on the adjacent continental slopes. This represents 29% of the total N discharged by the rivers (Table 1) and leaves $292 \cdot 10^9$ moles N y^{-1} as a potentially active contribution to the continental shelf.

There have also been few measurements of total P in deltaic sediments. Unfortunately, while there is a relationship between organic P and organic C in sediments that appears to vary in a regular way with sedimentation rate (Ingall & Van Cappellen 1990), a large fraction of the river borne P is in inorganic particulate form (Jahnke 1992). For example, Richey & Victoria (1993) found no correlation between organic C and total P in particulate matter carried by the Amazon. As a result, we estimated P burial in a different way than we did N burial. Following Froelich (1988), who concluded that "... about 25% of the total phosphorus eroded from the continents is activated and enters the ocean in a biologically available form," we assume that 75% of the total P discharge (Table 1) will be buried in the large river deltas and on the

Table 2. Estimated burial of nitrogen and phosphorus in major river deltas on the North Atlantic continental shelf. We assume that an equal amount of N and P is buried on the continental slope.

	Sediment discharge[1] 10^{12} g y^{-1}	Delta area[2] 10^3 km^2	10^9 moles y^{-1} N	P
Mississippi-Atchafalaya	210	30	12	1.3
Grijalva-Usumacinta	96	17	3	0.02
Magdelena	220	1.7	7	0.2
Orinoco	150	21	6	0.9
Amazon	1200	100	33	17.6
Total	1876	170	61	20

[1] From various sources summarized by Milliman & Syvitski (1992). The Grijalva-Usumacinta sediment discharge was calculated from Milliman and Syvitski's regression relating sediment discharge to runoff for high rivers using mean water discharge (1980–1983) and watershed data for the Usumacinta at Boca del Cerro, Mexico (UNESCO 1992). The resulting flux of 69 · 10^{12} g y^{-1} was increased by a factor of 1.39 to account for the higher water flow at the mouth of the combined Grijalva-Usumacinta river system (Czaya 1981). No estimate of combined watershed area was available.

[2] Amazon and Mississippi deltas from Czaya (1981), others from Mulholland (1981). The latter reports delta areas of 29 · 10^3 km^2 and 467 · 10^3 for the Mississippi and the Amazon, respectively.

continental slope. If this is correct, about 20 · 10^9 moles P y^{-1} from the very large rivers will be buried in the deltas, 20 · 10^9 moles y^{-1} will be buried on the continental slope, and 13 · 10^9 moles y^{-1} will be an active input to the shelf.

It follows from these assumptions that the average P concentration on the buried sediments from the very large rivers should be 40 · 10^9 moles P y^{-1} divided by 1876 · 10^{12} g y^{-1} of sediment or 0.02 mmol P g^{-1}. This is equivalent to 0.66 mg P g^{-1}, a value consistent with recent measurements of total P in marine sediments summarized by Howarth et al. (1995).

Aside from the five very large rivers, all of the remaining riverine fluxes of N and P from land must pass through estuaries before they reach the shelf. During the passage through these intensely metabolic systems, there is great potential for modification.

Estuaries

Considering that the first attempt to develop a quantitative mass balance or budget for nitrogen in a marine system was carried out almost 90 years ago in the North Sea (Johnstone 1908), there have been surprisingly few efforts to describe the overall balance of total N and P in estuarine embayments around the North Atlantic. Fortunately, for this exercise we were able to find or assemble relatively complete mass balances for nine impressively diverse systems. They vary widely in size (24 km^2 to 375.10^3 km^2), mean depth (1 m to 55 m), water residence time (3 days to 20 years), and nutrient loading rate (230 mmol N m^{-2} y^{-1} to 13,400 mmol N m^{-2} y^{-1}). The systems represented include a coastal lagoon (the Guadalupe Estuary), drowned river estuaries (e.g., Delaware Bay), coastal embayments (Boston Harbor), and an inland sea (the Baltic). The sample also includes relatively pristine systems where non-point sources of N and P dominate (e.g., Ochlockonee Bay) as well as intensively developed estuaries where point source sewage discharges provide large amounts of nutrients (e.g., the Scheldt Estuary). Some of the systems experience hypoxic or anoxic conditions (e.g., Chesapeake Bay, Potomac Estuary, Scheldt, the Baltic), while others are relatively well mixed and characterized by oxygenated bottom water (e.g., Delaware Bay, Narragansett Bay). Some have extensive areas of intertidal salt marsh (e.g., Delaware Bay), while others do not (e.g., Narragansett Bay, Boston Harbor).

Despite the considerable efforts expended by many investigators, we can not claim to have a rigorously closed budget for N and P in any of these diverse systems. In no case have all of the inputs, outputs, and changes in internal storage of N or P been measured independently and contemporaneously (see discussion in Nixon et al. 1995). In most cases, the flux in which we are most interested, the net export of N and P to the shelf, has been calculated by difference. The Baltic Sea and the Scheldt are notable exceptions in which that flux has been measured or modeled and denitrification has been calculated by difference. The estuarine-shelf exchanges at the mouth of Chesapeake Bay have also been modeled (Cerco & Cole 1992) and agree well with the exchanges calculated by difference (Boynton et al. 1995).

In order to facilitate comparison among the systems, we have expressed all of the terms in each budget as a flux per unit area (Tables 3–11). It is important to emphasize that our sample only includes studies in which all of the various forms of N and P have been measured. There are many more estuaries in which the behavior of dissolved inorganic N and P has been observed, but these can not be taken as revealing anything about the exchange of total N or P. For example, it has been well established that dissolved reactive phosphorus accounts for only a small fraction of the total P flux in unpolluted rivers (Jahnke 1992; Howarth et al. 1995).

It is apparent from an inspection of the budgets that our sample of only nine coastal systems includes a wide range of biogeochemical behavior, from almost complete transfer of N and P between land and the shelf (e.g., Ochlockonee Bay and Boston Harbor) to almost complete interception of the flux from land (e.g., the Baltic Sea). Chesapeake Bay and the estuary of its tributary, the

Potomac River, appear to import P from the shelf (Boynton et al. 1995). In seeking to relate such behavior to various characteristics of the systems, we turned to limnology, where it has been shown that the retention of N and P in many lakes is largely a function of flushing rate, though the exact form of the relationship is described differently by various authors (Kirchner & Dillon 1975; Kelly et al. 1987; see brief overviews by Rigler & Peters 1995; Howarth et al., this volume). It is more difficult to estimate the effective flushing rate in estuaries than in lakes, but we were able to obtain at least a rough estimate of the mean fresh water replacement time for each system in our sample. Linear regressions of the fractional net transport of nitrogen and phosphorus from land to the coastal ocean as a function the log mean residence time suggest that this feature is also very important in influencing the net balance of N and P in estuaries (Figures 1 and 2).

This result was not necessarily expected because there are important differences between the behavior of N and P in fresh and salt water and in lake and marine sediments (Froelich 1988; Seitzinger 1988b; Nixon 1988; Caraco et al. 1990; Howarth et al. 1995). However, our conclusion is further supported (at least for N) by a recent independent study of the N balance in a Danish fjord by Nielsen et al. (1995), who found that 95% of the N entering that system from land was exported. Their published report described the fjord as having a water residence time ranging from 1.5 to 13 days, but they have provided us with a calculation of the time-averaged fresh water replacement rate of five days during the year of their study. Their export is thus consistent with the other systems shown in Figure 1. The behavior of N in the northern Adriatic Sea also fits with the pattern shown by estuaries and lakes in our sample (Degobbis et al. 1986).

Because the companion article by Howarth et al. (in this volume) discusses N retention in lakes as a function of the ratio of mean depth to water residence time according to the model of Kelly et al. (1987), we have also analyzed our data in that way. While the result (Figure 3) confirms that the N retention of estuaries is generally consistent with that of lakes and rivers (see Howarth et al., this volume), the inclusion of water depth did not result in a regression with a higher r^2.

The relationship between flushing rate and export is less convincing for P (Figure 2). We excluded Chesapeake Bay and its tributary estuary, the Potomac, from the P export regression because they both appear to have a net input of P from the coastal ocean (Tables 4 and 8). This exceptional behavior appears to be due, at least in part, to very rapid sedimentation rates in these systems that are supported equally by sediment inputs from shoreline erosion and from rivers (Boynton et al. 1995). High rates of sediment deposition also make Chesapeake Bay and the Potomac appear to bury a larger fraction of

Table 3. The annual mass balance of total nitrogen and phosphorus in the Baltic Sea.[1] Data are from Elmgren (1984), HELCOM (1993), Larsson et al. (1985), Shaffer & Rönner (1984), Wulff & Stigebrandt (1989), and Wulff et al. (1994). In some cases values have been recalculated to exclude the Danish Straits area. Units are mmol m^{-2} y^{-1}.

	N	P
Inputs		
Direct atmospheric deposition[2]	68	0.2
Nitrogen fixation[3]	26	–
Rivers[4]	109	3.1
Direct urban areas and industries[5]	14	0.8
	217	4.1
Changes in storage		
Net accumulation in water[6]	16	0.3
Net accumulation in sediment[7]	11	3.0
Outputs		
Denitrification[8]	161	–
Loss as gaseous N_2O[9]	4	–
Fish landings[10]	5	0.3
Coastal ocean, net exchange[11]	20	0.5

[1] Inside the Danish Straits. Area = 374 600 km^2, mean depth = 55 m (Wulff & Stigebrandt 1989).
[2] Swedish Environment Protection Agency (1993: 142–143).
[3] Larsson et al. (1985).
[4] Swedish Environment Protection Agency (1993), minus direct loads from urban areas and industries.
[5] HELCOM (1993).
[6] Change between 1982–1986 and 1987–1991, Wulff et al. (1994).
[7] N value from Larsson et al. (1985); P value calculated by difference.
[8] Calculated by difference.
[9] Shaffer & Rönner (1984).
[10] Finfish only (Elmgren 1984), content of nutrient elements calculated as in Table 14.
[11] Wulff & Stigebrandt (1989).

their N input than calculated for the other estuaries. With the exception of the upper end of the estimated range for N burial in Narragansett Bay, all of the other estimates of N burial are less than 10% of the input, while N burial in the Potomac is reported to be 35% of the input and for the Chesapeake as a whole, it is almost 40% of the input (Boynton et al. 1995) (see Tables 3–11).

In a previous comparison of denitrification rates and the inputs of dissolved inorganic nitrogen (DIN) to several estuaries, Seitzinger (1988b) found that denitrification removed a relatively constant 45% of the DIN input. If we express denitrification as a fraction of the total nitrogen input, however, it

Table 4. The annual mass balance of total nitrogen and phosphorus in Chesapeake Bay, Maryland-Virginia.[1] during 1985–1986. Data from Boynton et al. (1995). Units are mmol m^{-2} y^{-1}.

	N	P
Inputs		
Direct atmospheric deposition[2]	113	2
Nitrogen fixation	?	0
Rivers (diffuse sources)	565	18
Direct sewage (point sources)	260	11
Coastal ocean[3]	0	10
	938	41
Changes in storage		
Accumulation in water	0	0
Accumulation in sediment	327	40
Outputs		
Denitrification	245	0
Fish landings	84	1.5
Coastal ocean[3]	282	0

[1] Area = 11 542 km^2, mean depth (MLW) = 6 m.
[2] Smullen et al. (1982); 1976–1981 mean TN and TP in wet deposition.
[3] P input and N output calculated by difference.

appears that N removal is largely a function of water residence time (Figure 4). Our finding that the fractional loss of N to denitrification increases with water residence time in estuaries is also consistent with the general behavior of five out of six Danish lakes studied by Andersen (1974) and with the unstratified lakes analyzed by Kelly et al. (1987).

Unfortunately, it is difficult to apply the relationships between flushing rate and N and P export in any exact way to the fluxes reported by Howarth et al. (this volume) because their results represent an aggregate flux through long stretches of coastline. Each region may contain numerous estuaries with highly variable flushing rates, many of which have never been estimated. As a practical expedient we have assumed that most estuaries are characterized by mean residence times of 0.5–12 months, and that they will acccordingly export some 35–70% of their total N input and about 45–90% of their total P input to the continental shelf. We have made an exception to this procedure for the Baltic Sea, where we reduced the N and P exported from the Baltic according to Table 3 in order to compute a net flux to the Northwest European Coast and North Sea. The consequence of these assumptions in combination with the burial of some of the N and P carried by the very large rivers is that the total riverine flux of N from land onto the total North Atlantic shelf is

Table 5. The annual mass balance of total nitrogen and phosphorus in Delaware Bay, Delaware-New Jersey[1] during the 1980s. Units are mmol m^{-2} y^{-1}.

	N	P
Inputs		
Direct atmospheric deposition[2]	100	3
Land drainage and sewage[3]	1800	155
	1900	158
Changes in storage		
Accumulation in water	0	0
Accumulation in sediment		
Salt marshes[4]	178	2
Subtidal[5]	77	21
	255	23
Outputs		
Denitrification[6]	825–1025	0
Fish landings	Small	Very small
Coastal ocean[7]	620–820	135

[1] Area = 1 989 km^2, mean depth = 9.7 m (Polis & Kupferman 1973).
[2] Direct wet DIN from Scudlark & Church (1994). They assumed dry deposition equal to wet. We assume DON equal to 50% of wet NO_3 and wet NO_3 equal to 50% of wet DIN based on Nowicki & Oviatt (1990). P based on recent regional AIRMoN data.
[3] From Jaworski & Howarth (in press). NOAA (1989) reported a total N input of 1630 mmol m^{-2} y^{-1} and a total P input of 190 mmol m^{-2} y^{-1}.
[4] N burial estimated assuming 10 mg N/g of marsh sediment based on various sources summarized by Nixon & Lee (1986) and 497 · 10^3 mt y^{-1} of marsh sediment accumulation (Biggs & Beasley 1988). P burial estimated from 300 μg P/g of marsh sediment (Strom 1976).
[5] Based on a sediment input of 1.6 · 10^6 mt accumulating in subtidal fine-grained deposits (Biggs & Beasley 1988) with a nitrogen content of 0.13% (Cifuentes 1991) for upper bay depositional area sediments (Weil 1977). P deposition based on 800 μg P g^{-1} inorganic sediment in the upper bay depositional area (Strom 1976).
[6] Calculated using denitrification rates in the Delaware River from (Seitzinger 1988a; 300 μmol m^{-2} h^{-1}) and Delaware Bay proper (Seitzinger 1988b; 100 μmol m^{-2} h^{-1}) and areas of 179 km^2 and 1819 km^2, respectively. Lower value obtained assuming the lower bay rate applied only to the area not classified by Biggs & Beasley (1988) as consisting of gravel or gravely sand. Lower bay rates measured in summer only.
[7] Calculated by difference.

reduced to 50–67% of the value calculated by Howarth et al. (this volume) and the flux of P is reduced to 29–39% of the calculated total discharge from land (Table 12). The reduction in P flux is probably even greater since some estuaries trap all of the P from land and import P from the shelf (e.g., Tables 4 and 8).

Table 6. The annual mass balance of total nitrogen and phosphorus in Narragansett Bay, Rhode Island.[1] Data were collected at various times during the 1980s and early 1990s. Units are mmol m^{-2} y^{-1}. From Nixon et al. (1995).

	N	P
Inputs		
Direct atmospheric deposition	90	0.4
Nitrogen fixation	~0	0
Rivers	1200	65
Direct sewage	550	40
Storm water runoff	20	12
	1960	115
Changes in storage		
Accumulation in water	~0	~0
Accumulation in sediment	140–335	17–38
Outputs		
Denitrification	260–520	0
Fish landings		
Hard clams	10	0.6
Fin fish	3	0.2
Coastal ocean[2]	765–1410	76–98

[1] Area = 328 km^2 including Mt. Hope Bay, mean depth (MLW) = 8.3 m (Pilson 1985).
[2] Calculated by difference. Each term rounded.

The N and P balance on the continential shelf

Deposition from the atmosphere

In addition to inputs of N and P from the large rivers that discharge on the shelf and the net export from estuaries, there is a significant flux of N to the continental shelf from atmospheric deposition (Duce et al. 1991; Paerl 1995). We have estimated this term for each region of the shelf around the North Atlantic by multiplying the shelf area estimates of Pilson & Seitzinger (this volume) by mean area-specific wet and dry N deposition intensities provided by three different atmospheric transport/chemistry models (Prospero et al., this volume). The result suggests that atmospheric deposition may provide an amount of N to continental shelf waters of the North Atlantic equal to 5–10% of the net N flux from large rivers and estuaries, depending on one's choice of the lower or upper range of probable transport from estuaries and one's choice of depth to define the outer edge of the shelf (Table 13). The relative importance of the atmosphere as a pathway for N input varies markedly

Table 7. The annual mass balance of total nitrogen and phosphorus in the Guadalupe Estuary, Texas[1] during a dry (1984) and a wet (1987) year. Data are from Brock et al. (1995) and Longley (1994). Units are mmol m^{-2} y^{-1}.

	N		P	
	1984	1987	1984	1987
Inputs				
Direct atmospheric deposition	40	40	3.6	3.6
Nitrogen fixation	48	48	0	0
Rivers	345	1900	49*	160*
Direct sewage	115	70	17	11
	548	2058	70	171
Changes in storage				
Accumulation in water	25	30	−7	2
Accumulation in sediment[2]	1	65	0.3	16
	26	95	−6.7	18
Outputs				
Denitrification	320	320	0	0
Fish landings[2]	35	85	2.2	5.3
Adjacent bays and gulf[3]	167	1558	75	148

[1] Area = 551 km^2, mean depth (MLW) = 1.4 m.
2 P burial reported by Longley (1994) adjusted to revised N burial reported by Brock et al. (1995) using an N/P ratio of 4 (D. A. Brock, Texas Water Development Board., personal communication. Fish landings adjusted assuming N/P (molar) of 16.
[3] Calculated by difference. Brock et al. (1995) used a numerical model to estimate a net N exchange for this term of 390 mmol m^{-2} y^{-1} in 1984 and 1500 mmol m^{-2} y^{-1} in 1987.
* River and sewage P data from D. A. Brock (personal communication).

among regions. For example, along the North Sea – Northwest European Coast, atmospheric deposition may provide an amount of N equal to 40–110% of the N supplied by the estuaries (Table 13). In fact, the input from the atmosphere is even larger because dissolved organic N is not included in these estimates, and recent evidence indicates that this may be a significant component of the total N deposition from the atmosphere (Cornell et al. 1995).

Denitrification

In a well-known review of estimates of nitrate uptake by continental shelf sediments, Christensen et al. (1987) concluded that denitrification in such sediments was an important term in the oceanic and global N cycle. While

Table 8. The annual mass balance of total nitrogen and phosphorus in the Potomac Estuary, Maryland-Virginia[1] during 1985–1986. Data from Boynton et al. (1995). Units are mmol m^{-2} y^{-1}.

	N	P
Inputs		
Direct atmospheric deposition[2]	113	2
Nitrogen fixation	?	0
Rivers (diffuse sources)	1277	5.6
Direct sewage (point sources)	705	20
Chesapeake Bay[3]	0	16
	2095	44
Changes in storage		
Accumulation in water	0	0
Accumulation in sediment	837	42
Outputs		
Denitrification	330	0
Fish landings	10	1.8
Chesapeake Bay[3]	918	0

[1] Area = 1210 km^2, mean depth (MLW) = 5.9 m
[2] Smullen et al. (1982); 1976–1981 mean TN and TP in wet deposition.
[3] P input and N output calculated by difference.

the summary compiled by Christensen et al. (1987) contains a number of errors and their calculations based on their own pore water nitrate profiles are not necessarily convincing, we agreed with their general conclusion that denitrification in shelf sediments was likely to be an important sink for N in the North Atlantic. For this reason a considerable effort was devoted to obtaining region-specific estimates of denitrification on the North Atlantic shelf. This was a considerable challenge because few measurements of denitrification have actually been made in the area. The details of the approach taken, the assumptions required, and the results obtained are presented elsewhere (Seitzinger & Giblin, this volume). Briefly, Seitzinger and Giblin first developed an empirical regression between primary production in the overlying water and sediment oxygen uptake. That regression was then linked to another regression relating sediment oxygen uptake to measurements of coupled nitrification and denitrification in shelf sediments. In this way, they were able to use relatively abundant primary production data to estimate region-specific benthic denitrification. We recognize that this model will underestimate total denitrification in areas where direct uptake of nitrate from the overlying water contributes significantly to denitrification in the sediments.

Table 9. The annual mass balance of total nitrogen in Ochlockonee Bay, Florida[1] during normal flow conditions. Data from Seitzinger (1987) and unpublished measurements. Units are mmol m^{-2} y^{-1}.

	N
Inputs	
Direct atmospheric deposition[2]	82
Nitrogen fixation	<3.5
Rivers[3]	5910
Direct sewage	0
	5995
Changes in storage	
Accumulation in water	0
Accumulation in sediment[4]	105–125
Outputs	
Denitrification	640
Fish landings	Very small
Coastal ocean[5]	5230–5250

[1] Area = 24 km^2, mean depth at low tide = 1 m.
[2] Hendry & Brezonik (1980), includes TON, NH_4, and NO_3 in bulk precipitation during 1976–1977.
[3] DIN input of Seitzinger (1987) increased by a factor of 5 based on TN:DIN ratios measured in June (4:1) and November (6:1).
[4] Based on ^{210}Pb profiles and measured N concentrations of sediments below 14 cm averaging 86 μmol N gdw^{-1}. Range for different assumptions regarding ^{210}Pb input.
[5] Calculated by difference.

The important conclusion for our effort in this paper is that denitrification appears to remove more nitrogen from every region on the North Atlantic shelf except the Amazon and, perhaps, the Gulf of Mexico, than the combined flux from rivers and the atmosphere can provide (Table 14). For the North Atlantic shelf as a whole, coupled nitrification-denitrification in sediments appears to exceed the effective N input from land and atmosphere by a factor of about 1.4–2.3. The net loss of N on the western side of the North Atlantic is about two-three times that on the eastern side, in spite of a much larger input of N from land on the west (Table 14).

In addition to denitrification, there are two additional fluxes that can remove N and P from biological circulation on the shelf – fisheries landings and long-term burial in sediments.

Fisheries harvest

The North Atlantic contains some of the world's most productive fishing grounds, including the North Sea, the Grand Banks, Georges Bank, and the

Table 10. The recent (1992–1994) annual mass balance of nitrogen and phosphorus in Boston Harbor, Massachusetts prior to sewage diversion.[1] Units are mmol m^{-2} y^{-1}.

	N	P
Inputs[2]		
Direct atmospheric deposition	200	0.3
Nitrogen fixation	0	0
Rivers	590	27
Groundwater	60	3
Storm water	85	4
Sewage	8160	625
	9095	660
Changes in storage[3]		
Accumulation in water	0	0
Accumulation in sediment	185	22
Outputs		
Denitrification[4]	910	0
Fish landings[3]	Very small	0
Dredging[5]	150	9
Coastal ocean[6]	7850	629

[1] Area = 108 km^2, mean depth = 5.8 m (Kelly 1991).
[2] Alber & Chan (1994).
[3] Kelly & Nowicki (1992); P storage calculated from N assuming an N/P ratio of 8.5 as in Narragansett Bay (Nixon et al. 1995).
[4] Giblin et al. (1993).
[5] Kelly & Nowicki (1992); P calculated from N assuming N/P ratio of 16.
[6] Calculated by difference.

Gulf of Mexico. We estimated the removal of N and P in fish by averaging the recent (1987–1991) annual commercial landings from the United Nations Food and Agricultural Organization (FAO) statistical areas that cover the North Atlantic. While finfish account for about 85% of the total North Atlantic catch, we separated the landings of other groups that may differ in their content of N and P (Table 15). If we assume that 90% of the catch is actually taken from the continental shelves, it appears that commercial fishing may remove from the shelf an amount of N equal to 3.6–6.4% of the total input from land and atmosphere (Table 15). Fishing may return to land some 4.1–10% of the total P flux from land onto the shelf (Tables 12 and 15). These calculations underestimate the N and P removal in fish because they do not include recreational landings from the shelf. On the other hand, they overestimate the removal because the FAO data include estuarine commercial landings as well as those from the shelf. It seems likely that inshore commercial landings

Table 11. The annual mass balance of total nitrogen and phosphorus in the western Scheldt Estuary.[1] N data are from Billen et al. (1985) for 1973–1983. P data are less certain but represent approximate conditions for alow flow year (1991). P data from Billen & Servais (1991). Units are mmol m^{-2} y^{-1}.

	N	P
Inputs		
Direct atmospheric deposition	?	?
Nitrogen fixation	?	0
Rivers	7220	770
Direct sewage	3350	260
Direct agriculture	2830	10
	13400	1040
Changes in storage		
Accumulation in water	?	0
Accumulation in sediment	1030	540*
	1030	
Outputs		
Denitrification*	5420	0
Fish landings	?	0
Coastal ocean	6950	500

[1] Area = 277 km^2, mean depth = 11.2 m (Billen, personal communication).
* Calculated by difference by original authors.

exceed the recreational catch on the shelf. In any case, the impact on the shelf N and P budgets is minor.

Burial exclusive of large river deltas

It is difficult to evaluate this term because of the complexity of sediment dynamics on continental shelves. We began by assuming that the burial of N and P only takes place in areas of sediment deposition, because of the general correlation between sedimentation rate and the burial rate of organic matter (Henrichs & Reeburgh 1987; Mayer 1994), and because sediments themselves also carry some inorganic N and P. The sediments being deposited may have been carried onto the shelf very recently by rivers, been eroded recently from headlands, or been eroded from much older deposits elsewhere on the shelf (Eisma 1981).

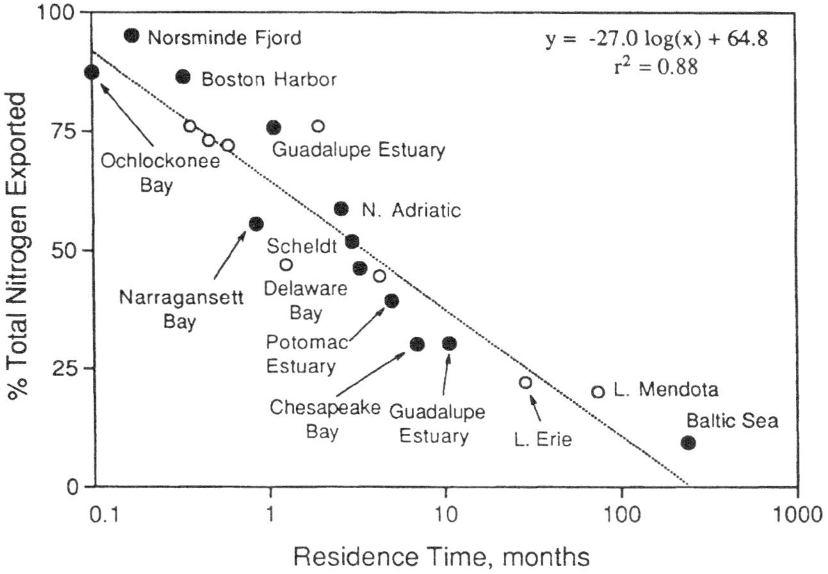

Figure 1. The percent of total nitrogen input from land and atmosphere that is exported from a sample of estuaries and lakes as a function of mean water residence time in the system. Estuarine data (solid points) from Tables 3–11. Norsminde Fjord export from Nielsen et al. (1995). Estuarine residence times from Seitzinger (1987) Ochlockonee Bay, FL; Signell & Butman (1992) Boston Harbor, MA; David Brock (personal communication) Guadalupe Estuary, TX in dry and wet years; Pilson (1985) Narragansett Bay, RI; Billen (personal communication) Scheldt Estuary; Polis & Kupferman (1973) Delaware Bay; W. Boicourt (personal communication) Potomac Estuary, VA-MD and Chesapeake Bay, VA-MD; C. Cerco, U.S. Army Corps of Engineers, recently provided a residence time estimate of 7.6 months for Chesapeake Bay based on a simulated dye study using a hydrodynamic model of the bay; Larsson (1986) Baltic Sea; K. Nielsen (personal communication) Norsminde Fjord. North Adriatic data from Degobbis et al. (1986). Lakes Erie and Mendota (open circles) from various sources summarized by Nixon (1988). Other lakes (unnamed open circles) from a study of six shallow Danish lakes by Andersen (1974). The regression line is for the combined data set. Separate regressions for estuarine and lake data were virtually identical.

Sediments carried by rivers

In their recent review of sediment discharge to the world ocean, Milliman & Syvitski (1992) provided a concise summary of our present state of knowledge:

> What is the sediment flux to the sea? This question really has two parts: how much sediment is carried by rivers, and how much escapes the present-day land/estuarine environment? The answer to both is more or less the same – we don't know.

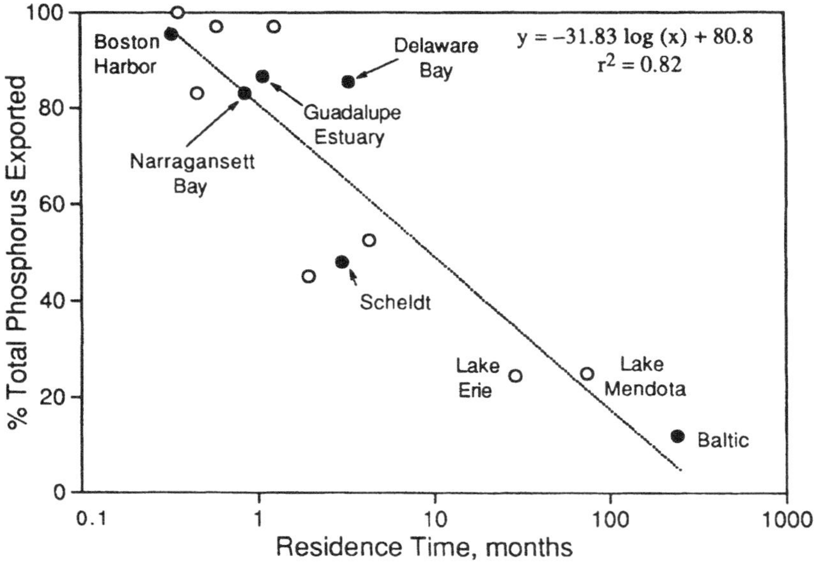

Figure 2. The percent of total phosphorus input from land that is exported from a sample of estuaries and lakes as a function of mean fresh water replacement time in the system. Data sources as in Figure 1. Regression line includes lakes and estuaries but excludes the Guadalupe in a low flow (long residence time) year.

Fortunately, the situation is not quite so difficult for the North Atlantic, where there appears to be good evidence that virtually all (90–95%) of the sediment carried by rivers on the east coast of the United States is deposited in estuaries rather than on the shelf (Meade 1972, 1982; Swift et al. 1981). The situation appears to be the same in the North Sea, where Eisma (1981) found that sediment deposition in estuaries (including the Wadden Sea and The Wash) was essentially equal to the input from rivers. While recognizing that some sediment escapes from the estuaries, especially during large floods, we assume that this flux is sufficiently small for the shelf as a whole that we can ignore it in subsequent calculations.

Sediments from coastal erosion

In the North Sea, this source appears to provide an amount of sediment equal to about 15% of that carried into the estuaries by rivers (Eisma 1988). We are not aware of any comprehensive estimate of this kind for the North Atlantic as a whole, and we are reluctant to extrapolate from the North Sea alone. As a practical matter, we simply acknowledge that coastal erosion may provide an unquantified, but almost certainly small, amount of sediment to the inner shelf. Moreover, even if the amount were known, we would still have to

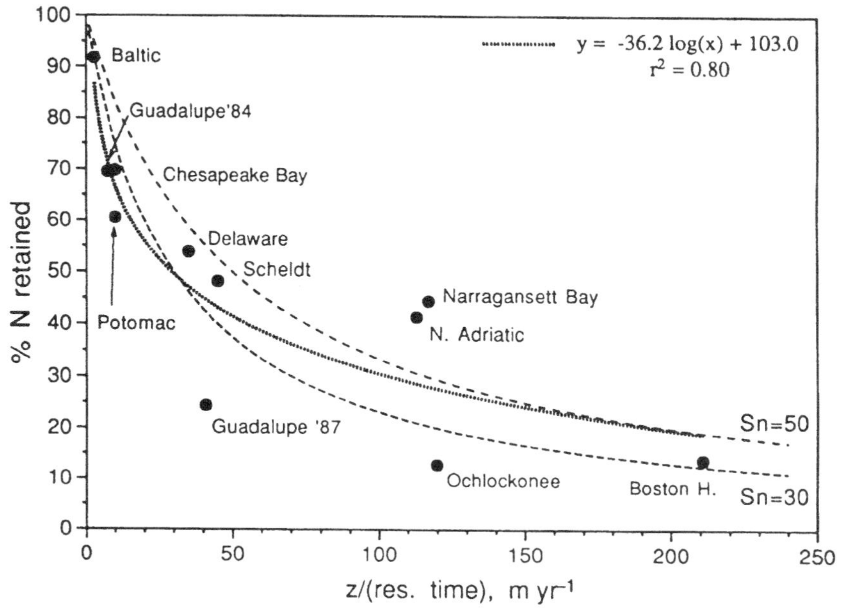

Figure 3. The percent of total nitrogen input from land and atmosphere that is "retained" (buried and denitrified) in various estuaries as a function of the ratio of mean depth (z, m) to fresh water residence time in the system. We have used retention rather than export and years rather than months for the residence time in this figure to be consistent with the discussion of Figure 6 in Howarth et al. (this volume). The broken lines were calculated using the model of Kelly et al. (1987) for mass transfer coefficients (Sn) of 30 and 50. The estuarine systems follow the pattern described for lakes, reservoirs and rivers.

determine the net increase (if any) in N and P concentration on the sediment between the time it was eroded and the time it was buried on the shelf. This would be necessary because Howarth et al. (this volume) did not include coastal erosion in their estimates of N and P flux from land.

Sediments resuspended from relic deposits

Again using the North Sea as an example, Eisma (1981, 1988) estimated that this source contributes as much or more sediment to depositional areas as do rivers. Overall, he calculated that some $14–24 \cdot 10^{12}$ g y^{-1} of fine-grained sediment are redeposited on the bottom of the North Sea outside the estuaries. Major depositional areas include the Skagerrak, the Kattegat, and the German Bight (van Weering 1981; Lohse et al. 1995).

It is difficult to know how much N and P is buried as older deposits are resuspended, transported, and deposited again. Sediments in the large depositional area of the Skagerrak contain 3–6 times more organic C than

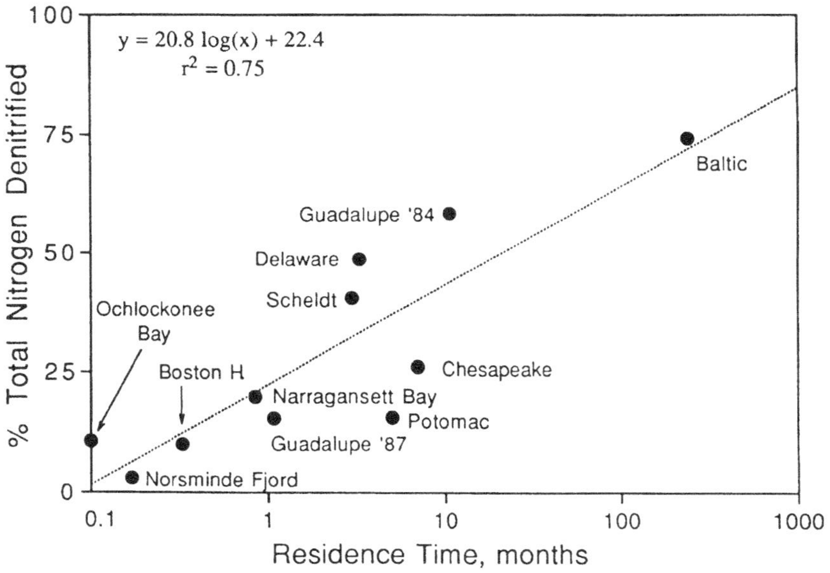

Figure 4. The fraction of total nitrogen input from land and atmosphere that is denitrified in various estuarine systems as a function of mean water residence time. Data sources as in Figure 1. The estimate for Delaware Bay is almost certainly too high because it is based on summer measurements only (see Table 5). Anoxic bottom waters over parts of Chesapeake Bay and the Potomac estuary may reduce the fraction of total N input that is denitrified.

adjacent sediments (van Weering 1981), but this may be due to grain size (Mayer 1994; Milliman 1994; Lohse et al. 1995) and we do not know if there was an increase in the concentrations of N and P on the sediments since they were eroded.

In the absence of specific information for all but a few of the various depositional areas on the North Atlantic shelf, we can only make a very rough beginning in estimating the potential magnitude of N and P burial. Since we are dealing with older sediment that may have been on the shelf for thousands of years (e.g., Bothner & Bacon 1987), the N and P of concern to us must enter the sediment from the deposition of organic matter formed in the water column above. In other words, in contrast to the active river delta sediments, we do not need to consider the mineral P associated with the redeposited sediments, since that P was on the shelf long before the time frame of our mass balance.

Deposition and burial of organic matter

As part of their effort to estimate denitrification on the North Atlantic shelf, Seitzinger & Giblin (this volume) reviewed and summarized numerous

Table 12. Estimated flux of total nitrogen and phosphorus from estuaries onto the continental shelf in various regions around the North Atlantic. The fluxes from large rivers that discharge directly on the shelf have also been included after correction for the estimated burial of N and P in large river deltas and on the continental slope. See text. Boundaries of the regions are given in Howarth et al. (this volume).

	10^9 moles y^{-1}	
	Total N	Total P
North Canadian Rivers	8–15	0.3–0.5
St. Lawrence Basin	16–33	0.3–0.6
Northeast Coast of U.S.[1]	13–26	1.0–1.9
Southeast Coast of U.S.	6–12	0.2–0.3
Gulf of Mexico[2]	136–159	1.1–1.3
Central America and Caribbean Islands[3]	33–43	1.4–2.2
Amazon Basin[4]	161–165	13–14
Total Western Atlantic	373–453	17.3–20.8
Northwest European Coast	11–22	0.4–0.8
North Sea	38–68	1.4–2.9
Southwest European Coast	5–10	0.8–1.6
Northwest Africa[5]	37–74	1.3–2.5
Total Eastern Atlantic	91–174	3.9–7.8
Total North Atlantic	464–627	21.2–28.6

[1] Includes Bay of Fundy and part of Nova Scotia and the U.S. coast to Cape Hatteras, North Carolina.
[2] Effective contribution from the Mississippi plus Grijalva-Usumacinta = $113 \cdot 10^9$ moles N and $0.91 \cdot 10^9$ moles P.
[3] Effective contribution from the Orinoco plus Magdelena = $22.5 \cdot 10^9$ moles N and $0.7 \cdot 10^9$ moles P.
[4] Effective contribution from the Amazon = $157 \cdot 10^9$ moles N and $11.8 \cdot 10^9$ moles P.
[5] Not including any net flux to the deep Atlantic from the Mediterranean (see Michaels et al. this volume).

measurements of primary production in the area. They also examined the relationship between primary production in the water and the consumption of organic matter by shelf sediments. While there is considerable variation, almost all of the data were consistent with the view that about 10–30% of the shelf primary production was consumed in or on the sediments. This is in agreement with more detailed studies of the U.S. shelf south of New England (Rowe et al. 1988) and the North Sea (Billen et al. 1990). As a working number, Seitzinger & Giblin (this volume) used 16% for their estimate of the fraction of primary production respired on the bottom. Given the scatter in the data, and considering that the deposition of organic matter on the bottom

Table 13. Estimated deposition of NO_y and NH_x (wet plus dry) from the atmosphere onto the continental shelf in various regions around the North Atlantic. Deposition rates are from Prospero et al. (this volume), shelf areas (0–100 m depth and 0–200 m depth) are from Pilson & Seitzinger (this volume). Boundaries of the regions are given in Howarth et al. (this volume).

	10^9 moles y^{-1}			
	NO_y		$NO_y + NH_x$	
	0–100 m	0–200 m	0–100 m	0–200 m
North Canadian Rivers[1]	3.5	6.4	4.9	9.0
St. Lawrence Basin[2]	4.7	7.8	6.6	11.0
Northeast Coast of U.S.[3]	8.8	14	11	18
Southeast Coast of U.S.[4]	2.7	3.0	3.9	4.4
Gulf of Mexico[5]	11	13	18	21
Central America and Caribbean Islands[6]	1.9	2.4	4.6	5.8
Amazon Basin[7]	1.8	1.9	5.4	5.7
Total Western Atlantic	34.4	48.5	54.4	74.9
Northwest European Coast[8]	20	29	38	54
Southwest European Coast[9]	0.7	1.1	1.1	1.8
Northwest Africa[10]	1.9	2.3		
Total Eastern Atlantic	41.1	58.2		

[1] 0–100 m = 708,616 km^2 0.200 m = 1,276,783 km^2
[2] 0–100 m = 315,770 km^2 0.200 m = 518,998 km^2
[3] 0–100 m = 230.780 km^2 0.200 m = 361,042 km^2
[4] 0–100 m = 123,338 km^2 0.200 m = 137,364 km^2
[5] 0–100 m = 500,926 km^2 0.200 m = 573,565 km^2
[6] 0–100 m = 381,070 km^2 0.200 m = 474,755 km^2
[7] 0–100 m = 359,579 km^2 0.200 m = 389,562 km^2
[8] 0–100 m = 786,236 km^2 0.200 m = 1,181,318 km^2
[9] 0–100 m = 33,986 km^2 0.200 m = 51,435 km^2
[10] 0–100 m = 319,325 km^2 0.200 m = 392,888 km^2
N.W. European coast includes Norway through France, excluding the Baltic.
S.W. European coast includes Spain and Portugal.
Cape Hatteras divides N.E. and S.E. coast of the U.S.

can not be much larger than the amount respired, we assumed that deposition on average was equal to 25% of primary production. This is lower than the 32% estimated by Wollast (1993) for the world coastal zone, but higher than the findings of Billen et al. (1990), who estimated an average deposition of 25 g C m^2 y^{-1} in the whole of the North Sea, a value equal to about 18% of the primary production according to Seitzinger and Giblin's review.

Unfortunately, there are few direct measurements of organic deposition on non-upwelling shelves, and even fewer measurements of net long-term

Table 14. Estimated balance of nitrogen on the continental shelf in various regions around the North Atlantic excluding exchanges with the open ocean. Boundaries of the regions are given in Howarth et al. (this volume). Fluxes from the land are from Table 12; rates of atmospheric deposition are from Prospero et al. (this volume) as calculated in Table 13; denitrification rates are from Seitzinger & Giblin (this volume). Low and high ranges in atmospheric deposition and denitrification apply to shelf areas with 0–100 m or 0–200 m depth contours, respectively (Pilson & Seitzinger, this volume).

	10^9 moles y^{-1}			
	Flux from land[1]	Atmospheric deposition	Denitrification loss	Net input or (loss)
North Canadian Rivers	8–15	4.9–9.0	49–99	(29–82)
St. Lawrence Basin	16–33	6.6–11	94–155	(54–128)
Northeast Coast of U.S.	13–26	11–18	114–178	(77–147)
Southeast Coast of U.S.	6–12	3.9–4.4	105–117	(89–107)
Gulf of Mexico	136–159	18–21	171–195	6-(38)
Central America and Caribbean Islands	33–43	4.6–5.8	101–122	(53–83)
Amazon Basin	161–165	5.4–5.7	77–84	83–93
Total Western Atlantic	373–453	54–75	711–950	(204–502)
Northwest European Coast	49–90	38–54	165–258	(37–155)
Southwest European Coast	5–10	1.1–1.8	19–29	(7.9–22)
Northwest Africa	37–74	1.9–2.3	98–171	(22–132)
Total Eastern Atlantic	91–174	41–58	282–458	(67–309)
Total North Atlantic[2]	464–627	95–133	993–1408	(271–811)

[1] Does not include the N and P from very large rivers that we consider to be buried in deltas and on the continental slope.

[2] Seitzinger & Giblin (this volume) also calculated a denitrification loss in the shallow waters around Greenland & Iceland that amounted to an additional $22 \cdot 10^9$ moles y^{-1}.

Table 15. Estimated removal of nitrogen and phosphorus in fisheries landings from the North Atlantic continental shelf. Values are based on mean annual landings from FAO statistical areas 21, 27, 31 and 34 during 1987–1991.

	Landings, metric tons[1]	Nitrogen and Phosphorus Removed[2] 10^9 moles y^{-1}	
		N	P
Oysters	334,400	0.37	0.012
Mussels	510,100	0.46	0.028
Scallops	324,900	0.51	0.027
Clams, etc.	498,200	0.96	0.024
Squids, octopuses	286,900	0.41	0.018
Lobsters	152,200	0.25	0.010
Crabs	182,600	0.30	0.012
Shrimp, prawns	480,600	1.20	0.034
Fin fish	15,887,700	22.7–34.0	1.02–2.05
Total	18,656,900	27.2–38.5	1.18–2.21

[1] From FAO (1993).
[2] N composition from Vinogradov (1953): mussels = 7.2% fresh weight, oysters = 7.7%, clams = 2.7%, scallops = 2.2%, squid and octopuses = 2% (Pilson 1970), lobster and crabs = 2.3%, shrimp = 3.5%, fin fish = 2–3%.
P composition from Vinogradov (1953): mussels = 0.17% fresh weight, oysters = 0.11%, clams = 0.15%, scallops = 0.26%, squid and octopuses = 0.2%, lobsters and crabs = 0.2%, shrimp = 0.22%, fin fish = 0.2–0.4%.

burial as a fraction of deposition. Based on the work of Canfield (1989) and Blackburn (1991), we have assumed a C preservation equal to 1–5% of deposition, or between 0.25 and 1.25% of primary production. The fact that this leaves a balance between our deposition estimates and the respiration estimates of Seitzinger & Giblin (this volume) is of no real importance given the variability of all of the data. It is probably also useful to note that a net accumulation of organic matter in shelf sediments is neither inconsistent with, nor supportive of, the view of Smith & Hollibaugh (1993) that continental shelves are essentially in metabolic balance, with organic production equalling consumption. Even in shelf areas such as the Georgia Bight off the southeast coast of the U.S., where Hopkinson (1985) found a great excess of respiration over production, it is possible that a fraction of the shelf primary production is buried while a large flux of organic matter from land supports an overall heterotrophic carbon balance.

Our estimate of a total North Atlantic shelf deposition of $19 \cdot 10^{12}$ moles C y^{-1} and burial of $0.2–1 \cdot 10^{12}$ moles C y^{-1} can be compared in a very rough way to global estimates made by others if we consider that the North

Atlantic (0–200 m) contains about 20% of the world shelf area. Wollast (1993) concluded that 183 · 10^{12} moles C y^{-1} (2.2 · 10^{15} g C) were sedimented onto the world continental shelves, about twice as much as a simple prorating of our estimate would give. However, his value presumably includes estuaries and active river deltas while ours does not, and a global estimate must also include very productive coastal upwelling areas with high rates of organic deposition and accumulation. Smith & Hollibaugh (1993) concluded that 9 · 10^{12} moles C y^{-1} were put into long-term burial on the "coastal bottom" of the world ocean, a value 1.8–9 times greater than ours. Again, however, their estimate presumably includes estuaries, river deltas, and major upwelling areas that are excluded from our value. All of these estimates must be considered as very rough approximations.

We next calculated N burial assuming a C/N (molar) ratio of 10. Walsh (1983) found such a value over large areas of the Gulf of Mexico shelf outside the Mississippi Delta, though there was also considerable local variation. Unfortunately, it was more difficult to decide on an approximate C/organic P ratio, since it is evident from a recent review that that ratio varies widely as a function of sedimentation rate (Ingall & Van Cappellen 1990). It is possible, of course, that a comparable review would show that our choice of a constant value for the C/N ratio was also an oversimplification.

There have been relatively few measurements of deposition rates on the North Atlantic shelf. One 13,000 km^2 area of fine-grained sediments on the southern New England shelf has been studied in some detail (Twichell et al. 1981), and geochemical evidence indicated a modern deposition rate of about 25 cm/1000 years (Bothner et al. 1981). This rate suggests a C/organic P ratio of about 500 (Ingall & Van Cappellen 1990). On the other hand, deposition rates in the German Bight and the Skagerrak appear to be much higher, perhaps on the order of 0.5 cm y^{-1} (Lohse et al. 1995). Eleven cores of fine-grained inner shelf sediments off the southeast U.S. coast dated using ^{210}Pb also showed high rates of deposition that averaged 0.9 cm y^{-1} (Pomeroy et al. 1993). Sediments accumulating at these higher rates are characterized by a lower C/organic P ratio of about 250 (Ingall & Van Cappellen 1990). In the absence of a convincing area-weighted sedimentation rate, we assumed a total shelf C/organic P range of 250–500.

The results suggest that the burial of N on the shelf outside of major river deltas is very small (Table 16) and amounts to only about 1–7% of the total denitrification estimate calculated by Seitzinger & Giblin (this volume). The removal of P may be more significant, however, accounting for 1–20% of the net flux onto the shelf from estuaries and very large rivers (Table 12). In areas away from the large rivers, the burial of organic P may be particularly important. For example, in the North Sea, we estimated that 1.4–2.9 · 10^9

Table 16. Estimated potential burial of N and P in North Atlantic continental shelf (0–200 m) sediments exclusive of major river deltas.[1] Burial calculated assuming 0.25–1.25% of the C fixed in the water column is preserved in long-term burial and that sediments accumulate with a C/N (molar) of 10 and C/organic P of 250–500.

	Primary production[2] 10^{12} moles C y^{-1}	Burial, 10^9 moles y^{-1}	
		N	P
North America	39.2	9.8–49	0.20–1.96
Central and South America	10.8	2.7–14	0.05–0.54
Europe	16.3	4.1–20	0.08–0.81
N. W. Africa	8.9	2.2–11	0.05–0.44
Greenland and Iceland	1.2	0.3–1.5	0.01–0.38
	76.4	19–95	0.39–4.13

1 These estimates also do not include 50% of the sediments discharged by the very large rivers that we assume are deposited outside the deltas on the shelf or over the shelf-slope break. See Table 17.
2 From Seitzinger & Giblin (this volume) who provide a more detailed geographical assessment. They used the shelf area within the 200 m depth contour.

moles of P were discharged from the estuaries (Table 12). Using Seitzinger & Giblin's (this volume) estimate for primary production of 32 mmol C m^{-2} d^{-1} (140 g C m^{-2} y^{-1}), we calculated a P burial rate on the North Sea shelf of 0.07–0.70 · 10^9 moles P y^{-1}, or 2.4 –50% of the input from land. It is obvious from the example, however, that large uncertainties very much limit our ability to develop a mass balance for the shelf.

Exchanges at the shelf break

If we summarize the estimates developed up to this point, it appears that there is a net flux of active P from the continental shelf to the open North Atlantic that is equal to 17.7–30.4 · 10^9 moles y^{-1}, or 24–41% of the value that Howarth et al. (this volume) estimate for the total P loss from land (Table 17). In addition there is an estimated 20 · 10^9 moles P y^{-1} buried on the continental slope with sediments from the large rivers. The situation with regard to N, however, is very different. The shelf budget developed thus far requires an additional large net source of N that is approximately equal to or greater than the combined input of N from large rivers, estuaries, and atmospheric deposition (Table 17).

There are two potential sources for the missing N – biological fixation directly on the shelf and the nutrient-rich deeper ocean waters off the shelf that Csanady (1990) termed the "slope sea." It is likely that the first is of little importance. Walsh (1988) estimated that N fixation on the entire shelf area of the world ocean was less than 100 · 10^9 moles N y^{-1}. Since the area of

Table 17. Comparison of the estimated fluxes of active N and P onto the North Atlantic continental shelf (0–200 m depth) from land and atmosphere with removal processes on the shelf. We estimate that an additional $61 \cdot 10^9$ moles y^{-1} of N and $20 \cdot 10^9$ moles y^{-1} of P are buried on the shelf in deltas of the very large rivers. We also assume that equal amounts of N and P associated with sediments carried by the very large rivers are deposited and buried directly on the continental slope.

	N	P
	Input 10^9 moles y^{-1}	
From direct atmospheric deposition	133	Very small
From very large rivers	292	13
From estuaries	172–335	11–19
Total	597–760	24–32
	Removal 10^9 moles y^{-1}	
Organic matter buried on the shelf	19–95	0.4–4.1
Fisheries harvest	27–38	1.2–2.2
Denitrification[1]	1408	0
Total	1454–1541	1.6–6.3

[1] Does not include shallow areas around Greenland and Iceland.

shelf in the North Atlantic (0–200 m) accounts for about 20% of the world total, biological fixation probably supplies less than $20 \cdot 10^9$ moles N y^{-1} compared to the apparent deficit of $690–940 \cdot 10^9$ moles N y^{-1} (Table 17). The largest single source of N for the North Atlantic shelf as a whole must be the onwelling of deeper nutrient-rich water from the continental slope.

Onwelling of N and P from the slope

As long ago as the 1930s, a few oceanographers noted that the physical circulation on at least some parts of the North Atlantic shelf was such that the slope waters could potentially make an important contribution to the nutrient supply and productivity of the shelf. As Iselin (1939) explained:

> The fact that the salinity is at a maximum near the bottom on the continental shelf shows conclusively that the coastal circulation has an off shore component at the surface and an in shore component beneath ... During the last few years physical oceanographers have begun to appreciate the important role which lateral turbulence plays in the sea ... Under the conditions of high stability often found in coastal waters, the isopicnal transport of nutrients should prove a most fruitful study ... The chemicals which are in this way pumped up from mid depths in the Sargasso Sea are transferred isopicnally across the Gulf Stream to the edge of the

continental shelf. Thus as the bottom water penetrates landward, it carries with it a good supply of nutrients ...

More recent work has shown that the physical exchange processes can be more complicated than this early picture (e.g., Walsh et al. 1978; Csanady 1990; Verity et al. 1993) and few attempts have been made to measure or calculate the fluxes of nitrate onto various regions of the North Atlantic shelf. The results that are available range by a factor of three, from about 1.9 mmol N m^{-1} of shelf break s^{-1} off Nova Scotia to about 5.7 mmol m^{-1} s^{-1} off the southeast coast of the U.S. (Table 18). When averaged over the entire width of the shelf in each area, these slope-shelf transports translate into N inputs amounting to 0.3 to 1.1 mol N m^{-2} y^{-1} between eastern Nova Scotia and Florida (Table 18). Georges Bank and, perhaps, a few other special areas are more intensively enriched.

If similar rates of onwelling characterize the entire North Atlantic shelf (0–200 m), this process would supply 1.7–6.2 $\cdot$ 10^{12} moles N y^{-1}, far more than enough to meet the deficit of 0.6–0.9 $\cdot$ 10^{12} moles N y^{-1} identified in Table 17. Unfortunately, the area of shelf between Nova Scotia and Florida where the most intensive effort has been made to quantify the slope-shelf exchange accounts for less than 10% of the total shelf area (0–200 m) in the North Atlantic (Pilson & Seitzinger, this volume).

We have also estimated the flux of nitrate onto the entire North Atlantic shelf using an empirical regression relating primary production (^{14}C uptake) to the supply of dissolved inorganic nitrogen (DIN) in a variety of phytoplankton-based marine systems (Figure 5). The plot shown modifies earlier versions (Nixon 1992; Nixon et al. 1994) in that we have only included systems where all of the DIN inputs have been considered. There are numerous estuarine sites for which primary production measurements and N input inventories from land and atmosphere are available, but there are few studies that also provide estimates of the gross flux of DIN into the estuary from the coastal ocean. This version also includes recent measurements of ^{14}C uptake from the Sargasso Sea and the North Pacific made using "clean techniques," as well as revised estimates of DIN input to those systems.

The review of primary production measurements given by Seitzinger & Giblin (this volume) provides an area-weighted mean value of 165 g C m^{-2} y^{-1} for the entire North Atlantic shelf. The relationship shown in Figure 5 suggests that this level of production would be associated with a DIN input of about 0.5 moles N m^{-2} y^{-1} over the total shelf area (0–200 m) of 5.6 $\cdot$ 10^6 km^2 (Pilson & Seitzinger, this volume). The total required flux of 2.9 $\cdot$ 10^{12} moles DIN y^{-1} might consist of 0.6–0.8 $\cdot$ 10^{12} moles N from land and atmosphere (if we ignore the fact that a significant but unknown fraction of the N flux from large rivers and estuaries is organic N) and 2.2–2.3 $\cdot$ 10^{12}

Table 18. Estimates of the flux of nitrate from slope waters onto various regions of the North Atlantic continental shelf

	mmol N m^{-1} s^{-1}	mol N m^{-2} y^{-1}
Scotian Shelf[1]	1.9	0.32
Georges Bank[2]		5.0
New York Shelf[3]		4.6
New York Shelf[4]	3.4	0.8
Mid Atlantic Shelf[5]	4.0	1.1
Outer Southeast U.S. Shelf[6]		5.6
Southeast U.S. Shelf[7]	5.7	1.0

[1] Houghton et al. (1978); flux per unit length of shelf at 50–100 m; area estimate for entire width of shelf.
[2] Walsh et al. (1987); estimate for entire area within 60 m isobath.
[3] Walsh et al. (1987). It is not clear how much of the entire shelf is included in this estimate of flux per unit area.
[4] Flux per unit length of shelf-slope break from Walsh et al. (1987); flux per unit area for entire shelf calculated using a mean width between Long Island and Cape May of 139 km (Stommel & Leetma 1972).
[5] Flux per unit length of 900 km long shelf from Csanady (1990); flux per unit area calculated for entire shelf area assuming a mean width between Nantucket shoals and Cape Henry of 113 km (Stommel & Leetmaa 1972).
[6] Flux per unit area calculated assuming slope waters only influence the outer 1/3 of the shelf area (Verity et al. 1993).
[7] Flux per unit length from Verity et al. (1993); flux per unit area calculated for entire width of shelf as given by Menzel et al. (1993).

moles of DIN from the slope. These values are consistent with the range of $1.7–6.2 \cdot 10^{12}$ moles DIN y^{-1} we calculated for the flux from the slope water by extrapolating from the measurements off Nova Scotia and the east coast of the U.S. Considering the uncertainty in all of these numbers, we accept a range of $2–6 \cdot 10^{12}$ moles DIN y^{-1} as a reasonable working estimate for the gross flux onto the shelf from the deeper ocean.

Since dissolved inorganic phosphorus (DIP) is present in the slope water with a DIN/DIP ratio of approximately 16, we estimate that the DIN flux of $2–6 \cdot 10^{12}$ moles y^{-1} would be accompanied by a DIP flux of $0.12–0.37 \cdot 10^{12}$ moles y^{-1}. This compares with the addition of only $0.02–0.03 \cdot 10^{12}$ moles of active P onto the shelf from large rivers and estuaries (Table 17).

The onwelled slope waters also contain particulate and dissolved organic N and P, but we do not know how to make a credible estimate of their flux onto the shelf. This may be of little consequence because we assume that the concentration of particulate organic matter in the deeper slope water is low and that the dissolved organic matter is sufficiently refractory that it is not metabolized to any significant degree during its relatively brief residence time on the shelf (e.g., about three months on the southeast U.S. shelf, Menzel et

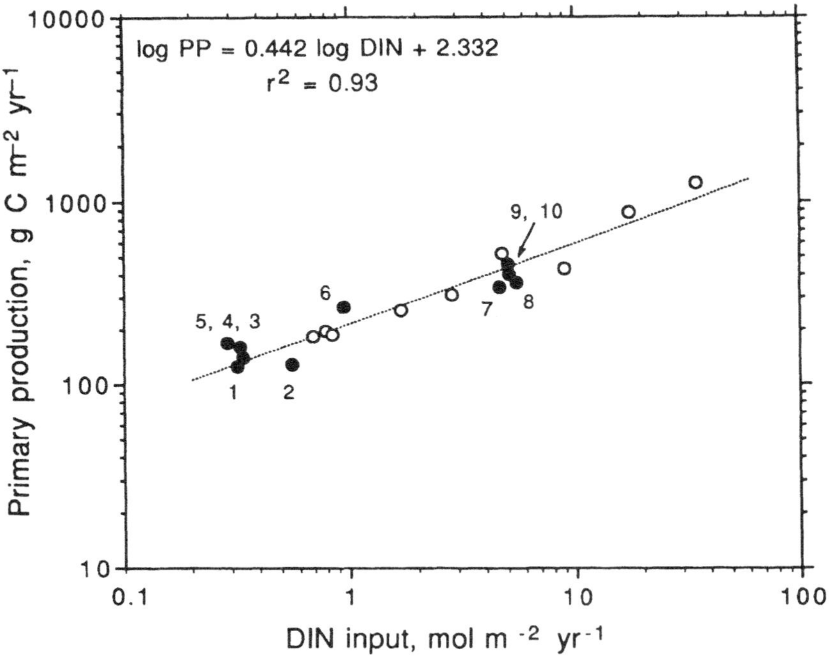

Figure 5. Primary production by phytoplankton (^{14}C uptake) as a function of the estimated rate of input of dissolved inorganic nitrogen (DIN) per unit area in a variety of marine ecosystems. The open circles are for large (13 m^3, 5 m deep) well-mixed mesocosm tanks at the Marine Ecosystems Research Laboratory (MERL) during a multi-year fertilization experiment (Nixon et al. 1986; Nixon 1992). Natural systems (solid circles) include (1) Scotian shelf – DIN from Houghton et al. (1978), production from Mills & Fournier (1979), (2) Sargasso Sea – DIN from Jenkins (1988), production from Lohrenz et al. (1992) mean of 1989 and 1990 values of 110 and 144 g C m^{-2} y^{-1}, (3) North Sea – DIN from Laane et al. (1993) assuming that the ratio of DIN/TN in the input from the Atlantic equals that in the Channel, production from Seitzinger & Giblin (this volume), (4) the Baltic Sea – DIN and production from Rönner (1985), including DIN flux across the halocline, (5) North Central Pacific – DIN from Platt et al. (1984), production from Tupas et al. (1993, 1994) mean of 1992 and 1993 values of 150 and 185 g C m^{-2} y^{-1}, (6) Tomales Bay, CA, DIN and production from Smith (1991), (7) Continental shelf off New York – DIN and production from Walsh et al. (1987), (8) Outer continental shelf off southeastern U.S., DIN and production from Verity et al. (1993), (9) Peru upwelling – DIN calculated from annual mean upwelling rate of 0.77 m d^{-1} (Guillén & Calienes 1981) and an initial 20 μM concentration of NO_3 in upwelled water (Walsh et al. 1980), production off Chimbote from Guillén & Calienes (1981), (10) Georges Bank – DIN from Walsh et al. (1987), production from O'Reilly et al. (1987). The equation is a functional regression.

al. 1993). As a practical matter, there may be essentially no net exchange of dissolved organic N or P between shelf and slope, and we ignore these fluxes in our budgets.

This discussion of the additions of N and P to the shelf from the slope leads directly to the last term in the shelf budget – the flux of N and P off the shelf. We have already established that the loss of N from the shelf in denitrification and burial exceeds the input from land and atmosphere (Table 17), but the onwelling of DIN from the slope is potentially large enough to meet the deficit and leave a large amount of excess N on the shelf. The onwelling of DIP is also much larger than the potential export of terrestrial P from the shelf (Table 17). Since the concentrations of N and P on the North Atlantic shelf as a whole are presumably not increasing (except perhaps in some local areas near shore), we can calculate this last term in the shelf N and P budgets by difference. Before proceeding to do so, however, it is useful to make some effort to estimate the amount of shelf to slope transport that consists of organic N and P in the form of phytoplankton detritus.

Export of organic N and P from the shelf

There has been a considerable recent effort along the Atlantic Coast of the U.S. to measure the loss of organic matter from the shelf and its possible accumulation in slope sediments. Much of the motivation for this work has come from the suggestion that such burial may be an important sink in the global carbon budget (e.g., Walsh et al. 1981, 1985; Walsh 1991). Unfortunately, it is difficult to measure the rate and extent of burial and, as noted by Yoder & Ishimaru (1989), "the magnitude of off-shelf phytoplankton carbon transport, the processes affecting the transport rate and the significance of such processes to global budgets are in dispute ..." For example, while Walsh et al. (1987) estimated that 2.2 moles N m^{-2} y^{-1} were exported to the slope from the shelf off New York (52% of the N consumed in annual primary production), a more recent budget developed for the same area by Rowe et al. (1988) concluded that there "... was little evidence that substantial organic matter escapes from the continental shelf to the upper continental slope." It is also apparent that the flux of organic matter from the shelf is not uniform. Even in an area thought to be an important export region, such as the slope off Cape Hatteras, NC (USA), measurements have shown highly variable rates of carbon and nitrogen burial over relatively small (150–180 km) distances (DeMaster et al. 1994).

Based on detailed studies of the carbon budget of the mid-slope deposition center of the Middle Atlantic Bight, Anderson et al. (1994) recently concluded that less than 2% of the shelf primary production could be buried in slope sediments. If we assume that an export of 1–2% of primary production might be realistic for the North Atlantic shelf as a whole, then the total primary production of $76.4 \cdot 10^{12}$ moles C y^{-1} estimated by Seitzinger & Giblin (this volume) would give a flux of $0.76–1.53 \cdot 10^{12}$ moles C y^{-1} from the shelf to the slope. If this is fresh organic matter with a Redfield C:N:P ratio of 106:16:1, there would be an associated loss of $115–230 \cdot 10^9$ moles of N and $7–14 \cdot 10^9$ moles of P. This would represent a return to the slope in organic form of about 2–12% of the onwelled DIN and DIP.

The Balance of Total N and P on the North Atlantic shelf

The various estimates summarized in Table 19 suggest a picture that is very different from textbook sketches that show N and P from land fertilizing the ocean. Denitrification in estuaries and on the continental shelf appears to return to the atmosphere an amount of N that exceeds the total N input from land and the atmosphere. If the estimates developed by Seitzinger & Giblin (this volume) and in this paper are approximately correct for the North Atlantic as a whole, denitrification also consumes between about 12 and 37% of the DIN brought onto the shelf from the deeper ocean. Overall, our budget suggests that the ocean provides a net flux of about $700–950 \cdot 10^9$ moles N y^{-1} to the North Atlantic shelf (Table 19), an amount approximately equal to the total N input from all other sources combined. If N associated with river sediments buried on the continental slope is included in the balance, the net loss from the ocean would be reduced by about $60 \cdot 10^9$ moles y^{-1}.

The situation with regard to P is quite different. About 70% of the total flux of P from the North Atlantic drainage basin is carried by the five very large rivers that discharge directly on the shelf. Even allowing for the burial of a large amount of this P in the deltas of the rivers and in riverine sediments deposited over the shelf-slope break, the addition of active P from these systems is 2–8 times greater than the removal of P from the shelf in fisheries and the burial of organic matter on the shelf (Table 19). The combination of $11–19 \cdot 10^9$ moles P that likely escape from estuaries, plus the $13 \cdot 10^9$ moles of active P from the large rivers provides for a net export of P from the shelf to the ocean that amounts to about $17.7–30.4 \cdot 10^9$ moles P y^{-1} (Table 19). An additional $20 \cdot 10^9$ moles P y^{-1} may reach the ocean but remain buried on the slope.

Since the shelf appears to be such a strong sink for N while allowing P to pass from land to the ocean, there could be a marked reduction in the N/P ratio of water during residence on the shelf. The estimates compiled in Table 19 suggest that the ratio of DIN/DIP exported from the shelf might be somewhere between 7 and 13. However, the large range again emphasizes the very preliminary stage of the shelf budget developed in Table 19. While it is clear that there is a great potential for biogeochemical processes in estuaries and on the shelf to modify greatly the exchange of N and P between land and the ocean, it is also clear that our ability to quantify that influence is still far from satisfying. The next order of business should be to obtain

Table 19. A preliminary assessment of the annual mass balance of active N and P on the continental shelf (0–200m) of the North Atlantic Ocean. Units are 10^9 moles y^{-1}. These estimates do not include an additional $122 \cdot 10^9$ moles y^{-1} of N and $40 \cdot 10^9$ moles y^{-1} of P that are carried by five very large rivers and buried with riverine sediments in deltas and on the continental slope. See text.

	N	P
Inputs		
Direct atmospheric deposition[1]	133	Very small
Biological nitrogen fixation[2]	~20	0
Estuaries[3]	172–335	11–19
Very large rivers[4]	292	13
Continental slope[5]	2000–6000	120–370
	2617–6780	144–402
Changes in storage		
Organic matter buried on shelf[6]	19–95	0.4–4.1
Outputs		
Fish landings[7]	27–38	1.2–2.2
Denitrification[8]	1430	0
Continental slope		
As organic detritus[9]	115–230	7–14
As DIN and DIP[10]	824–5189	124–393

[1] From Table 13.
[2] Prorated on the basis of world shelf rate from Walsh (1988).
[3] From Table 12 minus footnote values.
[4] From Table 1, minus burial in deltas and on the slope (see Table 2).
[5] See Table 18 and Text.
[6] From Table 16.
[7] From Table 15.
[8] From Table 14; includes shallow areas around Greenland and Iceland.
[9] Assuming 1–2% of primary production exported. See text.
[10] Calculated by difference. Ranges from maximum input minus minimum losses and vice versa. Thus, total flux to the slope = 1054–5304 $\cdot 10^9$ moles N y^{-1} and 138–400 $\cdot 10^9$ moles P y^{-1}.

direct measurements of denitrification in a variety of shelf environments, to improve our estimates of P burial, and to extend our knowledge of shelf-slope exchanges to additional areas. There is also great room for improvement in our understanding of the amount and fate of N and P associated with the sediments carried by the large rivers.

Acknowledgements

The authors are grateful to David Brock, Kurt Nielsen, and Carl Cerco for providing unpublished data and calculations and to Mike Kemp, Bjorn Sundby, and Ivan Valiela for helpful review comments. Preparation of the manuscript was supported by the R. I. Sea Grant College Program.

References

Alber M & Chan AB (1994) Sources of contaminants to Boston Harbor: Revised loading estimates. Report to Massachusetts Water Resources Authority, Boston, MA. Tech. Rept. No. 94–1

Aller JY & Aller RC (1986) General characteristics of benthic faunas on the Amazon inner continental shelf with comparison to the shelf off the Changjiang River, East China Sea. Cont. Shelf Res. 6: 291–310

Andersen JM (1974) Nitrogen and phosphorus budgets and the role of sediments in six shallow Danish lakes. Arch. Hydrobiol. 74(4): 528–550

Anderson RF, Rowe GT, Kemp PF, Trumbore S & Biscaye PE (1994) Carbon budget for the mid-slope depocenter of the Middle Atlantic Bight. Deep-Sea Res. II 41(2/3): 669–703

Balls PW (1994) Nutrient inputs to estuaries from nine Scottish East coast Rivers; influence of estuarine processes on inputs to the North Sea. Est. Coast. Shelf Sci. 39(4): 329–352

Berner RA (1982) Burial of organic carbon and pyrite sulfur in the modern ocean: Its geochemical and environmental significance. Amer. J. of Sci. 282: 451–473

Biggs RB & Beasley EL (1988) Bottom and suspended sediments in the Delaware River and Estuary. In: Majumdar SK, Miller EW & Sage LE (Eds) Ecology and Restoration of the Delaware River Basin (pp 116–131). The Penn. Acad. of Sci., Philadelphia, PA

Billen G & Servais P (1991) Modélisation du transport de pollutants par l'estuaire de l'Escant, Cas du Phosphore. In Unité de Gestion du Modèle Mathématique de la Mer du Nord et de L'Estuaire de 'Escaut, Ministère de la Santé Publique et de l'Environment. Buxelles, Belgique. 51 p

Billen G, Somville M, De Becker E & Servais P (1985) A nitrogen budget of the Scheldt hydrographical basin. Neth. J. Sea Res. 19(3/4): 223–230

Billen G, DeBecker E, Servais P, Lancelot C & Rousseau V (1990) Modèle mathématique de l'estuaire de l'Escaut. Développement d'un algorithme fournissant les conditions-limites amont (Modèle STRAHLER) au mod le de dégradation de la matière organique dans l'estuaire (Modèle MODESTY). Int. Rep. Unité de Gestion du Modèle Mathématique de la Mer du Nord et de l'Estuaire de l'Escaut. Réf. BH/88/26

Blackburn TH (1991) Accumulation and regeneration: Processes at the benthic boundary layer. In: Mantoura RFC, Martin JM & Wollast R (Eds) Ocean Margin Processes in Global Change (pp 181–195). Dahlem Workshop Reports, Wiley & Sons

Bothner MH & Bacon MP (1987) Natural radionuclides. In: Backus RH & Bourne DW (Eds) Georges Bank (pp 185–189). MIT Press, Cambridge, MA

Bothner MH, Spiker EC, Johnson PP, Rendigs RR & Aruscavage PJ (1981) Geochemical evidence for modern sediment accumulation on the continental shelf off southern New England. J. Sed. Petr. 51(1): 281–292

Boynton WR, Garber JH, Summers R & Kemp WM (1995) Inputs, transformations and transport of nitrogen and phosphorus in Chesapeake Bay and selected tributaries. Estuaries 18(1B): 285–314

Brock DA, Matsumoto J & Boyd N (1995) A nitrogen budget for the Guadalupe Estuary, Texas, over a range of freshwater inflow. Submitted to Contrib. in Mar. Sci.

Bugden GL (1981) Salt and heat budgets for the Gulf of St. Lawrence. Can. J. Fish. Aquat. Sci. 38: 1153–1167

Canfield DE (1989) Sulfate reduction and oxic respiration in marine sediments: implications for organic carbon preservation in euxinic environments. Deep Sea Res. 36: 121–138

Caraco N, Cole J & Likens GE (1990) A comparison of phosphorus immobilization in sediments of freshwater and coastal marine systems. Biogeochem. 9: 277–290

Cerco C & Cole T (1992) Application of the three-dimensional eutrophication model. CE-QUAL-ICM to Chesapeake Bay. Draft Technical Report. U.S. Army Engineer Waterways Experiment Station, Vicksburg, MI

Christensen JP, Murray JW, Devol AH & Codispoti LA (1987) Denitrification in continental shelf sediments has major impact on the oceanic nitrogen budget. Global Biogeochem. Cycles 1(2): 97–116

Cifuentes LA (1991) Spatial and temporal variations in terrestrially-derived organic matter from sediments of the Delaware Estuary. Estuaries 14(4): 414–429

Cornell S, Rendell A & Jickells T (1995) Atmospheric inputs of dissolved organic nitrogen to the oceans. Nature 376: 243–246

Csanady GT (1990) Physical basis of coastal productivity: The SEEP and MASAR experiments. EOS 71(36): 1060–1065

Czaya E (1981) Rivers of the World. Van Nostrand Reinhold Company

Degobbis D, Gilmartin M & Revelante N (1986) An annotated nitrogen budget calculation for the northern Adriatic Sea. Mar. Chem. 20: 159–177

DeMaster DJ, Pope RH, Levin LA & Blair NE (1994) Biological mixing intensity and rates of organic carbon accumulation in North Carolina slope sediments. Deep-Sea Res. II 41(4–6): 735–753

Devol AH, Richey JE & Forsberg BR (1991) Phosphorus in the Amazon mainstem: Concentrations, forms and transport to the ocean. In: Tiessen H, Lopez-Hernandez D & Salcedo IH (Eds) Phosphorus Cycles in Terrestrial and Aquatic Systems. Regional workshop 3: South and Central America. Proceedings of a workshop arranged by SCOPE and UNEP. Saskatchewan Institute of Pedology, Saskatoon, Saskatchewan

Duce RA, Liss PS, Merrill JT, Atlas EL, Buat-Menard P, Hicks BB, Miller JM, Prospero JM, Arimoto R, Church TM, Ellis W, Galloway JN, Hansen L, Jickells TD, Knap AH, Reinhardt KH, Schneider B, Soudine A, Tokos JJ, Tsunogai S, Wollast R & Zhou M (1991) The atmospheric input of trace species to the world ocean. Global Biogeochem. Cycles 5: 193–259

Edmond JM, Boyle EA, Grant B & Stallard RF (1981) The chemical mass balance in the Amazon plume I: The nutrients. Deep-Sea Res. 28A(11): 1339–1374

Eisma D (1981) Supply and deposition of suspended matter in the North Sea. In: Nio S, Shüttenhelm R & van Weering Tj (Eds) Holocene Marine Sedimentation in the North Sea Basin. Spec. Pub. Int. Ass. Sedimentologists, No. 5 (pp 415–428). Blackwell Scientific Publishers., Oxford, UK

Eisma D (1988) An introduction to the geology of continental shelves. In: Postma H & Zijlstra JJ (Eds) Continental Shelves. Ecosystems of the World, Vol. 27 (pp 39–89). Elsevier, NY

Elmgren R (1984) Trophic dynamics in the enclosed, brackish Baltic Sea. Rapp. P.-v. Reun. Cons. int. Explor. Mer. 183: 152–169

Emery KO, Orr WL & Rittenberg SC (1955) Nutrient budgets in the ocean. In: Essays in the Natural Sciences (pp 299–309). Univ. of Southern California Press, Los Angeles

FAO (1993) FAO Yearbook 1991. Fishery Statistics: Catches and Landings, Vol. 72. FAO Fisheries Series No. 40. FAO, Rome. 653 p

Froelich PN (1988) Kinetic control of dissolved phosphate in natural rivers and estuaries: A primer on the phosphate buffer mechanism. Limnol. Oceanogr. 33(4, part 2): 649–668

(GESAMP) (1987) Land/sea boundary flux of contaminants: Contributions from rivers. Reports and Studies No. 32

Giblin AE, Hopkinson C & Tucker J (1993) Metabolism, nutrient cycling and denitrification in Boston Harbor and Massachusetts Bay sediments. Report to Massachusetts Water Resource Authority, Boston, MA

Guillén O & Calienes R (1981) Upwelling off Chimbote. In: Richards FA (Ed) Coastal Upwelling. Coastal and Estuarine Sciences 1 (pp 312–316). American Geophysical Union, Washington, DC

HELCOM (Baltic Marine Environment Protection Commission – Helsinki Comission) (1993) Second Baltic Sea pollution load compilation. Baltic Sea Environment Proceedings 45: 1–161

Hendry CD & Brezonik PL (1980) Chemistry of precipitation at Gainesville, Florida. Environ. Sci. Technol. 14(7): 843–849

Henrichs SM & Reeburgh WS (1987) Anaerobic mineralization of marine sediment organic matter: Rates and the role of anaerobic processes in the oceanic carbon economy. Geomicrobiol. J. 5(3–4): 191–237

Hopkinson CS (1985) Shallow-water benthic and pelagic metabolism: Evidence for heterotrophy in the nearshore Georgia Bight, Mar. Biol. 87: 19–32

Houghton RW, Smith PC & Fournier RO (1978) A simple model for cross-shelf mixing on the Scotian Shelf. J. Fish. Res. Board Can. 35: 414–421

Howarth RW, Jensen HS, Marino R & Postma H (1995) Transport to and processing of P in near-shore and oceanic waters. In: Tiessen H (Ed) Phosphorus in the Global Environment: Transfers, Cycles and Management (pp 323–345). SCOPE 54. Wiley & Sons, Chichester, UK

Howarth RW, Billen G, Swaney D, Townsend A, Jaworski N, Downing JA, Elmgren R, Caraco N & Lajtha K (1995) Regional nitrogen budgets and riverine N & P fluxes for the drainages to the North Atlantic Ocean: Natural and human influences. (This volume)

Ingall ED & Van Cappellen P (1990) Relation between sedimentation rate and burial of organic phosphorus and organic carbon in marine sediments. Geochim. et. Cosmochim. Acta 54: 373–386

Iselin CO'D (1939) Some physical factors which may influence the productivity of New England's coastal waters. J. Mar. Res. 2: 75–85

Jahnke RA (1992) The phosphorus cycle. In: Butcher SS, Charlson RJ, Orians GH & Wolfe GV (Eds) Global Biogeochem. Cycles. (pp 301–313). Academic Press

Jaworski N & Howarth RW (In press) Preliminary estimates of the pollutant loads and fluxes into the northeast shelf LME. In: Sherman K (Ed) The Northeast Shelf Large Marine Ecosystem. NOAA

Jenkins WJ (1988) Nitrate flux into the eutrophic zone near Bermuda. Nature 331: 521–523

Johnstone J (1908) Conditions of Life in the Sea. Cambridge University Press. Reprinted by Arno Press, New York. 1977. 332 p

Kelly CA, Rudd JM, Hesslein RH, Schindler DW, Dillon PJ, Driscoll CT, Gherini SA & Hecky RE (1987) Prediction of biological acid neutralization in acid-sensitive lakes. Biogeochem. 3: 129–140

Kelly JR (1991) Nutrients and Massachusetts Bay: A synthesis of eutrophication issues. Report prepared for Massachusetts Water Resources Authority, Boston, MA

Kelly JR & Nowicki BL (1992) Sediment denitrification in Boston Harbor. Report to Massachusetts Water Resource Authority, Boston, MA

Kirchner WB & Dillon PJ (1975) An empirical method of estimating the retention of phosphorus in lakes. Wat. Resour. Res. 11: 1881–182

Kuehl SA, DeMaster DJ & Nittrouer CA (1986) Nature of sediment accumulation on the Amazon continental shelf. Contin. Shelf Res. 6: 209–225

Laane RWPM, Groeneveld G, deVries A, van Bennekom J & Sydow S (1993) Nutrients (P, N, Si) in the Channel and the Dover Strait: seasonal and year-to-year variation and fluxes to the North Sea. Oceanologica Acta 16(5–6): 607–616

Larsson U (1986) The Baltic Sea. In: Rosenberg R (Ed) Eutrophication in Marine Waters Surrounding Sweden – A Review (pp 16–70). The National Swedish Environmental Protection Board. Report 3054, Solna, Sweden
Larsson U, Elmgren R & Wulff F (1985) Eutrophication and the Baltic Sea – causes and consequences. Ambio 14: 9–14
Lewis Jr WM & Saunders III JF (1989) Concentration and transport of dissolved and suspended substances in the Orinoco River. Biogeochem. 7: 203–240
Lohrenz SE, Knauer GA, Asper VL, Tuel M, Michaels AF & Knap AH (1992) Seasonal and interannual variability in primary production and particle flux in the northwestern Sargasso Sea: U.S. JGOFS Bermuda Atlantic Time-series Study. Deep-Sea Res. 39: 1373–1391
Lohse L, Malschaert JFP, Slomp CP, Helder W & van Raaphorst W (1995) Sediment-water fluxes of inorganic nitrogen compounds along the transport route of organic matter in the North Sea. Ophelia 41: 173–197
Longley WL (Ed) (1994) Freshwater inflows to Texas Bays and estuaries: Ecological relationships and methods for determination of needs. Texas Water Development Board and Texas Parks and Wildlife Department, Austin, TX. 386 p
Mayer LM (1994) Surface area control of organic carbon accumulation in continental shelf sediments. Geochem. Cosmochem. Acta 58: 1271–1284
Meade RH (1972) Sources and sinks of suspended matter on continental shelves. In: Swift D, Duane D & Pilkey O (Eds) Shelf Sediment Transport: Process and Pattern (pp 249–260). Dowden, Hutchinson & Ross Publishers, Stroudsburg, PA
Menzel DW, Pomeroy LR, Lee TN, Blanton JO & Alexander CR (1993) Introduction. In: Menzel DW (Ed) Ocean Processes: U.S. Southeast Continental Shelf (pp 1–8). U.S. Department of Energy, Office of Scientific and Technical Information, Oak Ridge, TN
Meybeck M (1982) Carbon, nitrogen, and phosphorus transport by world rivers. Amer. J. Sci. 282: 401–450
Michaels A, Olson D, Sarmiento J, Ammerman J, Fanning K, Jahnke R, Knap A, Lipschultz F & Prospero J (1995) Inputs, losses and transformations of nitrogen in the deep North Atlantic Ocean. (This volume)
Milliman JD (1994) Organic matter content in U.S. Atlantic continental slope sediments: decoupling the grain size factor. Deep-Sea Res. II, 4: 797–808
Milliman JD & Syvitski JPM (1992) Geomorphic/tectonic control of sediment discharge to the ocean: The importance of small mountainous rivers. The Journ. of Geol. 100: 525–544
Mills EL & Fournier RO (1979) Fish production and the marine ecosystem of the Scotian Shelf, eastern Canada. Mar. Biol. 54: 101–108
Mulholland PJ (1981) Deposition of riverborne organic carbon in floodplain wetlands and deltas. In: Carbon Dioxide Effects Research and Assessment Program: Flux of Organic Carbon by Rivers to the Oceans, Report of a Workshop. United States Department of Energy
Nielsen K, Nielsen LP & Rasmussen P (1995) Estuarine nitrogen retention independently estimated by the denitrification and mass balance methods: a study of Norsminde Fjord, Denmark. Mar. Ecol. Prog. Ser. 119: 275–283
Nixon SW (1988) Physical energy inputs and the comparative ecology of lake and marine ecosystems. Limnol. Oceanogr. 33(4, part 2): 1005–1025
Nixon SW (1992) Quantifying the relationship between nitrogen input and the productivity of marine ecosystems. Proc. Adv. Mar. Tech. Conf. 5: 57–83
Nixon SW & Lee V (1986) Wetlands and water quality: A regional review of recent research in the United States on the role of freshwater and saltwater wetlands as sources, sinks, and transformers of nitrogen, phosphorus, and various heavy metals. U.S. Army Engineer Waterways Experiment Station, Vicksburg, MS. Tech. Report No. Y–86–2. 229 p
Nixon SW, Granger SL & Nowicki BL (1995) An assessment of the annual mass balance of carbon, nitrogen, and phosphorus in Narragansett Bay. Biogeochemistry 31: 15–61
Nixon SW, Oviatt CA, Frithsen J & Sullivan B (1986) Nutrients and the productivity of estuarine and coastal marine ecosystems. J. Limnol. Soc. sth. Afr. 12(1/2): 43–71

Nixon SW, Granger SL, Taylor DI, Johnson PW & Buckley BA (1994) Subtidal volume fluxes, nutrient inputs and the Brown Tide – an alternate hypothesis. Est. Coast. Shelf Sci. 39: 303–312

NOAA/EPA (Quinn H, Tolson JP, Klein CJ, Paul Orlando S & Alexander C (1989) Strategic assessment of near coastal waters. Susceptibility of East Coast estuaries to nutrient discharges: Passamaquoddy Bay to Chesapeake Bay. Summary Report, Strategic Assessment Branch, NOS/NOAA, Rockville, MD. 37 p

Nowicki BL & Oviatt CA (1990) Are estuaries traps for anthropogenic nutrients? Evidence from estuarine mesocosms. Mar. Ecol. Prog. Ser. 66: 131–146

O'Reilly JE, Evans-Zetlin C & Busch DA (1987) Primary production. In: Backus RH & Bourne DW (Eds) Georges Bank (pp 220–233). MIT Press, Cambridge, MA

Paerl HW (1995) Coastal eutrophication in relation to atmospheric nitrogen deposition: Current perspectives. Ophelia 41: 237–259

Pilson MEQ (1970) Water balance in California sea lions. Physiol. Zool. 43(4): 257–269

Pilson MEQ (1985) On the residence time of water in Narragansett Bay. Estuaries 8: 2–14

Pilson MEQ & Seitzinger S (1995) Areas of shallow water in the North Atlantic (This volume)

Platt T, Lewis M & Geider R (1984) Thermodynamics of the pelagic ecosystem: Elementary closure conditions for biological production in the open ocean. In: Fasham MJR (Ed) Flows of Energy and Materials in Marine Ecosystems: Theory and Practice (pp 49–84). Plenum Press, NY

Polis DF & Kupferman SL (1973) Physical characteristics of the Delaware Estuary. Trans. Del. Acad. of Sci. – 1970/1971: 163–180. Del. Acad. of Sci., Newark, DE

Pomeroy LR, Blanton JO, Poffenhofer G-A, Von Damm KL, Verity PG, Windom HL & Lee TN (1993) Chapter 2, inner shelf processes. In: Menzel DW (Ed) Ocean Processes: U.S. Southeast Continental Shelf (pp 9–43). U.S. Department of Energy, Office of Scientific and Technical Information, Oak Ridge, TN

Prospero J et al. (1995) Atmospheric deposition of nutrients to the North Atlantic Basin (This volume)

Richey JE & Victoria RL (1993) C, N, and P export dynamics in the Amazon River. In: Wollast R, Mackenzie FT & Chou L (Eds) Interactions of C, N, P and S Biogeochemical Cycles and Global Change. NATO ASI Series 1, Vol. 4 (pp 123–139). Springer-Verlag, Berlin Heidelberg

Rigler FH & Peters RH (1995) Science and Limnology, vol. 6 In Kinne O (Ed) Excellence in Ecology, Ecology Institute (p 239). Oldendorf/Luhe, Germany

Rönner U (1985) Nitrogen transformations in the Baltic Proper: Denitrification counteracts eutrophication. AMBIO 14: 134–138

Rowe GT, Theroux R, Phoel W, Quinby H, Wilke R, Koschoreck D, Whitledge TE, Falkowski PG & Fray C (1988) Benthic carbon budgets for the continental shelf south of New England. Cont. Shelf Res. 8(5–7): 511–527

Scudlark JR & Church TM (1994) Atmospheric input ofinorganic nitrogen to Delaware Bay. Estuaries 16(4): 747–759

Seitzinger SP (1987) Nitrogen biogeochemistry in an unpolluted estuary: The importance of benthic denitrification. Mar. Ecol. Prog. Ser. 37: 65–73

Seitzinger SP (1988a) Benthic nutrient cycling and oxygen consumption in the Delaware Estuary. In: Majumdar SK, Miller EW & Sage LE (Eds) Ecology and Restoration of the Delaware River Basin (pp 132–147). Penn. Acad. of Sci.

Seitzinger SP (1988b) Denitrification in freshwater and coastal marine ecosystems: Ecological and geochemical significance. Limnol. Oceanogr. 33(4, part 2): 702–724

Seitzinger SP & Giblin AE (1995) Estimating denitrification in North Atlantic continental shelf sediments (This volume)

Shaffer G & Rönner U (1984) Denitrification in the Baltic proper deep water. Deep-Sea Res. 31: 197–220

Showers WJ & Angle DG (1986) Stable isotopic characterization of organic carbon accumulation on the Amazon continental shelf. Cont. Shelf Res. 6: 227–244

Signell RP & Butman B (1992) Modeling tidal exchange and dispersion in Boston Harbor. J. Geophys. Res. 97(C10): 15,591–15,606

Smith SV (1991) Stoichiometry of C:N:P fluxes in shallow-water marine ecosystems. In: Cole J, Lovett G & Findlay S (Eds) Comparative Analyses of Ecosystems – Patterns, Mechanisms, and Theories (pp 259–286). Springer-Verlag, New York

Smith SV & Hollibaugh JT (1993) Coastal metabolism and the oceanic organic carbon balance. Rev. Geophysics 31(1): 75–89

Smullen JT, Taft J & Macknis J (1982) Nutrient and sediment loads to the tidal Chesapeake Bay system. In: Macalaster EG, Barker DA & Kasper M (Eds) Chesapeake Bay Program Technical Studies: A Synthesis (pp 147–262). U.S. E. P. A., Washington, DC

Solati E, Sylvester-Bradley R & Victoria RL (1982) Regional gains and losses of nitrogen in the Amazon basin. Plant and Soil 67: 367–376

Stommel H & Leetmaa A (1972) Circulation on the continental shelf. Proc. Nat. Acad. Sci. 69(11): 3380–3384

Strom RN (1976) Phosphorus fractionation in estuarine and marsh sediments. Ph.D. Thesis in Geochemistry. U. of Del. 554 p

Swedish Environment Protection Agency (1993) Eutrophication of soil, freshwater and sea. Swedish Environment Protection Agency, Report 4134: 1–199 (In Swedish)

Swift DJP, Young RA, Clarke TL, Vincent CE, Niedoroda A & Lesht B (1981) Sediment transport in the Middle Atlantic Bight of North America: Synopsis of recent observations. Spec. Publs. Int. Ass. Sediment 5: 361–383

Trefry JH, Metz S, Nelsen TA, Trocine RP & Eadie BJ (1994) Transport of particulate organic carbon by the Mississippi River and its fate in the Gulf of Mexico. Estuaries:17: 839–849

Tupas L, Santiago-Mandujano F, Hebel D, Lukas R, Karl D & Firing E (1993) Hawaii Ocean Time-series Data Report 4, 1992. SOEST Technical Report 93–14, University of Hawaii, Honolulu, HI

Tupas L, Santiago-Mandujano F, Hebel D, Firing E, Bingham F, Lukas R & Karl D (1994) Hawaii Ocean Time-series Data Report 5, 1993. SOEST Technical Report 94–5, University of Hawaii, Honolulu, HI

Turner RE & Rabalais NN (1991) Changes in Mississippi River water quality this century. BioSci. 41(3): 140–147

Twichell DC, McClennen CE & Butman B (1981) Morphology and processes associated with the accumulation of the fine-grained sediment deposit on the southern New England shelf. J. Sed. Petrology 51(1): 269–280

UNESCO (1992) Discharge of Selected Rivers of the World, 1980–1984, vol. 3, part 5. UNESCO, Paris

van Andel TH & Postma H (1954) Recent sediments of the Gulf of Paria: Verh. Konik. Nedesl. Akad. van Wetenschappen, D. Natuurkunde, Series I(20): 1–240

van Weering TjCE (1981) Recent sediments and sediment transport in the northern North Sea: surface sediments of the Skagerrak. In: Nio S, Shüttenhelm R & van Weering Tj (Eds) Halocene Marine Sedimentation in the North Sea Basin. Spec. Pub. Int. Ass. Sedimentologists, No. 5 (pp 335–359). Blackwell Sci. Pub., Oxford, UK

Verity PG, Lee TN, Yoder JA, Paffenhofer GA, Blanton JO & Alexander CR (1993) Chapter 3, outer shelf processes. In: Menzel DW (Ed) Ocean Processes: U.S. Southeast Continental Shelf (pp 45–74). U.S. Dep. of Energy, Office of Sci. Tech. Inf., 11674

Vinogradov AP (1953) The elementary chemical composition of marine organisms. Memoir Sears Foundation for Marine Research No. 11. New Haven: Yale Univ. Press. 677 p

Walsh JJ (1983) Death in the sea: Enigmatic phytoplankton losses. Prog. Oceanogr. 12: 1–86

Walsh JJ (1988) On the Nature of Continental Shelves. Acad. Press, Inc. 520 p

Walsh JJ (1991) Importance of continental margins in the marine biogeochemical cycling of carbon and nitrogen. Nature 350: 53–55

Walsh JJ, Whitledge TE, Esaias WE, Smith RL, Huntsman SA, Santander H & de Mendiola BR (1980) The spawning habitat of the Peruvian anchovy *Engraulis ringens*. Deep-Sea Res. 27: 1–27

Walsh JJ, Rowe GT, Iverson RL & McRoy CP (1981) Biological export of shelf carbon is a neglected sink of the global CO_2 cycle. Nature 291: 196–201

Walsh JJ, Whitledge TE, O'Reilly JE, Phoel WC & Draxler AF (1987) Nitrogen cycling on Georges Bank and the New York Shelf: A comparison between well-mixed and seasonally stratified waters. In: Backus RH (Ed) Georges Bank (pp 234–246). MIT Press, Cambridge, MA

Walsh JJ, Whitledge TE, Barvenik FW, Wirick CD, Howe SO, Esaias WE & Scott JT (1978) Wind events and food chain dynamics within the New York Bight. Limnol. Oceanogr. 23: 659–683

Walsh JJ, Premuzic ET, Gaffney JS, Rowe GT, Balsam W, Harbottle G, Stoenner RW, Betzer PR & Macko SA (1985) Storage of CO_2 as organic carbon on the continental slopes off the mid-Atlantic Bight, southeastern Bering Sea, and the Peru coast. Deep-Sea Res. 32: 853–883

Weil CB (1977) Sediments, structural framework, and evolution of Delaware Bay, a transgressive estuarine delta. U. of Del. Sea Grant Tech. Rep., DEL-SG-4-77, 199 p

Wollast R (1993) Interactions of carbon and nitrogen cycles in the coastal zone. In: Wollast R, Mackenzie FT & Chou L (Eds) Interactions of C, N, P and S Biogeochemical Cycles and Global Change. NATO ASI Series 1: Global Environmental Change, Vol. 4 (pp 195–210). Springer-Verlag, Berlin Heidelberg

Wulff F & Stigebrandt A (1989) A time-dependent budget model for nutrients in the Baltic Sea. Global Biogeochem. Cycles 3: 63–78

Wulff F, Rahm L & Rodriguez-Medina M (1994) Long-term and regional variations of nutrients in the Baltic Sea: 1972–1991. Finnish Mar. Res. 262: 35–50

Yoder JA & Ishimaru T (1989) Phytoplankton advection off the southeastern United States continental shelf. Cont. Shelf Res. 9(6): 547–553

Biogeochemistry **35**: 181–226, 1996.

Inputs, losses and transformations of nitrogen and phosphorus in the pelagic North Atlantic Ocean

A.F. MICHAELS[1], D. OLSON[2], J.L. SARMIENTO[3], J.W. AMMERMAN[4], K. FANNING[5], R. JAHNKE[6], A.H. KNAP[1], F. LIPSCHULTZ[1] & J.M. PROSPERO[2]

[1]*Bermuda Biological Station for Research, Ferry Reach, Bermuda;* [2]*University of Miami;* [3]*Princeton University;* [4]*Texas A and M University;* [5]*University of South Florida;* [6]*Skidaway Insitute of Oceanography*

Received 24 August 1995; accepted 18 March 1996

Abstract. The North Atlantic Ocean receives the largest allochthonous supplies of nitrogen of any ocean basin because of the close proximity of industrialized nations. In this paper, we describe the major standing stocks, fluxes and transformations of nitrogen (N) and phosphorus (P) in the pelagic regions of the North Atlantic, as one part of a larger effort to understand the entire N and P budgets in the North Atlantic Ocean, its watersheds and overlying atmosphere. The primary focus is on nitrogen, however, we consider both nitrogen and phosphorus because of the close inter-relationship between the N and P cycles in the ocean. The oceanic standing stocks of N and P are orders of magnitude larger than the annual amount transported off continents or deposited from the atmosphere. Atmospheric deposition can have an impact on oceanic nitrogen cycling at locations near the coasts where atmospheric sources are large, or in the centers of the highly stratified gyres where little nitrate is supplied to the surface by vertical mixing of the ocean. All of the reactive nitrogen transported to the coasts in rivers is denitrified or buried in the estuaries or on the continental shelves and an oceanic source of nitrate of 0.7–0.95 $\times$ 10^{12} moles NO_3^- y^{-1} is required to supply the remainder of the shelf denitrification (Nixon et al., this volume). The horizontal fluxes of nitrate caused by the ocean circulation are both large and uncertain. Even the sign of the transport across the equator is uncertain and this precludes a conclusion on whether the North Atlantic Ocean as a whole is a net source or sink of nitrate. We identify a source of nitrate of 3.7–6.4 $\times$ 10^{12} moles NO_3^- y^{-1} within the main thermocline of the Sargasso Sea that we infer is caused by nitrogen fixation. This nitrate source may explain the nitrate divergence observed by Rintoul & Wunsch (1991) in the mid-latitude gyre. The magnitude of nitrogen fixation inferred from this nitrate source would exceed previous estimates of global nitrogen fixation. Nitrogen fixation requires substantial quantities of iron as a micro-nutrient and the calculated iron requirement is comparable to the rates supplied by the deposition of iron associated with Saharan dust. Interannual variability in dust inputs is large and could cause comparable signals in the nitrogen fixation rate. The balance of the fluxes across the basin boundaries suggest that the total stocks of nitrate and phosphate in the North Atlantic may be increasing on time-scales of centuries. Some of the imbalance is related to the inferred nitrogen fixation in the gyre and the atmospheric deposition of nitrogen, both of which may be influenced by human activities. However, the fluxes of dissolved organic nutrients are almost completely unknown and they have the potential to alter our perception of the overall mass balance of the North Atlantic Ocean.

Introduction

The North Atlantic Ocean is one of the smallest of the ocean basins and it is surrounded by some of the most heavily industrialized countries in the world. Human activities at the boundaries of the North Atlantic modify the nitrogen (N) and phosphorus (P) dynamics of the ocean, watersheds and atmosphere. It is also one of the most heavily studied ocean basins and may be most amenable to a synthesis. In May 1994, a meeting was held on Block Island, R.I. (Galloway et al., this volume) to evaluate our understanding of the N and P cycles in the North Atlantic Ocean, its surrounding watersheds and overlying atmosphere. This question was addressed from four perspectives, the atmosphere (Prospero et al., this volume), watersheds (Howarth et al., this volume), coastal zone (Nixon et al., this volume) and open oceans (this paper). From the perspective of the open oceans group, the goal was stated succinctly in the question: "What are the current sources and sinks of nitrogen and phosphorus within the North Atlantic Ocean, what are the exchanges across its boundaries and how do the magnitudes of these exchanges, sources and sinks compare with recent changes from natural background levels as a result of human activity?" To address this question required specific information about both N and P in the oceanic realm: 1) The standing stocks of each form of each element, 2) the fluxes of N and P into and out of the ocean, and 3) the fluxes and transformations of N and P within the ocean and between important sub-sections of the ocean, particularly in areas where the biological transformations of N have resulted in depletion of reactive N with respect to the normal requirements of N and P for plant growth. This paper draws heavily on the estimates from the atmosphere and coastal groups for data on boundary exchanges. Prospero et al. (this volume) provided the N, P and Fe atmospheric deposition estimates (Prospero et al., this volume) and we relied on an analysis by Quinn et al. (this volume) to discuss the ammonia flux out of the North Atlantic. The net exchanges of N and P with neritic coastal waters were estimated by Nixon et al. (this volume).

We begin with assessments of the standing stocks of each form of N and P and a discussion of the vertical and horizontal flux processes that affect each stock. We then focus on two topics which we feel are important for understanding the whole-basin elemental cycles in the North Atlantic. Both topics involve the overall mass balance of nitrate and a comparison of nitrate fluxes with phosphate fluxes. 1) We evaluate a variety of estimates of the meridional fluxes of nitrate and phosphate at latitudes from 8° N to 65° N, across the Straits of Gibralter and the exchange with the Caribbean Sea. These analyses can be used to bound the likely sources and sinks of nitrate and phosphate within different subregions of the North Atlantic. 2) We reconsider the previously described anomalies in the ratios of nitrate to phosphate in the upper 1000 m of the Sargasso Sea (Fanning 1989, 1992; Michaels et al. 1994) and estimate a net nitrate creation rate for the gyre of 3.7–6.4 $\times 10^{12}$ moles y^{-1} that we infer is caused by nitrogen fixation. Finally, we discuss the overall balance of N and P in the pelagic North Atlantic.

Standing stocks and fluxes of nitrogen and phosphorus

A. *The boundaries of the North Atlantic Ocean*

Our choice for the horizontal boundaries of the Atlantic was determined primarily by the availability of data. For the southern extent of the Atlantic, we chose the Equator; however, for the meridional transport estimates at the southern boundary, we used data from 12° N as a proxy. We assumed that the outflow from the Amazon River, which lies directly on the equator, is part of the North Atlantic system because most of the offshore transport is to the north. We ignored the Panama Canal. We chose a boundary at the straits of Gibraltar and characterized all of the coastal and riverine inputs into the Mediterranean by the net flow through the straits. In the north, we chose boundaries at 60° N and 65° N, again based on the available zonal sections. Within the Atlantic, we further subdivided the basin at 24° N and 36° N, in part because of the earlier meridional flux estimates at these latitudes (Rintoul & Wunsch 1991; Martel & Wunsch 1993). The boundary with the coastal zone was chosen operationally as the 200 m bathymetric contour. In practice, the coastal group (Nixon et al., this volume) determined the net N and P fluxes into the shelf sediments. For nitrogen, there was frequently a nitrate deficit, and the required amount was supplied from the main thermocline of the adjacent ocean. For phosphate, net river outflow exceeded the estimates of burial and the residual was supplied to the oceans (Nixon et al., this volume).

B. *Standing stocks of nitrogen and phosphorus in the North Atlantic Ocean*

There are six major forms of nitrogen and three forms of phosphorus (some operationally defined) that are common in the ocean. Some of the trace nitrogen gases may also be useful as tracers of process, even if they do not contribute to the total nitrogen balance. The forms of nitrogen are nitrogen gas, nitrate, nitrite, ammonium (and ammonia), dissolved and particulate organic nitrogen (Table 1). Nitrogen gas is biologically unavailable to all but a few specialized organisms. The other species of nitrogen, which are readily converted by biological and chemical processes, can be called "reactive" nitrogen species. The principle forms of phosphorus are inorganic phosphate and dissolved and particulate organic phosphorous (Table 1). For each of these forms of N and P, we estimate the total standing stock for the basin and, for most, we compare these estimates with the fluxes of N and P in and out the basin. These fluxes are then discussed in further detail in the next section.

Nitrogen gas (N_2) is the most prevalent form of nitrogen in the ocean (Table 1), present at concentrations of approximately 450–600 μmoles kg^{-1} (Weiss 1970). It is generally biologically inert except for its utilization by nitrogen-fixing organisms and its production in denitrification. Although both of those transformations may affect the concentration of dinitrogen, the effects are likely to be local and must be small compared to the size of the standing stocks. The factor with the largest effect on N_2 concentration is the temperature dependence of N_2 solubility coupled to air-sea exchange.

Nitrate (NO_3^-) is the second most abundant form of nitrogen in the ocean and the most abundant form of reactive nitrogen. Concentrations below 1000 m are typically near 15–25 μmoles kg^{-1} in the North Atlantic. Concentrations near the surface are usually below the standard analytical detection limits (0.05 μmoles kg^{-1}) with seasonal exceptions that are very important for the biology of the upper ocean. In the permanent thermocline, there is a regular transition between these two extremes, determined both by mixing and remineralization. The spatial distribution of nitrate, particularly at 500–2000 m depth is strongly affected by midwater circulation. South Atlantic thermocline and Antarctic Intermediate water, with elevated nitrate concentrations, intrude into the North Atlantic from the south and create a zone of elevated nitrate at the base of the main thermocline.

We estimate the total standing stock of nitrate in the North Atlantic (Table 1) using mean nitrate concentrations from the TTO and GEOSECS datasets and the areal extent of different subregions of the North Atlantic (Platt et al. 1991). We estimate that there are 2,800 $\times$ 10^{12} moles NO_3^- in the North Atlantic, of which only 0.01% is on the continental shelves. Thus, the nitrate pool in the North Atlantic is 3–4 orders of magnitude larger than the nitrogen stocks described for the adjoining coastal areas, atmosphere and watersheds (Howarth et al., this volume; Nixon et al., this volume; Prospero et al., this volume). Even trivial uncertainties in the total standing stock of nitrate in the North Atlantic will preclude any attempt to infer boundary exchanges or the overall mass balance from sequential observations of the total nitrate standing stock. For example, if we naively assume that the accuracy of our estimate of the total nitrate standing stock is similar to the analytical accuracy of the measurement (as good as 0.5%) instead of the larger uncertainties inherent in the sparse data coverage, we would estimate an uncertainty in the total standing stock of 14 $\times$ 10^{12} moles NO_3^- (Table 1). This is larger than the annual nitrogen exchanged at the boundaries (Table 2), such as the 0.7–0.95 $\times$ 10^{12} moles N y^{-1} estimate of oceanic nitrate supply onto the shelves to

Table 1. Significant forms of nitrogen and phosphorus in the North Atlantic Ocean. Uncertainties for nitrogen are minimal values estimated from analytical accuracy and do not take into account the uncertainties that may arise when the sampling is inadequate to estimate the total stock. Uncertainties for the phosphorus species estimates are assumed to be the same as the corresponding nitrogen species. * In the case of dissolved organic N and P, the number of measurements is very small and very local, thus these estimates are actually much more uncertain than indicated in the table.

	Concentrations (μmoles/kg)	Stock in North Atlantic ($\times 10^{12}$ moles)	Uncertainty (%)	Uncertainty ($\times 10^{12}$ moles)
Nitrogen				
Dinitrogen gas (N_2)	450–600	112,000		
Nitrate (NO_3^-)	0–25	2,800	0.5	14
Nitrite (NO_2^-)	~0.01–0.2	2	1.0	<0.1
Ammonium (NH_4^+)	~0.01–0.5	5	1.0	<0.1
Dissolved Organic N (DON)	*3–5*	300–500	>5.0	>15–25
Particulate Organic N (PON)	0.02–1.0	13	>1.0	0.1
Phosphorus				
Phosphate (PO_4^{3-})	0.01–1.5	180	0.5	0.9
Dissolved Organic Phosphorus (DOP)	~0.2	37	>5.0	>1.9
Particulate Phosphorus (PP)	0.01–0.05	1	>1.0	<0.1

support denitrification and burial (Nixon et al., this volume), the 0.5–1.0 × 10^{12} moles N y^{-1} for atmospheric deposition of all reactive nitrogen species (Prospero et al., this volume) and the 0.9 × 10^{12} moles N y^{-1} for riverine fluxes of nitrogen to the coasts (Howarth et al., this volume).

Nitrite (NO_2^-) is generally present in low concentrations with the exception of a primary nitrite maximum that is present at the base of the euphotic zone in many areas (Herbland & Voituriez 1979; Olson 1981; Lipschultz et al., in press). Nitrite is an intermediate in the nitrification and denitrification pathways. Its standing stocks are likely unimportant in the overall budget of nitrogen (Table 1), but it can be an important marker for transformation processes. Nitrous oxide (N_2O) is a trace compound and is not a significant standing stock of nitrogen in the oceans. It is produced as a byproduct during nitrification and can be consumed during denitrification.

Ammonium (NH_4^+) and ammonia (NH_3) are trace nitrogen species in most areas. They are biologically very labile and are intermediates in many biologically-mediated transformations. As with NO_2^- and N_2O, ammonia/ammonium do not constitute a major standing stock of nitrogen. However, efflux of ammonia gas may be a significant source of nitrogen into the atmosphere, particularly in comparison to some of the atmospheric deposition fluxes (Quinn et al. 1988, 1990). Quinn et al. (this volume) compile the existing ammonia/ammonium data to attempt to estimate the air-sea flux of ammonia for the North Atlantic. The data are sparse, show both air-to-sea and sea-to-air fluxes of up to 100 μmoles m^{-2} d^{-1} and no conclusion was possible about the average flux (Quinn et al., this volume). If the mean net flux was only 1 μmole m^{-2} d^{-1} it would account for 16 × 10^9 moles y^{-1}, a small flux compared to the North Atlantic nitrogen budget, but large compared to some of the riverine and air-sea fluxes.

Dissolved organic nitrogen (DON) is a significant standing stock of nitrogen (Table 1) and likely the second largest pool of reactive nitrogen. However, there are very few measurements in the Atlantic Ocean. Typical concentrations are thought to be in the range of 3–5 μmoles kg^{-1} with only small variations with depth (Sharp 1983; Dennis Hansell, unpublished data). At these concentrations, the total standing stock of nitrogen in DON would be on the order of 300–500 × 10^{12} moles. There has been a great deal of controversy in recent years about the measurement of DON (Suzuki et al. 1985; Walsh 1989; Hansell 1993; Hopkinson et al. 1993), with suggestions in the early paper by Suzuki et al., that a new technique yielded higher concentrations and larger surface-deep sea gradients (Suzuki et al. 1985). However, these reports of higher concentrations have not been supported by subsequent research (Walsh 1989). It now appears that the traditional concentrations are reasonable and that vertical gradients are relatively small. Urea is

Table 2. Summary of nutrient fluxes across the boundaries of the North Atlantic Ocean. All fluxes are in units of 10^{12} moles y^{-1}.

Process and location	N	P
Atmospheric deposition		
0–10° N	0.13–0.26*	0.003
10–40° N	0.22–0.44*	0.003
40–70° N	0.18–0.36*	~0
Total atmospheric deposition	0.5–1.0	0.006
Ammonia air-sea exchange	?	–
Exchange with continental shelves **		
Western boundary, 0–10° N (Amazon)	0.11	0.014
Western boundary, 10–40° N	−0.26	0.004
Western boundary, 40–70° N	−0.41	0.003
Eastern boundary, 0–10° N	−0.15	0.003
Eastern boundary, 10–40° N	−0.03	0.002
Eastern boundary, 40–70° N	−0.18	0.004
Total exchange with shelves	−0.92	0.03
Burial in sediments	0.05	0.002
Nitrogen fixation		
0–10° N	?	–
10–40° N	3.7 to 6.4	–
Total nitrogen fixation	>3.7 to 6.4	–
Physical transport (negative is southward or westward)		
Equator***	−1.4 to 4.6	−0.07 to 0.31
Arctic Sea****	−0.3 to −2.8	−0.16 to −0.3
Mediterranean*****	−0.05 to −0.2	−0.003 to −0.005
Total North Atlantic Ocean	1.9–13.8	0.13–0.65

* The lower value in each range is the $NO_y + NH_x$ flux as calculated by Prospero et al. (this volume). These estimates exclude the potential contribution of DON which Prospero et al. estimate would approximately double the flux. Thus, the upper bound is estimated as twice the $NO_y + NH_x$ flux.

** The fluxes of nitrate and phosphate from the ocean onto the shelves are derived from the estimates of Nixon et al. (this volume) at the 200 m isobath. The partitioning of the nitrate fluxes by coastline is derived from the net balance of atmospheric deposition, riverine input and denitrification for each coastal region (Table 14 in Nixon et al., this volume). Nixon et al. also modify the total nitrate flux estimates by the inclusion of burial in shelf sediments and a loss to fisheries. This increases the net nitrogen deficit by approximately 16%. We modify each of the regional fluxes in their Table 14 upwards by 16% to approximate the small modifications from burial and fisheries. For phosphate, the regional data are restricted to riverflow and estuarian flux estimates (Table 12 in Nixon et al., this volume). Atmospheric fluxes, shelf burial and fisheries losses are small and ignored.

*** From Table 4. Range of estimate by Martel & Wunsch (1993) for 12 ° N

**** From Table 4. Full range of estimates by Martel & Wunsch (1993) for 60° N and of Anderson & Dyrssen (1981, 1983).

***** From Table 3. Range of estimates excluding the case where an inflow is calculated.

a biologically labile form of dissolved organic nitrogen and has been studied separately in many cases (Antia et al. 1991). The uncertainties in the size of the total DON pool are also large compared to the annual amount of nitrogen exchanged at the boundaries of the North Atlantic. The accuracy of the historical estimates is a combination of the 5–10% analytical uncertainty and the very small number of measurements. Even at 5%, the total DON pool is uncertain by 15–25 $\times$ 10^{12} moles (Table 1) and the true uncertainty is much higher.

Particulate organic nitrogen (PON) is an important standing stock of nitrogen, even though the total standing stock is only a fraction that of nitrate or DON. Concentrations range from 0.5–10 μmoles kg^{-1} in the surface waters and are generally less than 0.1 μmoles kg^{-1} in the deep water (e.g. Sharp 1983; Martin et al. 1993; Michaels et al. 1994b). In the upper water column, PON is comprised of phytoplankton, bacteria, zooplankton and detritus and it is these organisms that are responsible for the transformation of most other forms of nitrogen. In the water column, autotrophic and heterotrophic bacteria likely account for a significant fraction of the total PON (order 10–80%). Phytoplankton (some of which are prokaryotes) are a highly variable fraction, they can dominate during blooms and in oligotrophic environments are 10–30% of the PON. Detritus is usually calculated as the difference between the total standing stock and the living standing stocks and has been determined as anywhere from 10–90% of the PON.

Inorganic phosphorus in aerobic marine ecosystems is almost entirely orthophosphate. Dissolved phosphate is often called soluble reactive phosphate (SRP). Molybdate complexation techniques are most commonly used to measure phosphate concentrations in seawater, and these techniques measure both the dissolved phosphate and any phosphate that is released from organic compounds in the acidic conditions of the analysis (the term phosphate will be used to denote SRP throughout this paper). Phosphate is typically present in deep waters at concentrations of 1.0–1.5 μmoles kg^{-1}. At the surface, concentrations are low (order 0.1–0.2 μmoles kg^{-1}), although usually measurable with colorometric techniques. In the center of the Sargasso Sea, phosphate concentrations are almost always below 0.03 μmoles kg^{-1}.

We estimate a total standing stock of phosphate in the North Atlantic of 181 $\times$ 10^{12} moles PO_4^{3-} using the same approach as for nitrate. Similarly, the annual boundary exchanges of phosphate are small compared to the total standing stock (Table 2). The boundary exchange estimates are 18–30 $\times$ 10^9 moles P y^{-1} net input to the ocean from the shelves (Nixon et al., this volume), 6 $\times$ 10^9 moles P y^{-1} for atmospheric deposition of phosphate (Prospero et al., this volume) and 7 $\times$ 10^{10} moles P y^{-1} for riverine fluxes of phosphate to the coasts (Howarth et al., this volume).

The vertical and horizontal patterns of phosphate typically mirror those of nitrate with a tight covariance near the Redfield ratio of 16 moles NO_3^-:1 mole PO_4^{3-}. The major exception to this pattern is an excess of nitrate or depletion of phosphate in the upper 1000 m of the western Sargasso Sea which results in NO_3^-:PO_4^{3-} ratios of between 20 and 50 moles NO_3^-:1 mole PO_4^{3-}. This local deviation from Redfield stoichiometry indicates that other processes are influencing the behavior of nitrate and/or phosphate in this area compared to the rest of the North Atlantic. Proposed processes include nitrogen fixation, nitrogen pollution or unique phosphate dynamics (Fanning 1989; Michaels et al. 1994a; Michaels et al. 1994b). We return to this topic later in the paper.

The fluxes of nitrate and phosphate in and out of the North Atlantic Ocean are also not at Redfield ratios. Riverine supplies of nutrients to the coasts occur at a ratio of 13 moles NO_3^-:1 mole PO_4^{3-} (Howarth et al., this volume). However in the estuaries and shelves, denitrification removes nitrate. Consequently, the shelves are a net sink for oceanic nitrate, while they provide a small source of phosphate to the open ocean (Nixon et al., this volume). Rates of atmospheric deposition are very low for phosphate compared to nitrate, with a basin average deposition ratio of 80–150 moles reactive N:1 mole PO_4^{3-} (Prospero et al., this volume).

Dissolved organic phosphorus (DOP) pools are typically on the order of 0.2 μmoles kg^{-1} (Ammerman et al. 1994), exceeding the inorganic phosphate concentration in many surface waters, but lower than phosphate concentrations below the surface. DOP is a severely under-sampled pool of P, even though its dynamics may influence the cycling of P in many oceanic ecosystems. Data from near Bermuda (Ammerman et al. 1994) suggest that dissolved DNA in the Sargasso Sea is only a small fraction of the DOP. Dissolved nucleic acids in general appear to be a small fraction of the DOP unless the concentration of dissolved RNA is very high (Karl & Bailiff 1989). Particulate organic phosphorus is usually present at relatively low concentrations, typically 10–20 nanomoles kg^{-1} in oligotrophic environments (Ammerman et al. 1994) and 1–2 orders of magnitude higher in productive zones. As with DOP, POP is under sampled and more poorly understood than particulate nitrogen. Ammerman's results suggest that much of the POP in Bermuda is bacterial DNA.

C. *Transformation and transport processes*

We divide the fluxes of N and P into vertical and horizontal fluxes and distinguish between physical and biogeochemical flux processes. We concentrated on estimating a subset of the relevant fluxes as they may be relevant to calculating the overall mass balance of N and P in the North Atlantic (Table 2). Horizontal and vertical fluxes of N and P from rivers and the atmosphere are subdivided geographically by reference to the appropriate companion papers (Nixon et al., this volume; Prospero et al., this volume).

The physical environment determines most of the patterns of N and P in the ocean. Large-scale transport creates the major oceanic provinces such as the subtropical gyre called the Sargasso Sea and provides the dominant mechanism for horizontal transport of nutrients. Mixing at many scales acts to smooth the strong vertical and horizontal gradients created by biogeochemical processes. The vertical gradients are greatest near the surface where biological activity causes a near-complete depletion of nutrients; here mixing becomes the major regulatory force that determines the resupply of nutrients to the surface and, thus, controls many of the subsequent biogeochemical transformation rates. At the surface, both gas exchange and atmospheric deposition can also change the concentrations of N and P species.

Biological and chemical processes transform nutrient elements from one species to another and transport them from one water mass to another. The biogenic transformation of nitrogen and phosphorus species is usually associated with the creation or destruction of the organic species (e.g. assimilation, remineralization) and with the energy yielding reactions that fuel microbial metabolism (e.g. nitrification and denitrification). Biological transport processes are restricted to the active movements of large organisms (such as vertical migration of zooplankton and nekton) and the passive settling of particles (primarily detritus, fecal wastes or recently moribund organisms). Transformations of nutrient elements within a parcel of water modify the gradients in the concentration of a particular form of the element, but cannot modify the total standing stock of N or P in that parcel. For example, the conversion of nitrate into organic nitrogen at the surface produces gradients in both nitrate (an increased concentration with depth) and PON (a decreased concentration with depth). However, the total standing stock of N at the surface (NO_3^-+PON) will remain unchanged in the absence of a transport process. Biological transport (e.g. the sinking of PON for the example above) can change both the total standing stock of a nutrient element (decrease the surface NO_3^-+PON) and the gradient of one particular form of the element (e.g. remove the surface PON maximum).

Vertical transport processes

The primary vertical pattern in the ocean is the surface depletion of inorganic nutrients by biological utilization in illuminated surface waters and the subsequent transport of the organic matter to depth. Vertical mixing within the ocean acts to homogenize these gradients and results in a net flux of nutrient to the surface. Where there are surface accumulations of organic matter (e.g.

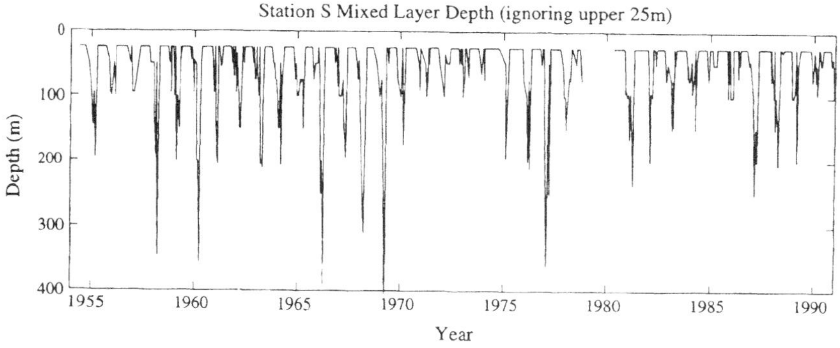

Figure 1. Time-series of mixed-layer depth at Hydrostation S (32.2° N 64.5° W) from 1954 to 1992. Modified from (Michaels et al. 1994b)

a plankton bloom or surface maximum in dissolved organic matter), this same mixing will act to transport the organic material to depth (Toggweiler 1989; Carlson et al. 1994). The balance between the processes that remove nitrate and phosphate from surface waters and the mixing that returns nutrients to the surface regulates both the surface concentrations and, for many ecosystems, the overall productivity. Thus, the near-surface nitrate or phosphate concentration provides an indication of the areas where uptake processes are inefficient or where vertical transport of nutrients by mixing or upwelling exceeds the biological uptake rate, at least over a part of the year. For example, upwelling off West Africa and the deep winter convection of the North Atlantic Ocean both cause elevated nutrient concentrations at the surface for part of the year.

Estimates of the actual vertical velocities or the net vertical mixing rates in a few regions have been determined by measurements of the microscale physical structure (Lewis et al. 1986; Gregg 1989), tracer release experiments (Ledwell et al. 1993), or models of the physical processes in an area (e.g. Sarmiento 1986). North of approximately 30° N latitude, there are strong seasonal patterns in the near-surface mixing. Cooling of the surface waters in the fall and winter causes a convectively driven mixing of the upper ocean. The magnitude of winter mixing increases with increasing latitude (Siegel et al. 1990) and there is significant interannual variability in this wintertime overturn (Figure 1).

The spatial patterns in vertical transport from wind-driven or eddy-driven upwelling, downwelling and wind or convection driven mixing, can be deduced from coupled circulation/biogeochemistry models (e.g. see Sarmiento et al. 1993) and from remote sensing. Direct estimates of wind-driven upwelling and downwelling velocities from satellite data show the upwelling centers off Western Africa and the downwelling of the central gyres (McClain & Firestone 1993). The circulation models give detailed maps of nitrate fluxes to the surface which, when coupled to a biological model, allow inferences of the biological uptake rates of nitrate (e.g Figure 2c). These types of models also illustrate the ventilation of mode water, the path of the Gulf Stream, the upwellings in the equator and off of the west coast of Africa, and the nitrate supply to the surface in the North Atlantic that causes the spring bloom at higher latitudes. These models have limitations in terms of the actual magnitude of the fluxes, due to uncertainties that are usually related to assumptions in the physical models. For instance, in Sarmiento et al. (1993), nitrate supply in the middle of the gyre is much lower than many of the direct measurements for these types of habitats.However, the models provide a good relative standard for comparison with the spatial patterns of other processes.

Nitrogen is added to or removed from the surface of the ocean by two processes, atmospheric deposition (both wet and dry) and gas exchange (for ammonia and N_2). Phosphorus lacks a significant gas phase and thus the air-sea exchange of P is only through deposition. Prospero et al. (this volume) have compiled and validated a set of measurements and model results on the spatial patterns and magnitude of atmospheric N and P deposition into the North Atlantic. They conclude that the total N deposition to the basin is approximately 1×10^{12} moles N yr^{-1}. About half of this deposition is due to a flux of dissolved organic nitrogen that is very uncertain (Prospero et al., this volume). The other half of the N deposition is approximately evenly divided between NO_y and NH_x species. We partition the total deposition of N into the three latitude bands (Table 2) using the deposition patterns for the sum of NO_y and NH_x. Approximately 25% of the atmospheric deposition of NO_y + NH_x occurs between 0 and 10° N and 33% occurs north of 40° N (Table 2). Atmospheric deposition of phosphate is approximately 6×10^9 moles N yr^{-1} and largely occurs south of 40° N (Prospero et al., this volume). This atmospheric transport of phosphate is largely associated with dust.

The atmospheric reactive nitrogen is produced by natural and human activities on the continents and much of this nitrogen is redeposited onto the surface after being transported relatively short distances. There is strong seasonality to most atmospheric deposition (Prospero et al., this volume). The patterns of deposition show strong maxima near the continental sources of North America, Europe and, to a lesser extent, Africa (Figure 2). Ammonia and nitrate wet deposition have very similar magnitudes and spatial patterns (Figures 2a and 2b). Much of the production of atmospheric reactive nitrogen is from human activity. The rates of deposition are certainly higher than before

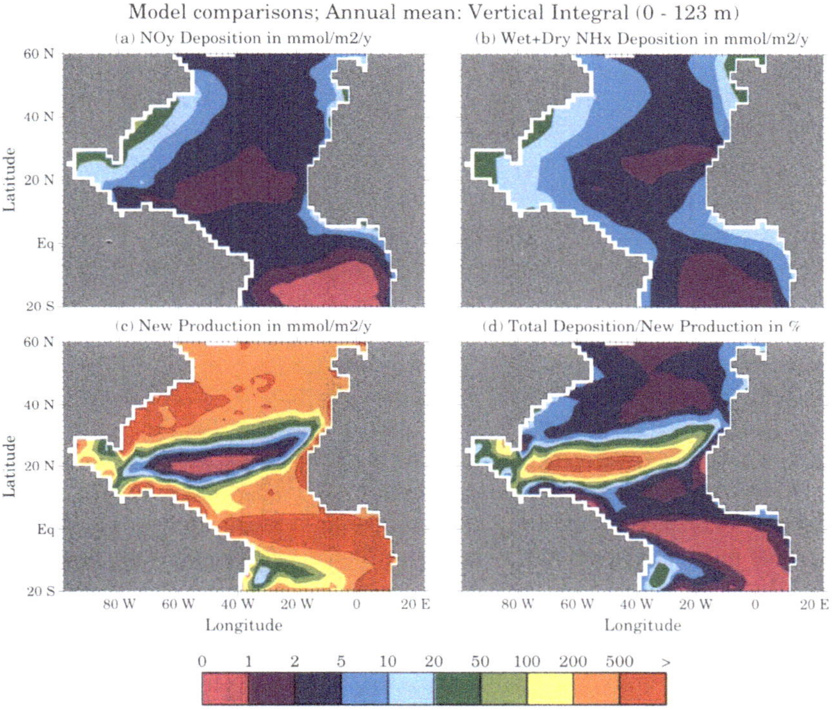

Figure 2. Comparisons of atmospheric deposition (courtesy of Chip Levy, see Prospero et al., this volume) of (a) NO_y and (b) NH_x and (c) new production from the Sarmiento et al. (1993) model. In (d) these are expressed as the sum of NO_y and NH_x as a percent of new production.

the industrial revolution and will likely continue to increase (Galloway et al. 1995; Prospero et al., this volume).

To assess the potential impacts of the current supply of reactive nitrogen by atmospheric deposition on marine ecosystem productivity, the deposition rates can be compared to the nitrate supplied to the surface by mixing as estimated with the Sarmiento et al. (1993) model. Since nitrate is usually completely depleted at the surface, the annual amount of nitrate used by phytoplankton will be approximately equal to the nitrate supplied by mixing (Figure 2c, keeping the caveats about the absolute magnitude of those fluxes in mind). Atmospheric deposition is a large source of N compared to the nitrate supply very near the coasts, where the deposition is high, and in the very center of the Sargasso Sea where the model predicts an extremely low nitrate utilization (Figure 2d). In intermediate locations, such as Bermuda, the combination of a modest nitrate supply and diminished atmospheric deposition make the atmospheric source unimportant (Knap et al. 1986; Michaels et al. 1993). Atmospheric deposition can be an important source of N in coastal waters (Paerl 1985). Few direct measurements of deposition or nitrate supply exist for the middle of the gyre (Knap et al. 1986; Owens et al. 1992; Michaels et al. 1993). Also, nitrate utilization in the the Sargasso Sea is certainly higher than these model estimates, perhaps as high as 0.2–0.7 moles N m^{-2} y^{-1} (Jenkins 1982; Jenkins & Goldman 1985; Sarmiento et al. 1990). Consequently, even though the deposition is very low this far from the sources, the low new production expected for these regions may make atmospheric deposition an important nitrogen source (Owens et al. 1992).

The atmospheric deposition of nitrogen in the open ocean is both seasonal and very episodic. During short periods of high deposition, it may be the dominant source of reactive N to the near-surface phytoplankton (Glover et al. 1988; Owens et al. 1992; Michaels et al. 1993). In the highly stratified waters that characterize the central gyre, most nitrate uptake occurs in the lower part of the euphotic zone (Goldman 1988). Thus, for the surface populations of phytoplankton, these atmospheric sources may be an important local nutrient source and in these stratified ecosystems, future changes in the anthropogenic production of atmospheric reactive nitrogen may have their largest effects. Furthermore, the stimulation of surface plankton blooms could cause a pigment signature that would be visible from ocean color satellites (Michaels et al. 1993).

The biological transport processes include the passive sinking of large particles (McCave 1975), the downward mixing of suspended particles and dissolved organic matter (Toggweiler 1989) and the vertical migration of zooplankton (Longhurst & Harrison 1988). The passive sinking of large particles has long been considered the major mechanism for the downward transport of organic material in the ocean (McCave 1975). Particle fluxes are highest near the surface and decline rapidly with depth (Martin et al. 1987; Bishop 1989) because of the remineralization or degradation of most sinking particles in the upper thermocline. Particulate fluxes near the sea floor are usually only a few percent of the fluxes leaving the euphotic zone. Recent measurements near Bermuda call into question our overall understanding of vertical flux processes. A comparison of all vertical fluxes with the changes in the standing stocks of carbon over a number of years indicates threefold discrepancies that can only be resolved if some of the vertical fluxes have been measured incorrectly, or if many of the observed patterns are due to horizontal advection (Michaels et al. 1994a). Further, seasonal measurements of dissolved organic carbon indicate that the downward mixing of dissolved organic matter may be an important transport process, at least for carbon (Carlson et al. 1994). Dissolved organic nitrogen and phosphorus dynamics are rarely measured, but have the potential to be important components of the large-scale N and P cycles (Najjar et al. 1992).

Burial of nitrogen and phosphorus in sediments is the ultimate link to elemental cycling on geological time-scales. However, even on the time-scales of anthropogenic production of reduced nitrogen, burial can be an important process. Jahnke & Jackson (1992) compare estimates of sediment accumulation rates of carbon with benthic respiration measurements. Using this dataset, and a C:N ratio of 10 for sediment organic matter (Nixon et al., this volume) we calculate a net burial of $1.1–1.7 \times 10^{10}$ moles N y^{-1} on the deep sea floor. Berner (1982) estimated that approximately 2.3 times as much organic carbon is buried in sediments under upwelling regions than in the deep sea for the world's oceans. This is exclusive of shallow carbonate, shelf and delta sediments (treated by Nixon et al., this volume). Assuming that a ratio of 2.3 applies to the North Atlantic, we estimate that $2.5–3.9 \times 10^{10}$ moles N y^{-1} are buried in slope sediments using the pelagic burial rates above. Similarly, we estimate a total phosphate burial of $2–3 \times 10^{9}$ moles P y^{-1} from the same data (Jahnke & Jackson 1992) using a C:P ratio of 200 (Table 2). These estimates are sensitive to both the accuracy of the carbon burial term and the C:N and C:P conversion factors. However, the fluxes are very small compared to other processes.

Nutrient transformation and cycling processes

The biological cycling of nitrogen in the upper ocean is usually interpreted in the context of the New Production Hypothesis (Dugdale & Goering 1967; Eppley & Peterson 1979). This hypothesis describes the nitrogen flows into and out of the euphotic zone. The euphotic zone is defined as the surface layer of the ocean which has adequate light for the growth of phytoplankton. It is usually delimited by the depth where the *in situ* irradiance is 1% of surface irradiance (typically depths of 30–120 m in different parts of the North Atlantic). New production is defined as phytoplankton growth on nutrients exogenous to the euphotic zone, primarily nitrate mixed into this layer from below. There are additional contributions from *in situ* nitrogen fixation and atmospheric deposition of various nitrogen species. Regenerated production is based on internal recycling within the euphotic zone and the total production is the sum of new and regenerated. The convenient division of nutrient species by their source (nitrate is largely an exogenous nutrient and ammonia is largely derived from regeneration) allows a partitioning of the total production based on the relative uptake of nitrate and ammonia (Dugdale & Goering 1967). When nitrogen fixation or atmospheric deposition are significant, the production of amines or deposition of ammonia will confound this simple distinction.

Nitrogen fixation and denitrification both occur in the water column of the North Atlantic. Nitrogen fixation is the process through which N_2 is converted to reactive nitrogen for use in the creation of organic nitrogen. This process is mediated by specialized bacteria and cyanobacteria, both free-living and symbiotic. The rates of N_2 fixation have traditionally been thought to be very low (although see below) and it has usually been ignored as a source of reactive N to the oceans (Carpenter 1983; Galloway et al. 1995). Denitrification is the bacterially-mediated conversion of nitrate to N_2 gas. It is largely confined to anoxic waters or microzones and is probably the major loss mechanism for reactive nitrogen; global estimates of marine denitrification are of the order $11–13 \times 10^{12}$ moles N y^{-1}, approximately evenly divided between the water column (primarily in the equatorial Pacific and the Indian Ocean) and the sediments (Galloway et al. 1995). Water column denitrification is largely restricted to areas with strong sub-surface oxygen minimum zones and anoxic basins. In the Atlantic, these conditions only occur in marginal seas (e.g. the Cariaco Trench), suggesting that water column denitrification is low in the rest of this basin. Sediment denitrification is predominantly found on continental shelves and the basin-wide rates have been estimated by Seitzinger and Giblin (Seitzinger & Giblin, in press) and summarized by Nixon et al. (this volume).

Phosphate uptake can also be conceptually separated into autochthonous and allochthonous sources, however, the lack of a unique P species associated with new versus regenerated production makes distinguishing the source of the production technically more difficult. There is also no analog of nitrogen fixation for phosphate; therefore, the ocean inventory of phosphate is tied to the combination of sedimentation and the hydrologic cycle. Atmospheric deposition rates for phosphate are small (Prospero et al., this volume). The dominant allochthonous source of phosphate to the phytoplankton communities in oceanic regimes will be the subsurface nutrient stocks.

Typically, organic biomass in the ocean has an N:P ratio of approximately 16:1 (the Redfield ratio; Redfield et al. 1963) and the concentrations of the two nutrients covary closely. Deep-ocean remineralization rates inferred by nutrient changes along neutral density surfaces are also at this ratio (Anderson & Sarmiento 1994). However, deviations do occur (Fanning 1989, 1992) and, when present, they are probably due to anomalies in the cycling of nitrogen, not phosphate. Denitrification removes nitrate and reduces the apparent ratio. Nitrogen fixation will add reactive nitrogen and should skew the ratio in a positive direction. The differential remineralization of DON compared to DOP will also affect the nutrient ratios. We take advantage of the non-Redfield nutrient stoichiometry later in this paper to infer a nitrogen fixation rate in the Sargasso Sea.

Horizontal nitrogen/phosphorus transport

The goal of this section is to consider broad scale nutrient transport, the problems in estimating it and the consequences of transport estimates for the nutrient balance in the North Atlantic basin. We address the flux of nutrients across the boundaries of the North Atlantic (i.e. the cross equatorial, Mediterranean and Arctic inflows) and the exchanges with the continental shelves. The flux divergence between different latitude bands based on recent calculations such as those found in Rintoul & Wunsch (1991) or Martel & Wunsch (1993) is considered in relationship to nutrient dynamics within the ocean. We assess the methods by which estimates of fluxes are calculated. These vary from very crude estimates of fluxes based on cross-sections of properties and educated guesses of transport such as the rates available from the Arctic, to nutrient balance arguments for basins such as the Mediterranean or Gulf of Mexico and Caribbean Sea, to sophisticated estimates of fluxes across trans-ocean sections using inverse techniques. The latter two methods are more rigorous, but both make heavy demands on data and assumptions. This is not meant to degrade the importance of the more ad hoc estimates such as those available for the Arctic. These often end up being approximately correct even though they involve intuition and a host of assumptions. For our estimates of nutrient exchange between the open ocean and the continental shelves, it proved easier to define the net balance on the shelves and then assume that any nutrient deficit was supplied from the ocean and any residual nutrient tranported offshore.

The large-scale currents in the North Atlantic Ocean cause horizontal transports of the dominant species of N and P. Surface currents are extremely pronounced, particularly the northerly flow of the Gulf Stream in the Western Atlantic. Thermohaline circulation at both intermediate and deeper depths also plays an important role in the meridional transport of nutrients with a net southerly flow in North Atlantic Deep Water and net northerly transport in Antarctic Bottom Water and Intermediate Water. Unfortunately, for making estimates of the appropriate meridional fluxes, we are constrained by the nature of the historical measurements. Zonal sections of nitrate and phosphate exist at a number of latitudes, but measurements of dissolved organic nitrogen and phosphorus are nearly non-existent. Nitrate and phosphate standing stocks are nearly an order of magnitude larger than the dissolved organic pools (Table 1) and the vertical patterns of the inorganic and organic species differ. Thus, the N and P flux by transport of DON and DOP is likely both smaller in magnitude and different in latitudinal pattern than the nitrate or phosphate fluxes. In the absense of any zonal sections of DON and DOP, this flux remains unknown. The unfortunate conclusion for a North Atlantic basin budget of N and P is that the uncertainties in the transport estimates are still very large and, in most cases, the ocean meridional transport uncertainties are larger than the total of riverine and atmospheric inputs to the North Atlantic basin.

Shelf boundary exchanges

The net loss of nitrate from the oceans required to satisfy the balance of nitrogen sources and sinks on the shelves is estimated as 0.7–0.95 $\times$ 10^{12} moles NO_3^- y^{-1} (Nixon et al., this volume; Table 2). This estimate arises from a detailed consideration of the riverine fluxes, atmospheric deposition, shelf denitrification, burial on shelves, deltas and estuaries and the losses through fish harvests in the coastal zone and continental shelves out to the 200 m isobath (Nixon et al., this volume). Shelf denitrification removes approximately 1.4 $\times$ 10^{12} moles NO_3^- y^{-1} from the overlying and *in situ* waters (Nixon et al., this volume). Denitrification and burial exceed the supply of nitrogen from the atmosphere and rivers; thus the remainder of the nitrate must come from water-column reactive nitrogen, ultimately nitrate. The shelf-ocean exchange is rapid and we assume that all of the net nitrate requirements of the shelves are met by a supply from the ocean.

We can partition the on-shelf nitrate flux geographically using the data in Nixon et al. (this volume, their Table 14). Losses of nitrate to the shelves are predominantly in the higher latitudes and in mid-latitude North America (Table 2). The Amazon provides a net source of nitrate to the open sea. The same dataset indicates a net flux of phosphate to the open ocean of 18–30 $\times$ 10^9 moles PO_4^{3-} y^{-1}, again dominated by the Amazon flux (Nixon et al., this volume). This combination of net sediment denitrification and phosphate transport should result in low water column NO_3^-:PO_4^{3-} ratios. For the water column N:P to remain near Redfield or higher, there must be a complimentary rate of nitrogen fixation in the water column or the oxidation of dissolved organic matter at very high N:P ratios (see below).

Ocean transport calculations – approaches

There are basically two approaches that can be used to estimate element or nutrient budgets in an ocean basin; (1) estimates of the flux divergence for a basin and (2) direct computations of a flux using velocity and concentration estimates for the element of interest. In the first approach, one estimates the flux divergence over the bounding surface area of a basin and then uses this in combination with available vertical fluxes and an assumption of steady state to estimate the lateral fluxes. This approach has been used to estimate heat fluxes in the Atlantic and world oceans (Vonder Haar & Ort 1973; Trenburth & Soloman 1994). A good example of this approach applied to a biogeochemical budget is the analysis of nutrient and dissolved oxygen budgets in the Mediterranean by Sarmiento et al. (1988). The second approach is to directly compute the fluxes by determining the horizontal velocity and the concentration of the desired element or property. This direct approach is thought to provide the best estimates of meridional heat flux (Hall & Bryden 1982; Rintoul & Wunsch 1991). Direct calculations have typically given lower heat fluxes than the basin-scale flux divergence estimates (although the most recent measurements of the latter are converging with the direct calculations; Trenburth & Soloman 1994). A direct approach using inverse models has been used to estimate nutrient fluxes in the Atlantic (Rintoul & Wunsch 1991; Martel & Wunsch 1993). Below, we summarize these results and also compute nutrient fluxes associated with the overturning cell, i.e. the zonally averaged circulation, for various inverse calculations of transports found in the literature. Since the actual completion of a full inverse with nitrate and phosphate included is beyond the scope of this paper, the contribution of the horizontal circulation to these meridional fluxes cannot be assessed fully. The degree to which these estimates are inadequate, however, can be addressed by comparing these calculations with the full inverses in Rintoul & Wunsch (1991).

Before going into the details of actually estimating the transport fluxes of nitrate and phosphate in the North Atlantic, we consider a conceptual steady-state framework for interpreting these fluxes. Sarmiento et al. (1988) divide basins into what they term "desert" basins and "nutrient trap" basins, building upon an earlier conceptual system described by Berger (1970). Desert basins have a combination of inputs and losses that result in low nutrient concentrations while nutrient trap basin dynamics result in high nutrient concentrations. The North Atlantic as a whole is a desert basin with low nutrients as compared to the nutrient trap basins in the Indian and Pacific. On a regional scale the Mediterranean is a desert basin (Sarmiento et al. 1988) while the Caribbean is a nutrient trap basin. The Arctic Sea is more complicated and seems to be a nutrient trap basin even though it exports deep water into the North Atlantic proper. The reason for this is a combination of inputs of nutrient rich waters from the Atlantic thermocline and the contribution made from the Pacific through the Bering Straits. Although there is no strict reason that these basin types would be constrained to import or export nitrate, it appears that desert basins export nutrients with their deep waters and "nutrient trap" basins accumulate nutrients brought in by deep flows. The basic difference in the two types of basins is that the desert basins are very efficient in removing nutrients through the formation of deep waters. Nutrient trap basins may also be more likely to lose nutrients to the sediments as the high nutrient concentrations would promote productivity and export to the sediments and the low oxygen concentrations would enhance burial.

The distribution of nitrate in the North Atlantic can be interpreted as a modification of nutrient contents in water masses as they proceed through the basin with the far North Atlantic being the most nutrient impoverished. It is not, however, possible to interpret this change directly in terms of fluxes or more correctly flux divergences of nutrients. While the property profiles suggest an overall pumping of nutrients out of the Atlantic, it is not possible to sort out the importance of recirculation and mixing from net advective fluxes. For example, much of the high nitrate water in the lower latitudes of the North Atlantic may simply be caught up in a northerly flow of Antarctic Intermediate Water, much or all of which circulates back to the south without contributing to any net flux. The only means of estimating the net intra-basin flux is to complete a calculation that explicitly conserves mass across some bounding surface. In other words, the observed property distributions say nothing about rates although they may be used to infer the sign of the fluxes under a steady state assumption. In order to demonstrate these methods and explore their capabilities the Mediterranean situation is considered in some detail.

Nutrient fluxes between the Mediterranean and Atlantic

The nutrient balance for the entire Mediterranean can be calculated using both the flux divergence method (see Sarmiento et al. 1988 for details) and the direct calculation method using a combination of outflow estimates (Baringer 1993; Price et al. 1993) and historical nutrient versus density relationships. Sarmiento et al. (1988) assume the sign of the fluxes based on the nature of the nutrient relations in the basin and they calculate a balance using the inputs from riverine and atmospheric deposition, sediment deposition and a two layer (surface and deep water) depiction of the fluxes across the sills into the basins. Their analysis assumes steady state and makes use of a closed salinity budget. The Sarmiento et al. (1988) water transport estimate for the Gibraltar outflow is 0.69×10^6 m^3 s^{-1} to the west (out of the Mediterranean). This is lower but comparable to the recent estimate of 0.79×10^6 m^3 s^{-1} by Bryden et al. (1989). Sarmiento et al. (1988) estimate the PO_4^{3-} balance to give an outward flux of phosphate of 2.96×10^9 moles y^{-1}. Using the canonical Redfield ratio of 16 moles NO_3^-:1 mole PO_4^{3-}, we can estimate a westward NO_3^- flux of 4.74×10^{10} moles y^{-1}. The Sarmiento et al. (1988) analysis is based on property end points for the inflowing and outgoing waters at Gibraltar (indicated by the S symbol in Figure 3). As shown these fit on the

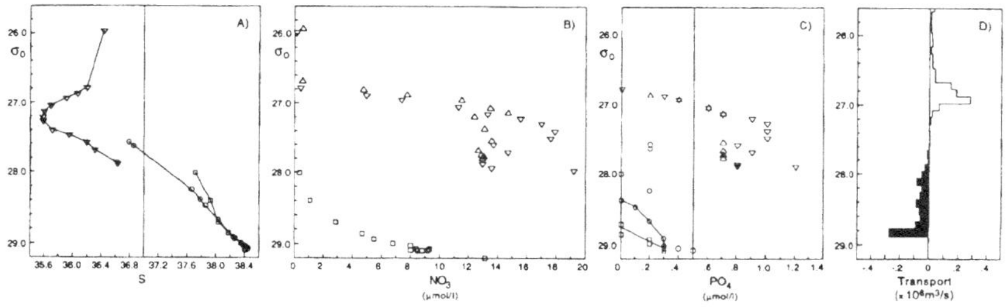

Figure 3. Properties for the flow through Gibralter as a function of potential density. Curves are based on historical station data from the Reid and Mantyla (per. comm.) data set (a.-c.) and Baringer's (1993) transport estimates. a) Salinity from either side of strait. Triangles indicate the Atlantic side while squares and circles are from the deep layers in the Mediterranean. Straight line indicates the salinity at the interface in the strait. b) Nitrate from outside (triangles) and inside the Mediterranean (squares and circles). c) Same but for phosphate. d) Transport estimates from the Atlantic side of the Straits.

mixing curves for the regions on either side of the sill and are in this sense reasonable.

The Mediterranean outflow nutrient fluxes can also be calculated using the property end points (Figure 3) and direct estimates of water transport at the sill (Bryden et al. 1989) or calculated transport rates just to the west of the Gibraltar sill (Baringer 1993; Price et al. 1993). As a first comparison, the westward salinity flux of 1.61×10^3 m^3 s^{-1} estimated by Bryden et al. (1989) is in good agreement with the salinity flux of 1.74×10^3 m^3 s^{-1} calculated using Sarmiento et al. (1988) end points and the Bryden et al. (1989) transport. Since Sarmiento et al. (1988) use a lower transport rate, they must be using larger differences in salinities. This makes sense given the high mixing rates over the sill region where the Bryden et al. calculation was made; i.e. mixing in the straits will reduce the contrast between the surface and deep layers.

Keeping the distinctions in end point values in mind, we can calculate westward fluxes of 21.8×10^{10} moles NO_3^- y^{-1} and 5.1×10^9 moles PO_4^{3-} y^{-1} using the Sarmiento et al. (1988) end points and the Bryden et al. (1989) transport. This estimate of the nitrate flux is greater than the estimate based on the Sarmiento et al. (1988) phosphate flux and the Redfield ratio. The nitrate:phosphate ratio of these fluxes is 42.5 and high nitrate:phosphate ratios are typical of certain parts of the Mediterranean (Krom et al. 1991). Clearly these calculations are sensitive to assumptions about the mixing curves in Figure 3 and nitrate:phosphate ratios. This sensitivity is obvious when the nutrient fluxes are calculated using historical profiles of nutrients from either side of the sill with the transport estimates made just to the west of the sill (Baringer 1993; Price et al. 1993; Figure 3d). With this combination, the calculated nutrient fluxes are in the opposite direction, eastward into the

Table 3. Estimates of the transport of nitrate and phosphate into or out of the Mediterranean through the Straits of Gibralter. Negative fluxes indicate a transport westward, out of the Mediterranean. All estimates are in units of 10^{10} moles y^{-1}.

Calculation approach	NO_3^-	PO_4^{3-}
Sarmiento et al. (1988)	−4.74	−0.30
Direct Baringer (1993) transports with historical end members.	9.59	0.19
Sarmiento end points with Bryden et al. (1989) transports	−21.78	−0.51
Sarmiento end points with Bryden et al. (1989) transports and accounting for changes in nutrient ratios	−10.4	−0.51

Mediterranean (Table 3), even though the salinity fluxes are in agreement with the other estimates. We can obtain yet another estimate of the flux of both NO_3^- and PO_4^{3-} using the Sarmiento et al. (1988) PO_4^{3-} end point, the Bryden et al. (1989) transport and by taking into account the differences in Redfield ratio between the incoming and outflowing waters. This calculation gives westward fluxes of 5.1×10^9 moles PO_4^{3-} y^{-1} (as above) and 10.4×10^{10} moles NO_3^- y^{-1}.

The diverse outcomes of these different methods for calculating nutrient fluxes for the Mediterranean illustrate several issues of general importance to the problem of estimating the nutrient fluxes caused by the ocean circulation. The results are sensitive to the shape of the property curves (e.g. Figure 3). The agreement between the salinity fluxes calculated by the different methods is due to the nearly linear salinity relationship in Figure 3. Thus the results from the direct estimates of Bryden et al. (1989) fall on basically the same portion of the mixing line as that assumed by Sarmiento et al. (1988). However, the nutrient curves (Figure 3b,c) are better approximated by a nonlinear S-shaped curve. The pronounced nutrient maximum associated with the Atlantic thermocline falls near the salinity of the interface between the upper and lower layers in the strait (discussed by Bryden et al. 1989, Figure 3a, vertical line) and therefore the zero in the transport versus density curve (Figure 3d). While the observations suggest that Sarmiento et al. (1988) are correct about the sign of the flux, the error bars on this estimate are likely to be large. It is hard to quantify the overall uncertainty in these calculations, given all of the assumptions, but with the uncertainties in the choice of end points and transports the variation is at least a factor of two. All of the methods could be improved with better observational data, primarily the concurrent measurement of transport and nutrients.

Finally these calculations all assume steady state conditions. Sarmiento et al. (1988) concluded that steady-state assumptions were reasonable in the Mediterranean, but they also pointed out that there have been major perturbations in the regional nutrient fluxes because of increased atmospheric and riverine fluxes of nutrients and the influence of dams on the Nile. These recent influences are expected to change the salt and nutrient property distributions in the Mediterranean and thus they could effect the exchange with the Atlantic since the fluxes across the sill are very sensitive to the property contrasts (Bryden & Kinder 1991). Secular trends in nutrient concentrations in the Mediterranean are not clear, but there have been changes in deep water salinity over the past few decades (Leaman 1994). Changes in nutrient concentrations and fluxes in the relatively small volume of the Mediterranean are proportionately greater than possible changes in the deep ocean and, for this reason, are easier to measure.

Nutrient fluxes in the Arctic, Caribbean and Gulf of Mexico

Estimation of net nutrient fluxes in the two other marginal sea complexes is even more difficult than in the Mediterranean. The Arctic, like the Mediterranean, is a basin that is dominated by deep-water formation. This occurs in the southern limb of the Arctic in the Greenland Sea. Deep-water formation would suggest that the Arctic is a desert basin (sensu Sarmiento et al. 1988) and possibly exports nutrients into the Atlantic Ocean. Indeed, based on the volume transports and nutrient concentrations in Anderson and Dyrssen (Anderson & Dyrssen 1981; Anderson et al. 1983) one obtains a southward flux of 30.0×10^{10} moles NO_3^- y^{-1} and 16.0×10^{10} moles PO_4^{3-} y^{-1}. These estimates have the same types of uncertainties as documented for the Mediterranean and the additional complexities associated with the various paths for water to enter and exit the basin.

Estimates of the nutrient fluxes through Caribbean Sea and Gulf of Mexico pose significant problems because of a very large volume flux and very small changes in nutrient concentrations (Figure 4). The flows in the Caribbean and Gulf of Mexico are dominated by the large volume transport of the southern limb of the North Atlantic subtropical gyre. There are also significant inputs of southern hemisphere waters (approximately 15×10^6 m^3 s^{-1}; Schmitz & Richardson 1991) and the input of two major rivers, the Orinoco and the Mississippi. These basins should be nutrient traps, consistent with the nutrient loading of the deep waters beneath the sill depth of the Caribbean passages. The calculations of Walsh et al. (1989) for the Gulf of Mexico suggest there is a net flux of NO_3^- into the Atlantic through the Florida Straits of $2.14 \times$

10^{10} moles y^{-1}. They suggest that an equal amount of nitrogen is exported as particulate organic nitrogen (PON) under current conditions. This estimate includes an increase of 0.71×10^{10} moles y^{-1} from anthropogenic influences, primarily changes in the Mississippi nitrogen flux. The PO_4^{3-} flux calculated by using the Walsh et al. (1989) computation and Redfield stoichiometry is an export of 1.34×10^9 moles PO_4^{3-} y^{-1} to the Atlantic Ocean. These estimates are based on a very coarse vertical resolution of the flow through the bounding straits (Yucatan and Florida, respectively). The nutrient cross sections seem to have considerable structure both in the horizontal and vertical and there are some questions about whether the data are accurate in time (Walsh et al. 1989).

The Walsh et al. (1989) physical transport estimates for the net nutrient balance in the Caribbean and Gulf of Mexico are quite different than the nutrient budgets implied by the estimates of riverine flows and shelf denitrification and burial (Nixon et al., this volume). Nixon et al. (this volume) estimate that denitrification exceeds the sum of atmospheric deposition and riverine flux for both basins. This net denitrification may be required if these basins are to stay nutrient traps as the large horizontal flows through a small basin could result in dramatically elevated deep nutrient concentrations in the absense of denitrification.

We attempted to estimate the flux of NO_3^- and PO_4^{3-} through the Caribean and Gulf of Mexico using the inverse calculation of water transport made by Roemmich (1981). This inverse calculation is well constrained and has reasonable resolution (five layers in the region of interest; Roemmich 1981). However, when combined with existing nutrient data, it is inadequate for computing the desired nutrient fluxes because of the large gradients in nutrient concentrations along the isopycnal surfaces used in the calculation of the water transport. Any brute force calculation that averages the data within layers, independent of position across the Caribbean or the Straits of Florida, would lead to a net inflow of both nitrate and phosphate into the basins. This arises because NO_3^- and PO_4^{3-} concentrations are actually higher on isopycnals in the central and southern Caribbean than they are in the Straits (Figure 4). The data are also sparse regionally. The northern Caribbean is under-represented and the Windward Passage had no nutrient data of sufficient quality.

One conclusion that can be drawn from the Roemich (1981) flows, as well as from those of other authors (Leaman et al. 1989) is that there is a net deposition of NO_3^- in the basin. The transports of waters with high NO_3^- and PO_4^{3-} into the Caribbean and the lower concentrations on the same isopycnal surface across the Florida Straits requires a net convergence throughout most of the thermocline in the Caribbean and Gulf of Mexico (Figure 4). This conclusion is at odds with that of Walsh et al. (1989) for the Gulf of Mexico.

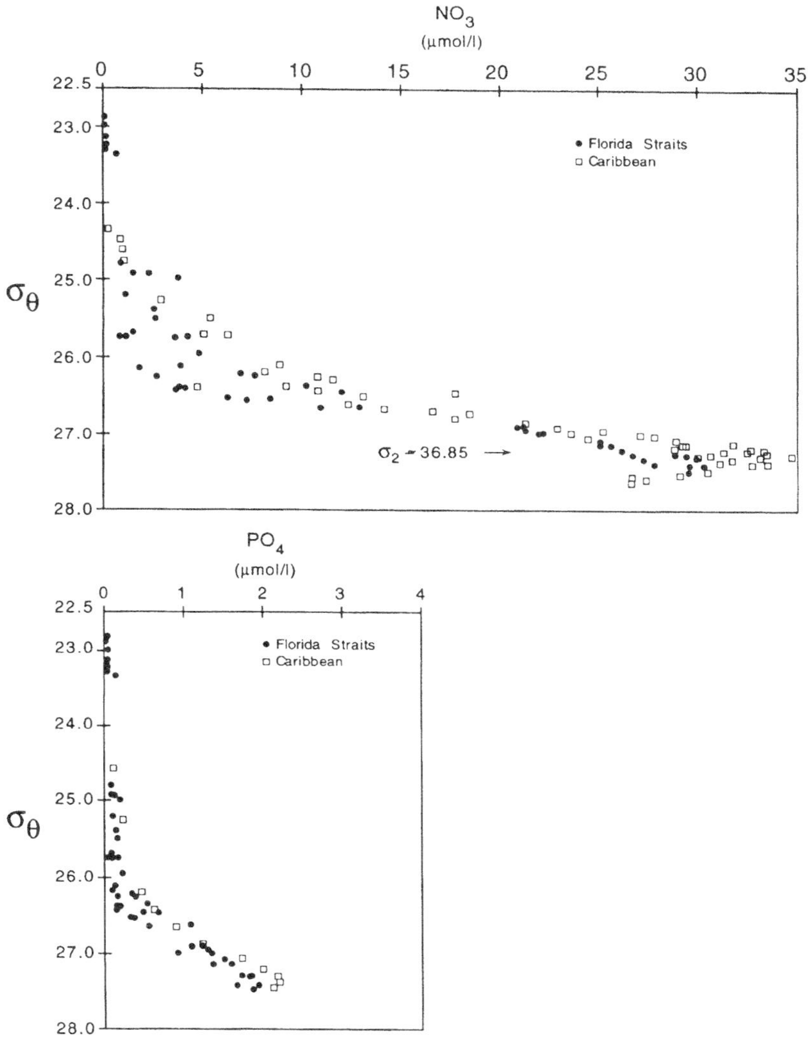

Figure 4. Nitrate and phosphate as a function of potential density in the Caribbean and Florida Straits. Notice that Florida Straits nutrients are lower on isopycnal surfaces suggesting a decline in nutrient levels in the upper pycnocline as the fluid traverses the basins.

The flow in the basins and the nutrient distributions (Figure 4) suggest that their treatment of the large scale oceanic transport is too coarse to define the oceanic flux. The current conclusion agrees with that of Nixon et al. (this issue) that overall, the Caribbean Sea and Gulf of Mexico complex should be a net sink for nitrate.

Meridional transport in the Atlantic Ocean

The north-south flux of nutrients within the North Atlantic can be estimated from either of the approaches outlined above. The Atlantic as a whole acts like a desert basin (sensu Sarmiento et al. 1988); it is depleted in nutrients at depth as compared to the other oceans. This depletion is marked by a strong north to south increase in NO_3^- and PO_4^{3-} at most depths in the Atlantic. If one assumes steady state conditions, there must be a balance between inputs into each latitudinal band and exports. Thus, the residual northward or southward flux in each latitudinal band must equal the balance between sources of reactive nitrogen (nitrogen fixation, riverine fluxes and atmospheric deposition) compared to the losses in that band (sedimentation and denitrification). Completing this balance using the flux convergence approach described for the Mediterranean is difficult because of the uncertainties in and relatively small size of many of the boundary exchanges (e.g. Nixon et al., this volume; Prospero et al., this volume). Thus most of the attempts to calculate meridional transport have come through direct calculations of volume fluxes and nutrient fluxes using oceanographic data.

We summarize three studies that apply inverse methods to oceanographic hydrographic station data in order to estimate nitrate and phosphate fluxes from meridional transects (Schlitzer 1988; Rintoul & Wunsch 1991; Martel & Wunsch 1993). We also discuss the earlier inverse estimates of meridional transports by Roemmich (1983) in an attempt to interpret the fluxes through the tropics into the North Atlantic. Of the four studies Schlitzer's (1988) calculations are based on highly smoothed historical data, those of Roemmich (1983) and Rintoul & Wunsch (1991) are based on cross-basin sections taken on single cruises and those of Martel & Wunsch (1993) are based on a set of data collected in the central Atlantic during the early 1980's. Martel & Wunsch (1993) make use of a rather coarse gridding of their data base and therefore their study is somewhat similar in resolution in space to that of Schlitzer (1988). All of these studies make use of mass, heat, salt and nutrients to estimate nutrient transport. While there are differences in the approaches used in these studies they are not faced with the uncertainties from the almost arbitrary choices of property values or transports for particular density strata as affected the previous calculations. Finally, these methods are capable of producing error estimates as an integral part of the calculation based on the statistics of the data used.

The results of the inverse estimates with the exception of Schlitzer's (Schlitzer 1988, 1989) are shown in Table 4. Schlitzer (1988, 1989) suggests that there is no flux of NO_3^- across lines of latitude in the Atlantic. The estimates by Martel & Wunsch (1993) contain Schlitzer's conclusion within their estimated uncertainty. Both of these studies utilize highly smoothed datasets either in terms of the actual data or in terms of the gridding of the data set for the finite difference analysis in Martel & Wunsch (1993). The Rintoul & Wunsch (1991) estimate makes use of data that resolve the eddy field along their sections. They find a significant northward flux of nitrate across 36° N and they argue that this becomes evident because they resolve the eddy field. Again, these estimates fall within the error bars and agree in sign with the estimates of Martel & Wunsch (1993). Unfortunately, the inverse methods, while quantitative, do little to constrain the nutrient budget for the North Atlantic because of the large uncertainty associated with each flux estimate.

In the absence of tightly constrained estimates of meridional fluxes, it is still useful to ask how the fluxes take place and what they imply about nutrient cycling in the Atlantic. One might question the relative role in nutrient fluxes of thermohaline circulation versus the locally semi-closed wind gyres. This distinction between a buoyancy driven vertical circulation (a zonally-averaged meridional and vertical circulation) and the horizontal gyre flows (with both meridional and zonal features) can be estimated by comparing the averaged fluxes from inverse computations (e.g. Roemmich 1983) with the full estimates.

The flux calculation for a substance carried in the meridional circulation has been worked out in detail by investigators interested in the oceanic heat flux (Bryan 1962, 1987; Hall & Bryden 1982). The basic computation involves decomposing the flux into zonally average contributions and correlations between the anomalies from this average. The flux of some quanity C is then

$$\overline{[F_c]} = \overline{[V]'[C]'} + \overline{[\bar{V}^*\bar{C}^*]} + \overline{[V^{*\prime}C^{*\prime}]}$$

where the brackets ([]) denote a zonal average, the asterix (*) the deviation from the zonal mean, the overbar a depth average, and the primes (′) the deviation from the depth average. The three terms on the right are the vertical overturning contribution, the depth averaged contribution of anomalies in the transport and concentrations in the horizontal, and the contribution due to deviations from both the zonal and depth averages. The full inverse calculations of Rintoul & Wunsch (1991) include the entire flux. Their published results, however, only give [V]'. Therefore the computations here only include the second term. This overturning contribution is typically thought of as a thermohaline contribution since there is no wind-driven contribution to this term.

To get meridional nutrient fluxes, the meridional overturning estimates from Roemmich's (1983) inversions are combined with historical mean nutrient profiles at these latitudes. The product of the average nutrient concentration for each density layer and Roemmich's transport estimate gives fluxes

Table 4. Nutrient fluxes in the North Atlantic. Unless otherwise noted fluxes are across lines of latitude with northward being positive. For a description of the estimation techniques and sources see text. The Arctic and Gulf of Mexico fluxes use the same nomenclature. The Mediterranean flux is negative denoting a flux out to the west. All estimates are in units of 10^{10} moles y^{-1}.

Latitude	NO_3^- flux	PO_4^{3-} flux
Martel & Wunsch (1993)		
12 N	160 ± 300	12 ± 19
24 N	−100 ± 250	−10 ± 16
36 N	50 ± 250	2 ± 16
43 N	−90 ± 275	−10 ± 19
48 N	−60 ± 175	−3 ± 10
55 N	−10 ± 150	−3 ± 10
60 N	−180 ± 100	−23 ± 7
Rintoul & Wunsch (1991)		
24 N	−25 ± 123	–
36 N	375 ± 119	–
Roemmich (1983) with historical nutrient profiles (see text)		
8 N	−665 to 610	−31 to 34
24 N	−805 to −400	−380 to −2
Mediterranean		
36 N	−5 to −22	−3 to −5
Arctic (Anderson & Dyrssen 1981, 1983)		
65 N	−30	−16
Gulf of Mexico (Walsh et al. 1989)		
25 N	2	–

at 8° and 24° N that can be compared to the full calculations of Rintoul & Wunsch (1991) and Martel & Wunsch (1993). This approach assumes that the Roemmich-mean nutrient profile calculation is indicative of the thermohaline overturning cell and the difference between this estimate and the full inversion is the flux in the horizontal or the along-isopycnal flux of nutrients. At 8° N, the flux has a very large uncertainty (Table 4, the range of values is discussed in Roemmich 1983). At 24° N both calculations indicate a southward flux of NO_3^- and PO_4^{3-}. A similar calculation done crudely from the figure for zonally averaged transport from Rintoul & Wunsch (1991) matches the result using Roemmich (1983) for 24° N and suggests a southward transport at 36° N by the thermohaline cell. Furthermore the range of nutrient estimates from the calculation using Roemmich (1983) transports falls below the error bars for both Rintoul & Wunsch (1991) and Martel & Wunsch (1993). If these crude computations are valid, they imply that the thermohaline cell may indeed be pumping nutrients southward, out of the North Atlantic. However, this is partially reversed in the subtropical gyre by the horizontal circulation. This conclusion is consistent with that of Rintoul & Wunsch (1991). The reversal in flux and the flux divergence between 24° N and 36° N calculated by Rintoul & Wunsch (1991) poses further questions concerning the nutrient budget in the central North Atlantic Ocean. We concur with the conclusion of Rintoul & Wunsch (1991) that the horizontal circulation in the gyres has a major impact on the flux of nutrients and that resolution in space is important in these calculations.

Rintoul & Wunsch (1991), in their interpretation of the analysis of hydrographic sections at 24° N and 36° N, invoke an unmeasured flux of dissolved organic nitrogen (DON) to the south at 36° N to balance the northward flux of nitrate. At 36° N, they estimate a net northward flux of $3.75 \pm 1.19 \times 10^{12}$ moles NO_3^- y^{-1} (119 ± 35 kmoles NO_3^- s^{-1}), causing a divergence between 24° and 36° N and a convergence north of 36° N. They suggest that there is a net southward transport of DON to balance this northward flow of nitrate, an amount equal to about 1/3 of the nitrate carried in the Gulf Stream. They then use these estimates to calculate a new production rate for the zone between 36° and 60° of 2.1 moles C m^{-2} y^{-1} over an area of 1.2×10^{13} m^2. Since there are no data on DON on these sections, their hypothesis cannot currently be tested. However with an appropriate section of DON data, this mechanism for balancing the apparent flux divergence of nitrate in the Sargasso Sea can be assessed. The estimates of excess nitrate creation in the Sargasso Sea described below are relevant to this question as the fluxes are similar to the northward nitrate flux across 36° N and they further constrain the source of the excess nitrate to be the center of the gyre. These constraints may indicate that nitrogen fixation is another logical (and unmeasured) source of the nitrate transported north across 36° N.

Excess nitrate and nitrogen fixation in the Sargasso Sea

The Sargasso Sea has elevated ratios of nitrate to phosphate in the main thermocline compared with surrounding waters and with other oceans (Fanning 1989, 1992; Michaels et al. 1994b). Nitrogen fixation and atmospheric deposition have previously been suggested to account for this feature

(Fanning 1989; Michaels et al. 1994b). The waters of the main thermocline (above 1000 m) where this nutrient ratio anomaly is most pronounced are ventilated on time-scales of decades or less (Jenkins 1980, 1988; Sarmiento 1983; Sarmiento et al. 1990). In the areas where these isopycnal surfaces outcrop, the concentrations of both nitrate and phosphate are often near zero. Thus nutrients that are present on these isopycnal surfaces in regions away from the ventillation areas come from remineralization of organic matter that has been transported from the overlying surface waters. If the organisms that comprise the bulk of the organic biomass in sinking material have N:P elemental concentrations near the Redfield ratio of 16, the remineralization of this biomass will lead to the production of 16 moles of nitrate for every mole of phosphate produced. If organisms such as nitrogen-fixing cyanobacteria create biomass that is richer in nitrogen (more correctly, phosphate-poor) this could yield the remineralization of more than 16 moles of nitrate for every mole of phosphate. Conversely, denitrification would remove nitrate with respect to phosphate and decrease the *in situ* elemental ratio. We may be able to use the changes in nutrient ratios on density surfaces along with some assumptions to estimate the annual supply of excess nitrate that would be required to maintain the observed patterns under the Sargasso Sea.

We define a parameter N^* as the concentration of nitrate in excess (or deficit) of that expected from the remineralization of phosphate at Redfield stoichiometries:

$$N* = [NO_3^-] - 16 * [PO_4^{3-}] + 2.72$$

$[NO_3^-]$ and $[PO_4^{3-}]$ are the concentrations of nitrate and phosphate respectively. We have arbitrarily added 2.72 to the nominal $([NO_3^-] - 16*[PO_4^{3-}])$ quantity so that the global average N^* as estimated from the GEOSECS dataset is zero (Gruber and Sarmiento in preparation). This implies that for the global ocean, denitrification exceeds nitrogen fixation. The addition of 2.72 will have no effect on the first-order flux estimates below as the 2.72 cancels out in the calculation of the gradient on a density surface.

In a mean profile of N^* at the Bermuda Atlantic Time-series Study (BATS) site, N^* is elevated throughout the upper 800 m and becomes less than 2.72 below this depth (Figure 5). A similar profile is observed at the TTO stations occupied in 1981 in the central Sargasso Sea, although the values are slightly lower than the BATS composite. GEOSECS nutrient profiles in this same area are similar to the BATS data. If we compare N^* at BATS or the TTO stations with N^* from the regions where each density surface outcrops during the winter ventilation, we can determine an N^* gradient. This gradient can be compared to the ventilation time-scale of each isopycnal surface to calculate the rate of supply of excess nitrate that would be required to maintain the

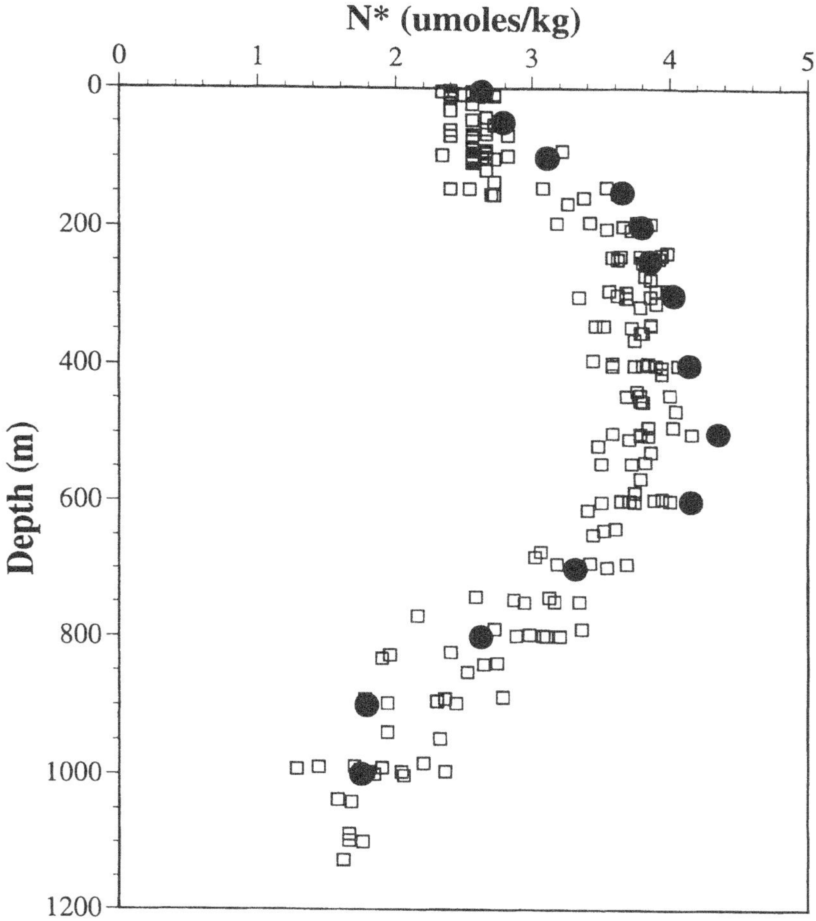

Figure 5. Mean profile of excess nitrate at BATS (solid circles) and at various TTO stations in the central Sargasso Sea (open squares).

gradient (using techniques described by Sarmiento 1983). This calculation is similar to the approach commonly used to estimate oxygen utilization rates (OUR) in the same region (Sarmiento et al. 1990). The outcrop N* (N*(O)) is estimated from the TTO stations near the outcrop regions (Sarmiento 1983; Table 5).

From these calculations, we estimate that 3.4–6.1 $\times$ 10^{12} moles of "excess" nitrate are created in the Sargasso Sea each year (Table 5). The range in the estimates is largely a function of the different residence time estimates for each density surface and less a function of the uncertainty in the magnitude of

Table 5. N* and excess nitrate creation-rate calculations. N* in units of μmoles kg^{-1}, volume in 10^{15} m^3, residence times τ in years with $\tau_{(J)}$ from Jenkins 1980 and $\tau_{(S)}$ from Sarmiento 1983. $N_{(J)}$ and $N_{(S)}$ are the rate of nitrate formation in 10^{11} moles y^{-1} estimated using $\tau_{(J)}$ and $\tau_{(S)}$ respectively. N* (B) data were collected at BATS and N* (O) data for outcrop regions from TTO. Data in parentheses are estimates made for this analysis that are not in the cited references. Density surfaces less than 26.2 ventillate near Bermuda every year, but do not ventillate just to the south. Exclusion of the data from the 26.0 and 26.1 isopycnals reduces the nitrate creation estimates by approximately 10%.

Density (sigma-theta)	N*(B)	N*(O)	ΔN*	Volume	$\tau_{(J)}$	$N_{(J)}$	$\tau_{(S)}$	$N_{(S)}$
26.0	3.06	2.22	0.84	(0.3)	(0.75)	3.38	(1)	2.53
26.1	3.31	2.20	1.11	(0.3)	(0.75)	4.44	(4)	0.83
26.2	3.69	2.24	1.41	0.36	(0.75)	6.95	7.2	0.72
26.3	3.80	2.22	1.58	0.48	(0.75)	10.08	7.3	1.04
26.4	3.99	1.42	2.57	0.64	1	16.46	7.5	2.19
26.5	4.14	1.45	2.70	0.98	7.4	3.57	8.2	3.22
26.6	4.30	0.26	4.04	1.2	10.8	4.49	9.3	5.22
26.7	4.26	0.94	3.32	1.1	11.5	3.18	11	3.32
26.8	4.06	1.06	3.00	1.2	16.5	2.18	12	3.00
26.9	3.56	0.64	2.92	1.3	21	1.81	14	2.71
27.0	3.16	0.56	2.60	1.6	28	1.49	15	2.78
27.1	2.84	0.69	2.15	1.9	39	1.05	16	2.56
27.2	2.50	0.48	2.02	2.2	49	0.91	19	2.34
27.3	2.08	0.98	1.10	2.4	61	0.43	29	0.91
27.4	1.78	0.91	0.87	2.6	71	0.32	39	0.58
Sum of nitrate creation rates from sigma-theta = 26.0–27.4:						60.72		33.95

the horizontal N* gradient. This rate is likely an underestimate of the gross rate of excess nitrate creation as other boundary exchanges could further modify the nitrate and phosphate concentrations in the upper layers of the Sargasso Sea. Nixon et al. (this volume) estimate that the continental shelves require 0.7–0.95 × 10^{12} moles NO_3^- y^{-1} from the open ocean to supply the excess of denitrification rates over riverine and atmospheric nitrogen supplies. At the same time, they calculate a net supply of phosphate to the open ocean of 1.8–3.0 × 10^{10} moles PO_4^{3-} y^{-1}. Some fraction of these fluxes would exchange with the density layers used to calculate the excess nitrate gradients and should act to lower N*. Thus the gross rate of excess nitrate production would have to be higher than the net rate to compensate for these and other sources and sinks. Most of the phosphate flux comes from the Amazon River (Nixon et al., this volume) and it is difficult to determine if it would exchange with the upper

Sargasso Sea isopycnal surfaces. We estimate that approximately one third of the net nitrate loss to the shelves may exchange with these surfaces (based on a comparison of the net nitrogen balance for each shelf section in Table 13 of Nixon et al., this volume), potentially enhancing the excess nitrate production rate by 0.3×10^{12} moles NO_3^- y^{-1}. This results in an estimate of total excess nitrate production of 3.7–6.4×10^{12} moles y^{-1} for the subtropical regions. There are two possible sources of the nitrate excess in the main thermocline, nitrogen fixation and/or differential fluxes or lability in dissolved organic nitrogen compared to dissolved organic phosphate. Unfortunately, neither process is well constrained by measurements.

Nitrogen fixation

Marine nitrogen fixation usually occurs in the nitrate and phosphate-depleted, illuminated waters near the surface. The link between diazotrophy at the surface and excess nitrate production in the mesopelagic requires that nitrogen-fixing organisms create biomass with a high N:P ratio (Karl 1992; Karl et al. 1995), either sink or become incorporated into large sinking particles and then that the organic matter be remineralized to nitrate and phosphate in the main thermocline. In oligotrophic ecosystems, most of the organic biomass is dominated by small particles and rapid recycling. Larger organisms play a disproportionate role in export processes dominated by sinking particles (Michaels & Silver 1988) and, thus, it seems likely that diazotrophs that could mediate the excess nitrate signal would have to be relatively large. In addition, these organisms would have to acquire an appropriate supply of phosphate in a part of the ocean that has been strongly depleted in phosphate to depths of 100–200 m.

The most conspicuous large diazotroph in the Sargasso Sea is the colonial cyanobacterium, *Trichodesmium sp.* . This organism has a number of attractive properties that may both mediate the transport of nitrogen to the mesopelagic and allow it access to the phosphate in the main thermocline (Karl 1992). It is capable of active bouyancy regulation (Villareal & Carpenter 1990) and can have vertical migration rates of up to 150 m d^{-1} through both buoyancy and sinking. It can store phosphate as polyphosphate when it is at depth (nitrogen storage is much less efficient compared to the growth requirements, although some organisms may also migrate to gain access to nitrogen). It uses a carbohydrate ballast to control buoyancy (Romans et al. 1994); the respiration of this ballast at depth would result in a net downward transfer of CO_2 in excess of its nominal biomass C:N ratio.

Carpenter (1983) estimated a global nitrogen fixation rate for *Trichodesmium sp.* of approximately 7.5×10^{11} moles N y^{-1}. Both specific growth rates and abundance tend to be low, and the areally weighted nitrogen fixation rates are usually very small compared to other parts of the nitrogen cycle (Lipschultz & Owens, this volume). Carpenter & Romans (1991) concluded that the rate of nitrogen fixation might be much higher than previously estimated for the Caribbean Sea. They found N fixation rates of 2.1 mmoles m^{-2} d^{-1}, comparable to total new production from other sources. This higher rate comes from a combination of using the higher estimates of the specific growth rates of *Trichodesmium sp.* , and from higher abundance estimates from a recent series of cruises in the area (Carpenter & Romans 1991). Interestingly, these new estimates completely change the global estimates of new production. At the estimated fixation rate of 2.1 mmoles m^{-2} d^{-1}, the approximately 7–19 $\times$ 10^{12} m^2 area in the Caribbean and Sargasso Seas, where one could logically extrapolate the measurements, would account for 5–14 $\times$ 10^{12} moles N y^{-1} of nitrogen fixation, 8–20 times the earlier estimates for the entire globe (Carpenter 1983). Thus, combining the higher rates with moderately higher abundance estimates indicates that nitrogen fixation could be significant in many parts of the Sargasso Sea. Lipschultz & Owens (this volume) summarize the existing literature on nitrogen fixation by *Trichodesmium sp.* They discount some assumptions in the calculations of high nitrogen fixation rates and estimate that *Trichodesmium sp.* could only account for about 1.1 $\times$ 10^{12} moles N y^{-1} in the Sargasso Sea. This is only a fraction of the excess nitrate signal. It remains possible that other, less well studied diazotrophs may play a role.

Dissolved organic nitrogen

The potential role of dissolved organic nitrogen (DON) in the creation of excess nitrate is more difficult to constrain, largely because DON is rarely measured. The net production of organic nitrogen must occur near the surface. DON can invade the thermocline by mixing along isopycnal surfaces or by horizontal transport into the Sargasso Sea at the surface, where it would then have to be used by surface biota and transferred into the thermocline by sinking particles. In both cases, the source of the putative DON would be the spring blooms in the outcrop zones at northern latitudes. Such a southward transport of DON was suggested to explain the measured northward transport of nitrate across 36° N latitude (Rintoul & Wunsch 1991). Both scenarios require a pool of DON that is very refractory for the first few years after creation, but quite labile on time-scales of a decade (it must create the N* maximum in the main thermocline at the center of the Sargasso Sea). If the remineralization time-scale of the DON was short, the N* signal would be strongest near the outcrop. If it was similar to the ventillation time scale of these layers, there would be no gradient as nitrate would be created relatively evenly over the density surface. Thus, the isopycnal mixing of DON would require an unlikely change in lability as it passed under the Sargasso Sea to produce the observed mid-gyre maximum. A surface transport of DON from the North would also require a similar change in lability after the water had moved south of 36° N.

As both the total concentration of nutrients and the excess nitrate increase towards the center of the gyre, a horizontal DON source would also require a DOP pool that has similar characteristics and a slightly higher lability on all time-scales. DOP pools are low (200 nmoles kg^{-1}) and fairly constant in the surface waters of the Sargasso Sea (Ammerman et al. 1994). Thus any uptake process that uses DON in the surface waters would face the phosphate limitations described earlier. It seems unlikely that DON and DOP with these characteristics could exist and not result in higher rates of remineralization at the source region than in the center of the gyre. Thus nitrogen fixation seems to be the most likely source of the elevated N^* in the Sargasso Sea. However, until adequate surveys of DON and DOP are performed, a dissolved organic source cannot be ruled out.

Midwater particle fluxes

We can attempt to further bound the mass balances in the main thermocline of the Sargasso Sea by examining the vertical transport of N and P within this layer. We estimate a particulate nitrogen flux across the 27.0 sigma-theta surface of 0.1–0.3 mmoles N m^{-2} d^{-1} (the mean depth of this surface is 430 m in the midst of the region of elevated NO_3^-:PO_4^{3-} ratios; Sarmiento 1983). This flux is comparable to the fluxes measured with sediment traps at these depths at the Bermuda Atlantic Time-series Study (BATS) site (0.05–0.3 at 31.7° N, 64.2° W; Lohrenz et al. 1992; Michaels & Knap 1996) and springtime fluxes at the North Atlantic Bloom Experiment site (0.3–0.6 at 42° N, 20° W; Martin et al. 1993). In addition, extrapolating the basin-wide benthic oxygen respiration rate (Jahnke 1992) to 430 m using the flux profile algorithm of Berger et al. (1987) yields a rate of 0.27 mmoles N m^{-2} d^{-1} at 430 m. The oceanic region between 10° N and 40° N encompasses approximately 27×10^{12} m^2 and the total vertical flux out of this region would be $1–3 \times 10^{12}$ moles y^{-1}. This estimate is likely to be uncertain by at least a factor of two because of the recognized uncertainties in the accuracy of sediment traps (Buesseler 1992; Michaels et al. 1994) and the limited number of measurements.

The downward flux of nitrogen is balanced by an upward mixing of nitrate. We can estimate this flux by taking the product of a gradient at 500 m of 0.03–0.04 mmols NO_3^- m^{-4} (calculated from the BATS and TTO datasets) and diapycnal eddy diffusivities of $1–4 \times 10^{-5}$ m^2 s^{-1}. Typical thermocline values of eddy diffusivity are thought to range between 10^{-5} and 10^{-4} (Munk 1966; Gregg 1989). A vertical diffusivity of 4×10^{-5} m^2 s^{-1} is calculated from the typical density profile following the parameterization suggested by Gargett (1984). This agrees well with estimates of 3.7×10^{-5} m^2 s^{-1} from microstructure measurements (Lewis et al. 1986). A more recent study using a tracer release found lower diapycnal diffusivities of 1×10^{-5} $m^2 s^{-1}$.(Ledwell et al. 1993) Using this range of parameters, we estimate an upward mixing of nitrate of 0.03–0.16 mmoles NO_3^- m^{-2} d^{-1} and 0.3–1.6 $\times$ 10^{12} moles NO_3^- y^{-1}. Thus the downward flux of nitrogen on particles appears to be larger than the upward nitrate flux, but the ranges in the two estimates overlap. A similar comparison is not possible for phosphate as there are no measurements of the flux of P on sinking particles. The phosphate gradients are 0.002–0.003 mmols PO_4^{3-} m^{-4}, indicating an upward flux of 2–12 $\times$ 10^{10} moles PO_4^{3-} y^{-1}.

Carbon cycle implications
Nitrogen fixation as a source of the excess nitrate could also help resolve some unanswered questions about the measured carbon imbalance in the Sargasso Sea. Each year, between April and December, approximately 3 moles m^{-2} of carbon disappear from the surface waters at BATS (Michaels et al. 1994a). This disappearance is a conundrum because it occurs in the absence of a supply of nitrate and because only 1 mole C m^{-2} can be accounted for in the balance of all of the vertical processes that modify the surface concentration of dissolved inorganic and organic carbon (DIC+DOC+POC). The imbalance is likely caused by one of two issues, either some of the measurements of vertical processes are in error or there is a summer/fall advection of low DIC waters into the region. Sediment traps seem to underestimate the sinking particle flux in this region and this error is probably responsible for at least part of the discrepancy between the observed disappearance and the flux measurements. Since latitudinal gradients in DIC are small relative to the seasonal signal, advection may just require that the anomalous carbon uptake occur elsewhere. This anomalous carbon uptake has a strong effect on the air-sea gas flux of CO_2 in this region (Bates et al. 1996); using a Redfield DIC:NO_3^- stoichiometry for the surface DIC balance would raise summer pCO_2 values by 60 μatm and turn the area at BATS from a weak sink into a strong source of CO_2 with respect to the atmosphere.

Nitrogen fixation would require the uptake of dissolved CO_2 as the new reduced nitrogen is converted into biomass. This may occur at the typical Redfield ratio of 6.6 moles C:1 mole N, or likely at a much higher ratio as the most conspicuous diazotroph, *Trichodesmium sp.* uses a carbohydrate ballasting process to regulate buoyancy . Using C:N ratios of both 6.6 and 15, we estimate the total carbon uptake in the surface waters overlying the Sargasso Sea as 24–96 $\times$ 10^{12} moles C y^{-1} (0.3–1.2 $\times$ 10^{15} g C y^{-1}). In the seasonal environment of BATS, this would likely be confined to the summer/fall period where oligotrophy is conducive to nitrogen fixation, but could occur year-round to the south. Approximately 50–80% of this carbon

Table 6. Wet and Dry deposition of iron to the Sargasso Sea. Calculated as 3.5% of the total dust deposition by weight. Each value is the total deposition of iron to a 10° × 10° square centered on each latitude and longitude in 10^9 moles Fe y^{-1}. Data modified from Prospero (this volume).

		Longitude								
		95° W	85° W	75° W	65° W	55° W	45° W	35° W	25° W	15° W
	35° N			0.60	0.99	0.54	0.16	0.15	0.29	0.91
Lat.	25° N	0.56	0.95	1.41	0.96	0.57	0.38	0.59	1.09	
	15° N		1.79	1.52	2.51	2.52	2.17	2.64	4.34	

would be sequestered for multi-year timescales, but little for time-scales of longer than a decade (Table 5). The area over which this nitrogen fixation likely occurs is approximately 20 × 10^{12} m^2 which gives an annual carbon fixation rate of 1.2–4.8 moles C m^{-2} y^{-1}. This rate is very similar to the observed summertime carbon drawdown at the BATS site (Michaels et al. 1994a).

Atmospheric deposition of iron

Nitrogen fixation requires a great deal of iron, approximately 100–1000 fold higher intracellular concentrations than required by organisms that utilize nitrate (Hutchins 1989; Rueter et al. 1992; Elardo 1994). The iron supply to the Sargasso Sea is primarily through the deposition of Saharan dust (Prospero et al., this volume). Model estimates of total dust deposition to the Sargasso Sea between 10 and 40° N latitude (excluding the areas near the African coast) are approximately 44 × 10^{12} g dry wt. y^{-1} (Prospero et al., this volume). Assuming 3.5% iron by weight, dust would supply 2.8 × 10^{10} moles Fe y^{-1} (Table 6). *Trichodesmium sp.* typically has an N:Fe ratio of the order 1000 moles N:1 mole Fe (range = 200–3000; Hutchins 1989; Rueter et al. 1992; Elardo 1994). The export of particulate biomass produced by nitrogen fixing organisms would likely transport the intracellular iron to depth as well. Thus, the total iron deposition could support an annual excess nitrate production rate from N_2 fixation of as much as 28 × 10^{12} moles N y^{-1}. Much of the atmospheric iron is insoluble and thus unavailable to nitrogen-fixing phytoplankton. The iron supply would support the estimated excess nitrate production if *Trichodesmium sp.* was able to utilize about 13–23% of the iron supply. Interestingly, the relatively close match between the nitrogen fixation rate and the rate supportable by the iron deposition may indicate that Fe supply limits the total amount of nitrogen fixation that can occur.

There is a large interannual variation in dust deposition (Figure 4 in Prospero et al., this volume). The Barbados record shows the typical annual pattern (summer maximum), the large interannual variability and a distinct trend towards higher rates of deposition over the past 3 decades. This long-term increase is the result of the large expansion of the Sahara in the early 1970s. A combination of the onset of the drought, perhaps coupled with changes in land use caused the area of the Sahara to expand four-fold. Prior to the 1970s, dust deposition to the Atlantic would have been approximately one third of today's rates and the levels of nitrogen fixation may have been correspondingly reduced. Thus the current nutrient anomalies in the Sargasso Sea may be the result of a recent change in iron supply. The N^* at BATS is slightly higher than the N^* from the TTO composite of a decade earlier, however, the few GEOSEC profiles in this area are similar to BATS profiles. If dust supply controls this process, we may even expect annual and decadal fluctuations in nitrogen fixation that could drive 0.3–1.2 Gt C y^{-1} fluctuations in the northern hemisphere atmospheric CO_2 stocks. However, the main thermocline ventillates on time-scales of 7–30 years. Thus, nitrogen fixation in the Sargasso Sea cannot result in the sequestration of large amounts of CO_2 on the longer time-scales associated with the anthropogenic signal in the atmosphere.

Mass balances in the pelagic North Atlantic Ocean

The estimates of the fluxes of nitrogen and phosphorus across the boundaries of the North Atlantic Ocean that we present here suggest that this basin is accumulating nitrogen. The range of possible accumulation rates from all of our estimates is 1.9–13.8 $\times$ 10^{12} moles N y^{-1} and 0.13–0.65 $\times$ 10^{12} moles P y^{-1} (Table 2). These net annual rates are approximately 0.06–0.4% of the stock of the nutrient in the basin and would suggest that the nutrients would increase by 10% on time-scales of 23–170 years. The relative similarity of nutrient profiles collected during the GEOSECS program in the early 1970s with the profiles in recent years suggests that the most rapid of these accumulation rates must be incorrect, however the uncertainties in the stocks and fluxes are large and they preclude a strong conclusion about the slower rates of accumulation. To further evaluate the inferred basin-wide balances, we look at the mass balances within various subsets of the basin. Just two processes account for the majority of the imbalance, the meridional transport estimates and the nitrogen fixation rate inferred from the excess nitrate calculation. We concentrate on these two processes and discuss three issues related to the observed patterns. 1. The predicted accumulation may be an artifact of errors in some of the estimated fluxes. 2. The basins are accumulating nutrients. 3. The nutrient fluxes are not constant on decade to century time-scales.

The range of meridional transport estimates in Table 2 is a reflection of the uncertainties in the transport estimates by each method. These transports only involve the inorganic nutrients and the inclusion of organic nitrogen or phosphorus could provide counter-balancing fluxes as the vertical profiles of organic matter are very different than those of nutrients. The flux across the equator as inferred from the 12° N section has a very wide range from -1.4 to 4.6×10^{12} moles NO_3^- y^{-1} and from -0.07 to 0.31×10^{12} moles PO_4^3 y^{-1}. The net accumulation from all other fluxes into the basin (i.e. ignoring the flux at 12° N) is still very large compared to the uncertainty in the equatorial flux: 3.3 to 9.2×10^{12} moles N y^{-1} and 0.2 to 0.34×10^{12} moles P y^{-1}. It would require a southward equatorial flux that was 2–6 times the estimate for the northward flux just to bring the basin into balance.

If we calculate the balance of fluxes for the North Atlantic north of 24° N we find that this subset of the basin has an estimated net accumulation rate that ranges from 0.6 to 7.0×10^{12} moles N y^{-1} (using the Rintoul and Wunsch fluxes) or from -1.4 to 7.5×10^{12} moles N y^{-1} (using the Martel and Wunsch fluxes, Table 7). The phosphate stock in this subregion could be in balance. The calaculation for the basin north of 36° N becomes less balanced as we include the large northward transport inferred by Rintoul and Wunsch (1991, Table 8), despite the lack of a nitrogen fixation signal. North of 43° N the atmospheric deposition, shelf losses and nitrogen fixation are all small and the various transport terms (Table 4) indicate similar southward transports. The divergence between the fluxes at 24° N and 36° N of 4×10^{12} moles NO_3^- y^{-1} and the divergence of 5.4×10^{12} moles NO_3^- y^{-1} between the fluxes at 12° N and 36° N is similar to the nitrogen fixation source of $3.7–6.4 \times 10^{12}$ moles NO_3^- y^{-1} estimated from the excess nitrate. Thus, the nitrogen balance in the North Atlantic seems to be most sensitive to the transport terms in the subtropical gyre and across the equator and to the large nitrogen fixation estimate. The fluxes are internally consistent with a source of nitrate in the mid-latitudes. However, there may still have to be an unmeasured southward flux of dissolved organic nitrogen at 36° N as inferred by Rintoul & Wunsch (1991).

On the time-scale of the observations that are used to estimate these fluxes, we must also consider the likelihood that the North Atlantic Ocean is not in steady state. This can either imply that there are changes in the concentrations of nitrogen and phosphorus stocks or that the fluxes themselves change with time. The dust deposition record already indicates that the putative nitrogen fixation rate may have been as much as 4 fold lower before 1970, a change that may have been exacerbated by human activity. Atmospheric deposition of reactive nitrogen was also lower before the industrial revolution. Removing the source of excess nitrate and the nitrogen deposition bring the entire

Table 7. Mass balances of N and P in the North Atlantic Ocean north of 24° N. Data are abstracted from Tables 2 and 4. All fluxes are in units of 10^{12} moles y^{-1}.

North of 24° N latitude	N	P
Atmospheric deposition		
24–40° N	0.22*	0.003
40–70° N	0.36	~0
Total atmospheric deposition	0.58	0.003
Exchange with continental shelves		
Western boundary, 24–70° N	−0.54	0.005
Eastern boundary, 24–70° N	−0.21	0.005
Total exchange with shelves	−0.75	0.01
Nitrogen fixation		
10–40° N*	1.9 to 3.2	–
Total nitrogen fixation	1.9 to 3.2	–
Physical Transport (negative is southward or westward)		
24° N (Rintoul & Wunsch 1991)	−1.48 to 0.98	–
24° N (Martel & Wunsch 1993)	−3.5 to 1.5	−0.26 to 0.06
Arctic Sea	−0.3 to −2.8	−0.16 to −0.3
Mediterranean	−0.05 to −0.2	−0.003 to −0.005
Estimated imbalance	0.6 to 7.0 (−1.4 to 7.5)**	−0.08 to 0.38**

* Approximately half of the 10–40° N estimate for total nitrogen deposition and nitrogen fixation given in Table 2.
** Using the Martel & Wunsch (1993) transport at 24° N.

basin quite close to balance. Further, there are indications from the Pacific Ocean that the rate of oceanic denitrification may change quite markedly on glacial-interglacial time-scales thus affecting the overall stock of nitrate (Ganeshram et al. 1995). Watermasses at different depths have residence times with respect to ventillation of 10s to 100s of years in the North Atlantic Ocean and any variability in nitrogen boundary fluxes on local scales will lead to heterogeneities in the oceanic transport rates as these signals propagate through the ocean.

Table 8. Mass balances of N and P in the North Atlantic Ocean north of approximately 36° N. Some estimates use a 40° N boundary. Data are abstracted from Tables 2 and 4. All fluxes are in units of 10^{12} moles y^{-1}.

North of 36° N–40° N latitude	N	P
Atmospheric deposition		
40–70° N	0.36	~0
Total atmospheric deposition	0.36	0
Exchange with continental shelves		
Western boundary, 40–70° N	−0.41	0.003
Eastern boundary, 40–70° N	−0.18	0.004
Total exchange with shelves	−0.59	0.007
Physical transport (negative is southward or westward)		
36° N (Rintoul & Wunsch 1991)	2.6 to 4.9	–
36° N (Martel & Wunsch 1993)	−2.0 to 3.5	−0.14 to 0.18
Arctic Sea	−0.3 to −2.8	−0.16 to −0.3
Estimated imbalance	2.7 to 7.5 (−1.9 to 6.1)*	0.03 to 0.49*

* Using the Martel & Wunsch (1993) transport at 36° N.

Summary

• The total standing stocks of nitrogen and phosphorus in the open ocean are large compared to the magnitude of the boundary exchanges in a single year. Changes in atmospheric deposition will only have local or ephemeral impacts on ocean ecosystems, either close to a pollution source or in the center of the gyre where nutrient depletion is stongest. Except for the Amazon outflow, nitrogen fluxes in rivers only meet a portion of the denitrification and burial fluxes on continental shelves; the remainder is supplied from oceanic nitrate pools. Increases in riverine fluxes of nitrogen could decrease the net on-shelf transport of nitrate; however, this would have little observable effect on the open ocean on time-scales of less than millenia.

• The ability to resolve nutrient transport in the ocean is still poor, in most cases because of limitations in data quality or quantity. Even the question of whether the North Atlantic exports nitrate to other basins cannot be adequately resolved at this time. The balance of flux estimates currently available indicates that there should be a net accumulation of nitrate in the North Atlantic Ocean and that this accumulation is related to processes that occur at low and mid-latitudes.

• We calculate that 3.7 – 6.4 × 10^{12} moles NO_3^- y^{-1} are created in the upper 1000 m of the Sargasso Sea. This nitrate comes from either nitrogen fixation or differential remineralization of dissolved organic nitrogen and phosphorus; we suggest that nitrogen fixation is the likely source. The rate of creation of nitrate is similar to a flux divergence observed by Rintoul & Wunsch (1991) and implies that the North Atlantic should either export nitrate to the south, have a much larger denitrification rate over the entire basin or be accumulating nitrogen. If nitrogen fixation occurs at the required rate, it requires an iron supply of the same order as the observed deposition. Thus interannual and decadal variations in dust deposition could play an important role in modifying the upper ocean nitrogen dynamics of the Sargasso Sea.

Acknowledgements

We thank the other participants at the Block Island Workshop on nitrogen in the North Atlantic Ocean for the stimulating discussion that lead to the contents of this paper. The workshop was supported by SCOPE and WMO. We thank Dave Siegel, Jim Galloway, Dave Karl, Claire Michaels and anonymous reviewers for critical reviews of the paper. The preparation of the paper and some of the original research reported herein was supported by NSF through grant OCE 9301950 to AFM and AHK. This is BBSR contribution number 1435 and JGOFS contribution number 232.

References

Ammerman JW, Bentzen E, Peele ER, Cotner JB & Jeffrey WH (1994) Phosphorus cycling at the Bermuda time-series station: An overview. EOS 75: 100

Anderson LA & Sarmiento JL (1994) Redfield ratios of remineralization determined by nutrient data analysis. Global Biogeochem. Cycles 8: 65–80

Anderson LG & Dyrssen DW (1981) Chemical constituents of the Arctic Ocean in the Svalbard area. Oceanol. Acta 4: 305–311

Anderson LG, Dyrssen DW, Jones EP & Lowings MG (1983) Inputs and outputs of salt, freshwater, alkalinity and silica in the Arctic Ocean. Deep-Sea Res. 30: 87–94

Antia NJ, Harrison PG & Oliveira L (1991) The role of dissolved organic nitrogen in phytoplankton nutrition, cell biology and ecology. Phycologia 30: 1–89

Baringer MO (1993) Mixing and dynamics of the Mediterranean outflow. M.I.T./WHOI, Ph.D. Dissertation

Bates NR, Michaels AF & Knap AH (1996) Seasonal and interannual variability of oceanic carbon dioxide species at the U.S. JGOFS Bermuda Atlantic Time-series Study Site. Deep-Sea Res. 43: 347–383.

Berger WH (1970) Biogenous deep-sea sediments: Fractionation by deep-sea circulation. Geol. Soc. Am. Bull. 81: 1385–1402

Berger WH, Fisher K, Lai C & Wu G (1987) Ocean carbon flux: Global maps of primary productivity and export production. Scripps Inst. Oceanog. Ref. 87-30: 1–67

Berner RA (1982) Burial of organic carbon and pyrite sulfur in the modern ocean: its geochemical and environmental signficance. A. J. Sci. 282: 451–475
Bishop JKB (1989) Regional extremes in particulate matter composition and flux: effects on the chemistry of the ocean interior. In: Berger WH, Smetacek VS & Wefer G (Eds) Productivity of the Oceans: Past and Present (pp 117–137). Wiley, New York
Bryan F (1987) Parameter sensitivity of primitive equation ocean general circulation models. J. Phys. Oceanogr. 17: 970–985
Bryan K (1962) Measurements of meridional heat transport by ocean currents. J. Geophys. Res. 67: 3403–3414
Bryden HL, Brady EC & Pillsbury RD (1989) Flow through the Strait off Gibralter. In: Almazan JP, Bryden H, Bryden T & Parrilla G (Eds) Seminario sobre la oceanografia fisca del estrecho de Gibralter, Madrid, 24–28 Octubre, 1988 (pp 166–194). SECEG, Madrid
Bryden HL & Kinder TH (1991) Steady two-layer exchange through the Strait of Gibralter. Deep-Sea Res. 38: S445–S463
Buesseler KO (1992) Do upper-ocean sediment traps provide an accurate record of particle flux? Nature 353: 420–423
Carlson CA, Ducklow HW & Michaels AF (1994) Annual flux of dissolved organic carbon from the euphotic zone in the northwest Sargasso Sea. Nature 371: 405–408
Carpenter EJ (1983) Nitrogen Fixation by Marine Oscillatoria (Trichodesmium) in the World's Oceans. Academic Press, Inc., New York
Carpenter EJ & Romans K (1991) Major role of the cyanobacterium Trichodesmium in nutrient cycling in the North Atlantic Ocean. Science 254: 1356–1358
Dugdale RC & Goering JJ (1967) Uptake of new and regenerated forms of nitrogen in primary productivity. Limnol. Oceanogr. 12: 196–206
Elardo KM (1994) Changes in proteins associated with nitrogen fixation and iron nutrition in the marine cyanobacterium Trichodesmium. Portland State University. 93 p
Eppley RW & Peterson BJ (1979) Particulate organic matter flux and planktonic new production in the deep ocean. Nature 282: 677–680
Fanning KA (1989) Influence of atmospheric pollution on nutrient limitation in the ocean. Nature 339: 460–463
Fanning KA (1992) Nutrient provinces in the sea: Concentration ratios, reaction rate ratios, and ideal covariation. J. Geophys. Res. C Oceans 97: 5693–5712
Galloway J, Howarth R, Michaels A, Nixon S, Prospero J & Dentener F (this volume) N and P budgets of the North Atlantic Ocean and its watershed. Biogeochem.
Galloway JN, Schlesinger WH, Levy II H, Michaels A & Schnoor JL (1995) Nitrogen fixation: Anthrogenic enhancement-environmental response. Global Biogeochecm. Cycles 9: 235–252
Ganeshram RS, Pedersen TF, Calvert SE & Murray JW (1995) Large increases in oceanic nutrient inventories from glacial to interglacial periods. Nature 376: 755–759
Gargett AE (1984) Vertical eddy diffusivity in the ocean interior. J. Mar. Res. 42: 359–393
Glover HE, Prezelin BB, Campbell L, Wyman M & Garside C (1988) A nitrate-dependent Synechococcus bloom in surface Sargasso Sea water. Nature 331: 161–163
Goldman JC (1988) Spatial and temporal discontinuities of biological processes in pelagic surface waters. In: Rothschild B (Ed) Towards a Theory on Biological-Physical Interactions (pp 273–296). Kluwer, New York
Gregg MC (1989) Scaling turbulent dissipation in the thermocline. J. Geophys. Res. 94: 9686–9698
Hall MM & Bryden HL (1982) Direct estimates and mechanisms of oceanic heat transport. Deep-Sea Res. 29: 339–360
Hansell DA (1993) Results and observations from the measnurement of DOC and DON in seawater using a high-temperature catalytic oxidation technique. Mar. Chem. 41: 195–202
Herbland A & Voituriez B (1979) Hydrological structure analysis for estimating the primary production of the tropical Atlantic Ocean. J. Mar. Res. 37: 87–101

Hopkinson C, Cifuentes L, Burdige D, Fitzwater S, Hansell D, Henrichs S, Kahler P, Koike I, Walsh T & Bergamaschi B (1993) DON subgroup report. Marine Chemistry 41: 23–36
Howarth RW, Billen G, Swaney D, Townsend A, Jaworski N, Downing JA, Elmgren R, Caraco N & Lajtha K (this volume) Regional nitrogen budgets and riverine nitrogen fluxes for the drainage systems of the North Atlantic Ocean: Natural and human influences. Biogeochemistry
Hutchins DA (1989) Nitrogen and iron interactions in filamentous cyanobacteria. Portland State University. 116 p
Jahnke RA (1992) The Spatial Distribution of Sea Floor Oxygen Consumption in the Atlantic and Pacific Oceans. Kluwer Academic Publishers, Netherlands
Jenkins WJ (1980) Tritium and 3He in the Sargasso Sea. J. Mar. Res. 38: 533–569
Jenkins WJ (1982) Oxygen utilization rates in North Atlantic subtropical gyre and primary production in oligotrophic systems. Nature 300: 246–248
Jenkins WJ (1988) The use of anthropogenic tritium and helium-3 to study subtropical gyre ventilation and circulation. Philos. Trans. R. Soc. London Ser. A 325: 43–61
Jenkins WJ & Goldman JC (1985) Seasonal oxygen cycling and primary production in the Sargasso Sea. J. Mar. Res. 43: 465–491
Karl DM (1992) Trichodesmium Blooms and New Nitrogen in the North Pacific Gyre. Kluwer
Karl DM & Bailiff MD (1989) The measurement and distribution of dissolved nucleic acids in aquatic environments. Limnol. Oceanog. 34: 543–558
Karl DM, Letelier R, Hebel D, Tupas L et al. (1995) Ecosystem changes in the North Pacific subtropical gyre attributed to the 1991–1992 El Niño. Nature 373: 230–234
Knap A, Jickells T, Pszenny A & Galloway J (1986) Significance of atmospheric-derived fixed nitrogen on productivity of the Sargasso Sea. Nature 320: 158–160
Krom MD, Kress N, Brenner S & Gordon LI (1991) Phosphorus limitation of primary productivity in the eastern Mediterranean Sea. Limnol. Oceanog. 36: 424–432
Leaman KD (1994) The formation of Western Mediterranean Deep Water. Coast. Est. Studies 46: 227–248
Leaman KD, Johns E & Rossby T (1989) The average distribution of volume transport and potential vorticity with temperature at three sections across the Gulf Stream. J. Phys. Oceanogr. 19: 36–51
Ledwell JR, Watson AJ & Law CS (1993) Evidence for slow mixing across the pycnocline from an open-ocean tracer-release experiment. Nature 364: 701–703
Lewis MR, Harrison WG, Oakey NS, Hebert D & Platt T (1986) Vertical nitrate fluxes in the oligotrophic ocean. Science 234: 870–873
Lipschultz F & Owens NJP (this volume) An assessment of nitrogen fixation as a source of nitrogen to the North Atlantic Ocean. Biogeochemistry
Lipschultz F, Zafiriou OC & Ball LA (1996) Seasonal fluctuations of nitrite concentrations inthe deep oligotrophic ocean. Deep-Sea Res. 43: 403–420
Lohrenz SE, Knauer GA, Asper VL, Tuel M, Michaels AF & Knap AH (1992) Seasonal and interannual variability in primary production and particle flux in the northwestern Sargasso Sea: U.S. JGOFS Bermuda Atlantic Time-series Study. Deep-Sea Res. 39: 1373–1391
Longhurst AR & Harrison WG (1988) Vertical nitrogen flux from the oceanic euphotic zone by diel migrant zooplankton and nekton. Deep-Sea Res. 35: 881–889
Martel F & Wunsch C (1993) The North Atlantic circulation in the early 1980's – an estimate from inversion of a finite-difference model. J. Phys. Oceanogr. 23: 898–924
Martin JH, Fitzwater SE, Gordon RM, Hunter CN & Tanner SJ (1993) Iron, primary production and carbon-nitrogen flux studies during the JGOFS North Atlantic Bloom Experiment. Deep-Sea Res., Pt. II 40: 115–134
Martin JH, Knauer GA, Karl DM & Broenkow WW (1987) VERTEX: carbon cycling in the northeast Pacific. Deep-Sea Research 34: 267–285
McCave IN (1975) Vertical flux of particles in the ocean. Deep-Sea Res. 22: 491–502
McClain CR & Firestone J (1993) An investigation of Ekman upwelling in the North Atlantic. J. Geophys. Res. 98: 12327–12339

Michaels AF, Bates NR, Buesseler KO, Carlson CA & Knap AH (1994a) Carbon-cycle imbalances in the Sargasso Sea. Nature 372: 537–540
Michaels AF & Knap AH (1996) An overview of the Bermuda Atlantic Time-series Study and Hydrostation S programs. Deep-Sea Res. 43: 157–198
Michaels AF, Knap AH, Dow RL, Gundersen K, Johnson RJ, Sorensen J, Close A, Knauer GA, Lohrenz SE, Asper VA, Tuel M & Bidigare RR (1994b) Seasonal patterns of ocean biogeochemistry at the U.S. JGOFS Bermuda Atlantic Time-series Study site. Deep-Sea Res. 41: 1013–1038
Michaels AF, Siegel DA, Johnson RJ, Knap AH & Galloway JN (1993) Episodic inputs of atmospheric nitrogen to the Sargasso Sea: Contributions to new production and phytoplankton blooms. Global Biogeochem. Cycles 7: 339–351
Michaels AF & Silver MW (1988) Primary production, sinking fluxes and the microbial food web. Deep-Sea Research 35: 473–490
Munk W (1966) Abyssal recipes. Deep-Sea Res. 13: 707–730
Najjar RG, Sarmiento JL & Toggweiler JR (1992) Downward transport and fate of organic matter in the ocean: Simulations with a general circulation model. Glogal Biogeochem. Cycles 6: 45–76
Nixon SW, Ammerman J, Atkinson L, Berounsky V, Billen G, Boynton BW, Church T, DiToro D, Elmgren R, Garber J, Giblin A, Jahnke R, Ownes N, Pilson MEQ & Seitzinger S (this volume) The fate of nitrogen and phosphorus at the land-sea margin of the North Atlantic Ocean. Biogeochemistry
Olson RJ (1981) 15N tracer studies of the primary nitrite maximum. J. Mar. Res. 39: 203–226
Owens NJP, Galloway JN & Duce RA (1992) Episodic atmospheric nitrogen deposition to oligotrophic oceans. Nature 357: 397–399
Paerl HW (1985) Enhancement of marine primary production by nitrogen-enriched acid rain. Nature 315: 747–749
Platt T, Caverhill C& Sathyendranath S (1991) Basin-scale estimates of oceanic primary production by remote sensing: The North Atlantic. J. Geophys. Res. 96: 15147–15159
Price JF, Barringer MO, Lueck RG, Johnson GC, Ambar I, Parrilla G, Cantos A, Kennelly MA & Sanford TB (1993) The Mediterranean Outflow. Science 259: 1277–1282
Prospero JM, a. Author & b. Author (this volume) Nitrogen dynamics of the North Atlantic Ocean: Atmospheric deposition of nutrients to the North Atlantic basin. Biogeochemistry
Quinn PK, Barrett KJ, Dentener FJ, Lipschultz F & Kurtz KD (this volume) Estimation of the air/sea exchange of ammonia for the North Atlantic Basin. Biogeochemistry
Quinn PK, Bates TS, Johnson JE, Covert DS & Charlson RJ (1990) Interactions between sulfur and reduced nitrogen cycles over the Central Pacific Ocean. J. Geophys. Res. 95: 16405–16416
Quinn PK, Charlson RJ & Bates TS (1988) Simultaneous observations of ammonia in the atmosphere and ocean. Nature 335: 336–338
Redfield AC, Ketchum BH & Richards FA (1963) The influence of organisms on the composition of sea water. In: Hill MN (Ed) The Sea (pp 26–77). Wiley, New York
Rintoul SR & Wunsch C (1991) Mass, heat, oxygen and nutrient fluxes and budgets in the North Atlantic Ocean. Deep-Sea Res. 38: S355–S377
Roemmich D (1981) Circulation of the Caribbean Sea: A well resolved inverse problem. J. Geophys. Res. 86: 7993–8005
Roemmich D (1983) The balance of geostrophic and Ekman transports in the tropical Atlantic Ocean. J. Phys. Oceanogr. 13: 1534–1539
Romans KM, Carpenter EJ & Bergman B (1994) Buoyancy regulation in the colonial diazotrophic cyanobacterium trichodesmium tenue – ultrastructure and storage of carbohydrate, polyphosphate and nitrogen. J. Phycol. 30: 935–942
Rueter JG, Hutchins DA, Smith RW & Unsworth NL (1992) Iron nutrition of Trichodesmium. In: Carpenter EJ, Capone DG & Rueter JG (Eds) Marine Pelagic Cyanobacteria: Trichodesmium and Other Diazotrophs. Kluwer Academic Publishers, Dordrecht

Sarmiento JL (1983) A tritium box model of the North Atlantic thermocline. J. Phys. Oceanog. 13: 1269–1274
Sarmiento JL (1986) On the north and tropical Atlantic heat balance. J. Geophys. Res. 91: 11677–11689
Sarmiento JL, Herbert T & Toggweiler JR (1988) Mediterranean nutrient balance and episodes of anoxia. Global Biogeochem. Cycles 2: 427–444
Sarmiento JL, Slater RD, Fasham MJR, Ducklow HW, Toggweiler JR & Evans GT (1993) A seasonal three-dimensional ecosystem model of nitrogen cycling in the North Atlantic euphotic zone. Global Biogeochem. Cycles 7: 417–450
Sarmiento JL, Thiele G, Key RM & Moore WS (1990) Oxygen and nitrate new production and remineralization in the North Atlantic subtropical gyre. J. Geophys. Res. 95: 18303–8315
Schlitzer R (1988) Modeling the nutrient and carbon cycles of the North Atlantic. 1. Circulation, mixing coefficients and heat fluxes. J. Geophys Res. 93: 10699–10723
Schlitzer R (1989) Modeling the nutrient and carbon cycles of the North Atlantic. 2. New production, particle fluxes, CO sub(2) gas exchange, and the role of organic nutrients. J. Geophys. Res. C Oceans. 94: 12781–794
Schmitz WJ & Richardson PL (1991) On the sources of the Florida Current. Deep-Sea Res. 38: S379–S409
Seitzinger SP & Giblin AE (in press) Estimating denitrification in North Atlantic continental shelf sediments. Biogeochem.
Sharp JH (1983) The distributions of inorganic nitrogen and dissolved and particulate organic nitrogen in the sea. In: Carpenter EG & Capone DG (Eds) Nitrogen in the Marine Environment (pp 1–35). Academic Press, New York
Siegel DA, Itturiaga R, Bidigare RR, Smith RC, Pak H, Dickey TD, Marra J & Baker KS (1990) Meridional variations of the spring-time phytoplankton community in the Sargasso Sea. J. Mar. Res. 48: 379–412
Suzuki Y, Sugimura Y & Itoh T (1985) A catalytic oxidation method for the determination of total nitrogen dissolved in seawater. Mar. Chem. 16: 83–97
Toggweiler JR (1989) Is the downward dissolved organic matter (DOM) flux important in carbon transport? In: Berger WH, Smetacek VS & Wefer G (Eds) Productivity of the Oceans: Past and Present (pp 65–83). Wiley, New York
Trenburth KE & Soloman A (1994) The global heat balance – Heat transports in the atmosphere and ocean. Climate Dynamics 10: 107–134
Villareal T & Carpenter EJ (1990) Diel buoyancy regulation in the marine diazotrophic cyanobacterium *Trichodesmium thiebautii*. Limnology and Oceanography 35: 1832–1837
Vonder Haar T & Ort AH (1973) New estimate of annual poleward energy transport by northern hemisphere oceans. J. Phys. Oceanogr. 2: 169–172
Walsh JJ, Dieterle DA, Meyers MB & Mueller-Karger FE (1989) Nitrogen exchange at the continental margin: A numerical study of the Gulf of Mexico. Prog. Oceanogr. 23: 245–301
Walsh TW (1989) Total dissolved nitrogen in seawater: A new high-temperature combustion method and a comparison with photo-oxidation. Mar. Chem. 26: 295–311
Weiss RF (1970) The solubility of nitrogen, oxygen, and argon in water and seawater. Deep Sea Res. 17: 721–735

Biogeochemistry **35:** 227–233, 1996.

Areas of shallow water in the North Atlantic

MICHAEL E.Q. PILSON[1] & SYBIL P. SEITZINGER[2]
[1]*Graduate School of Oceanography, University of Rhode Island;* [2]*Institute of Marine and Coastal Science, Rutgers/NOAA CMER Program, Rutgers University*

Received 27 September 1995; accepted 14 February 1996

Key words: area, continental shelves, North Atlantic, shallow water

Abstract. We report here estimates of the areas of water that are between 1 and 100 m and between 1 and 200 m deep in the North Atlantic, including the Caribbean Sea, the Gulf of Mexico, Hudson Bay and the North Sea, but excluding estuaries. The total areas within these depths, from the equator to 70° N, are 3.91 × 10^6 km^2 and 5.66 × 10^6 km^2, respectively. We also report the respective areas by selected geographic regions.

Introduction

While many factors influence the rates of biogeochemical processes in seawater and at the seafloor, characteristically the depth of the water is an important variable. In shallow water the seafloor plays an important role in the sum of biogeochemical processes (e.g. Seitzinger & Giblin, *this volume*). In order to calculate the rates of processes on the continental shelves, the areas of selected portions of the shelves are needed. A previous estimate (Levitus 1982; Levitus & Boyer 1994) was derived from chart measurements made by Smith et al. (1966). The latter measurements were made by summing visual estimates of the average depths of one-degree squares (60 nautical miles on a side at the Equator); for shallow regions the results are therefore somewhat uncertain.

Methods

The source of data for this compilation was the file called ETOPO5, supplied by the National Geophysical Data Center (NGDA 1993) on CD-ROM. This global data set is presented as elevations and depths, in meters, at the intersections of each 5-minute line of latitude and longitude.

Our procedure was to acquire the data in blocks of one to five degree squares or rectangles for all shallow water regions in the North Atlantic and adjacent marine waters (excluding the Mediterranean and the Baltic), carry the data into Excel, eliminate overlaps, and count the number of values falling between -1 and -100 m, and between -1 and -200 m. An area of 25 square minutes was assigned to each count, and the area corrected according to latitude. The respective areas were accumulated according to latitude band and also according to various regions.

An estimate of the quality of the data source was obtained by selecting regions on several charts at appropriate scales, picking off the depths at the intersections of the five-minute latitude and longitude lines, converting to meters where necessary, and comparing these with the ETOPO5 data.

Results

a. *Data quality.* As can be seen from Figures 1 and 2, there are numerous discrepancies between the depths obtained at specific locations using the ETOPO5 data and using navigational charts. The larger features of the ocean depths are certainly represented reasonably well (Figure 1). On a smaller scale, and in shallow water, there are many startling discrepancies (Figure 2). Nevertheless, the ETOPO5 values tend to be rather randomly distributed around those from the navigational charts, a conclusion drawn from viewing about 20 graphs such as those in Figures 1 and 2. These figures were selected to show the overall general trend on the large scale, and examples of the poorer fits on the small scale in that limited data set. There is, of course, some error in picking depths off the navigational charts, but from a detailed comparison it is evident to us that these cannot be as large as appear in the figures presented. It is not our purpose here to perform an exhaustive analysis of the problems with ETOPO5, as we hope that it will soon be superseded by a much more accurate data set. Figures 1 and 2 indicate that the general trend in the larger scale and the errors on the smaller scale are such that the derived values of the areas of the continental shelves and other shallow water regions, averaged over the selected regions, will not be grossly distorted.

An additional check on the accuracy of the data was carried out by comparing the areas obtained between 1 and 200 m water depth and between 30° N and 35° N on the SE coast of the United States. Planimetry on Defense Mapping Agency Chart 108 for this region, excluding Pamlico Sound, yielded a value of 85.7×10^3 km^2. Evaluation of the area from a detailed USGS data set gave a value of 81.9×10^3 km^2 (Michael Crowley, pers. comm., areas calculated from a compilation of USGS bathymetry data obtained from John Breckenridge, NRL South, Stennis Space Center), while the ETOPO5 data set yielded a value of 84.0×10^3 km^2. This comparison provides some sense of the accuracy with which the individual areas are known. We did not attempt a detailed point-by-point comparison to isolate the discrepancies.

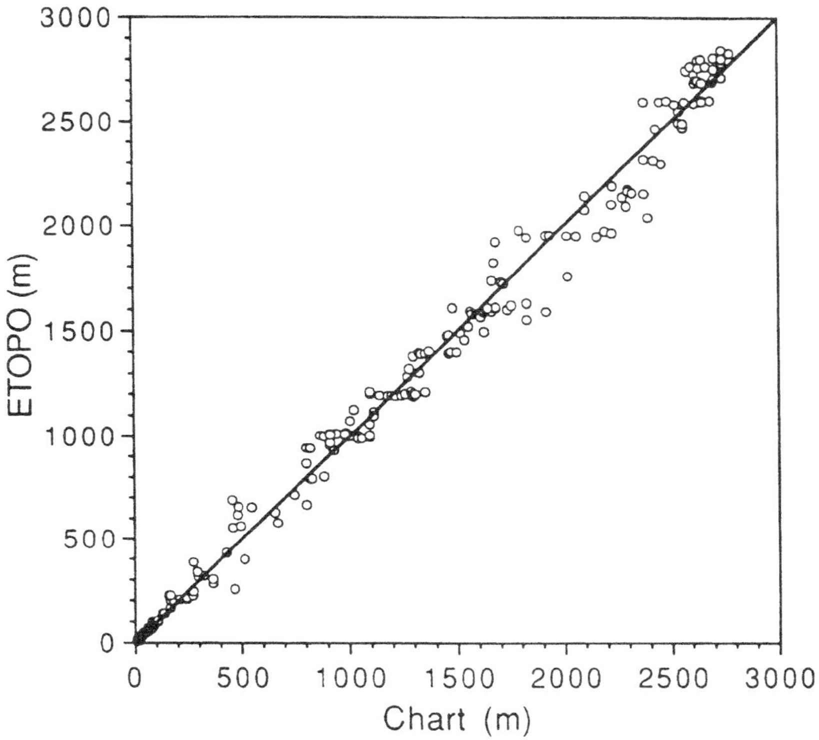

Figure 1. Comparison of depths at the intersections of five-minute latitude and longitude lines obtained using data from ETOPO5 with data from navigational charts. Here data are combined from three one-degree squares on the following charts: NOAA Chart No. 12300, Approaches to New York, fathoms, 40N 71W–39N 70W; NOAA Chart No. 11300, Gulf Coast, Galveston to Rio Grande, fathoms, 28N 97W–27N 96W, 27N 96W–26N 95W.

After the comparisons noted above were completed, we came across a paper (Smith 1993) wherein is included a comparison of ETOPO5 data with shipboard depth data. The evaluation deals mostly with the deep sea, but two cruise tracks on the southern Argentine shelf show evident discrepancies with ETOPO5, and the general evaluation is consistent with our conclusions here.

b. *Areas of the shelves.* The areas of shallow water in the North Atlantic were cumulated in five-degree latitude bands for the west and east sides of the N. Atlantic and totalled (Table 1). The total area between 1 and 100 m depth is 3.91 million square kilometers, and between 1 and 200 m depth it is 5.66 million square kilometers. These values are 88 and 64 % greater, respectively, than those listed in Levitus (1982) and Levitus & Boyer (1994).

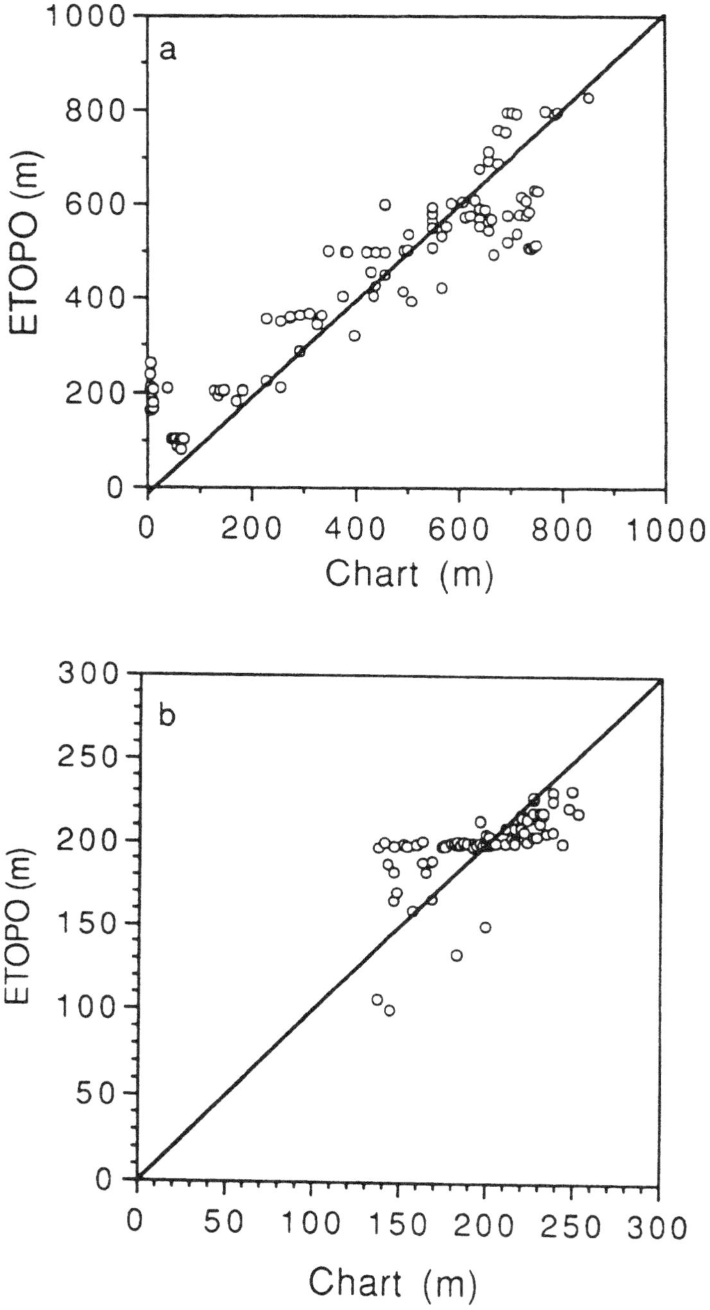

Figure 2. Comparison of depths at the intersections of five-minute latitude and longitude lines obtained from ETOPO5 with equivalent data from navigational charts. a) NOAA Chart No. 11460, Florida Cape Canaveral to Key West, fathoms, 28N 80W–27N 79W. b) NOAA Chart 13260, Bay of Fundy to Cape Cod, fathoms, 44N 68W–43N 67W. Several other charts gave similar results.

Table 1. Summary of the areas of shallow water in the North Atlantic, by five-degree latitude band, extracted from the ETOPO5 data set. Areas are listed both from 1 to 100 m depth and from 1 to 200 m depth. Values are given in units of 1000 km^2. Estuaries and small bays are not included (e.g. Chesapeake Bay is not included, neither is the Baltic, but Hudson Bay, the Bay of Fundy, and the Gulf of St. Lawrence up to Quebec City are included). For comparison, equivalent data from Levitus (1982) and Levitus and Boyer (1994) are also listed. Areas are listed from 10° S. latitude to 70° N. latitude for the west and east sides of the Atlantic. Island areas are allocated to either the west side or east side of the basin (see Table 2). The totals are given only for the region from the Equator north to 70° N. latitude.

	From ETOPO5						From Levitus	
	West side, to		East side, to		Total, E+W		Total, E+W	
	100 m	200 m	100 m	200 m	100 m	200 m	100 m	200 m
	Areas in each band (10^3 km^2)							
10S to 5S	21.87	28.31	19.92	29.41	41.79	57.72	20	20
5S to Eq.	130.57	138.77	30.23	38.26	160.80	177.03	120	130
Eq. to 5N	176.26	186.68	63.02	80.78	239.28	267.46	150	220
5N to 10N	196.14	214.45	80.41	94.43	276.56	308.88	220	240
10N to 15N	157.31	186.04	68.22	74.32	225.53	260.35	50	100
15N to 20N	133.20	178.01	19.83	27.10	153.03	205.11	30	110
20N to 25N	265.68	314.26	43.04	50.48	308.72	364.73	100	210
25N to 30N	326.11	377.82	30.72	42.50	356.83	420.32	150	270
30N to 35N	98.35	105.73	14.11	23.09	112.46	128.82	30	40
35N to 40N	65.70	71.83	13.29	22.49	78.99	94.32	60	60
40N to 45N	220.98	355.29	23.08	33.99	244.06	389.27	110	300
45N to 50N	242.76	408.67	88.31	195.93	331.07	604.60	260	510
50N to 55N	115.51	168.88	310.62	384.04	426.13	552.92	260	280
55N to 60N	283.59	541.53	337.19	467.30	620.78	1,008.83	210	440
60N to 65N	185.80	455.57	71.08	174.39	256.88	629.97	60	150
65N to 70N	211.02	280.37	65.19	142.35	276.21	422.73	390	530
	Totals (10^3 km^2)							
Eq. to 70N	2,678	3,845	1,228	1,813	3,907	5,658	2,080	3,460

For the purpose of assessing biological and geochemical processes in various shelf regions and other shallow water areas (e.g. Seitzinger and Giblin, *this volume*; Nixon et al., *this volume*), we have calculated the areas of various geographical or biotic regions (Table 2).

Discussion

The earlier data presented in Levitus (1982) and Levitus & Boyer (1994) were derived by visual inspection and assigning a single depth to one-degree

Table 2. Areas of shallow marine water in the North Atlantic, arranged by region. The criteria used for selecting areas to measure are given in the text and in Table 1. In this compilation, the northern limit of measurement is 70° N, and the southern limit is the Equator. Values are given in units of 1000 km^2.

Region	Areas down to a depth of	
	100 m	200 m
Greenland	88.92	159.90
Hudson Bay	402.49	780.18
Labrador, Foxe Basin, Baffin Island, Hudson Strait, Ungava Bay	306.13	496.31
Gulf of St. Lawrence & Grand Banks	315.77	519.00
Nova Scotia to Cape Hatteras	230.78	361.04
Cape Hatteras to S. Florida	123.34	137.36
West Florida	126.24	145.13
Northern Gulf of Mexico	178.17	204.40
West and South Gulf of Mexico	196.52	224.03
Caribbean Islands	142.99	194.96
Caribbean Shelf, Central & South America	238.08	279.80
Amazon Shelf	359.58	389.56
Faroes, Azores, Rockall Bank, Madiera, Cape Verdes	11.32	33.73
African Shelf (Equator to Strait of Gibralter)	319.33	392.89
Spain and Portugal (to 45° N)	33.99	51.44
France, Great Britain, West. Europe, North Sea, (from 45° N to 60° N)	741.53	1,087.43
Norway Shelf (N from 60° N)	44.71	93.88
Iceland	46.69	107.27
Totals	3,907	5,658

squares. This procedure probably produces a reasonably accurate estimate of the volume of the ocean basins, but is not suitable for accurately assessing the areas of shallow water, where many squares may overlap dry land, shallow water and deep water. The ETOPO5 data set is certainly suspect; however, it has coverage at relatively dense five minute intervals. Accordingly, we believe that it provides values for the selected areas that are more accurate than those previously available. Until a better global data set is available, and the areas extracted, the values derived here will have to do.

Acknowledgements

We thank Heeson Jeon, Scott Glenn and Michael Crowley for help in assessing the quality of the data. Financial support was provided by the Rhode Island Sea Grant Program. We also wish to acknowledge and express our appreciation for the funding of this project from the UNEP and the Mellon Foundation to support the efforts of the SCOPE Nitrogen Project, and from the World Meteorological Organization to investigate the interaction between the N cycles of the atmosphere and the NAO&W.

References

Levitus S (1982) Climatological Atlas of the World Ocean. NOAA Professional Paper 13. U.S. Dept. of Commerce, National Oceanic and Atmospheric Administration, Rockville, MD. 173 pp

Levitus S & Boyer TP (1994) World Ocean Atlas 1994, Vol. 4: Temperature. U.S. Dept. of Commerce, National Oceanic and Atmospheric Administration Rockville, MD. 117 pp

National Geophysical Data Center (1993) Global Relief CD-ROM. National Oceanographic and Atmospheric Administration, E/CG4 Boulder CO

Nixon SW et al. (1996) The fate of nitrogen and phosphorus at the land-sea margin of the North Atlantic Ocean. Biogeochemistry *(this volume)*

Seitzinger S & Giblin A (1996) Estimating denitrification in North Atlantic continental shelf sediments. Biogeochemistry *(this volume)*

Smith SM, Menard HW & Sharman G (1966) World-wide ocean depths and continental elevations averaged for areas approximating one degree squares of latitude and longitude. Technical Report, SIO Ref. 65–8. Scripps Institution of Oceanography, La Jolla, CA

Smith WHF (1993) On the accuracy of digital bathymetric data. J. Geophys. Res. 98: 9591–9603

Biogeochemistry **35**: 235–260, 1996.

Estimating denitrification in North Atlantic continental shelf sediments

SYBIL P. SEITZINGER[1] & ANNE E. GIBLIN[2]
[1]*Institute of Marine and Coastal Sciences, Rutgers/NOAA CMER Program, Rutgers University, New Brunswick, NJ 08903-0231 (Fax: 908-932-1792);* [2]*Ecosystems Center, Marine Biological Laboratory, Woods Hole, MA 02543 (Fax: 508-457-1548)*

Received 23 February 1995; accepted 6 March 1996

Key words: benthic mineralization, continental shelf, denitrification, global N cycle, nitrogen, North Atlantic, nutrients, onwelling, phytoplankton, sediments

Abstract. A model of coupled nitrification/denitrification was developed for continental shelf sediments to estimate the spatial distribution of denitrification throughout shelf regions in the North Atlantic basin. Using data from a wide range of continental shelf regions, we found a linear relationship between denitrification and sediment oxygen uptake. This relationship was applied to specific continental shelf regions by combining it with a second regression relating sediment oxygen uptake to primary production in the overlying water. The combined equation was: denitrification (mmol N m^{-2} d^{-1}) = 0.019* phytoplankton production (mmol C m^{-2} d^{-1}). This relationship suggests that approximately 13% of the N incorporated into phytoplankton in shelf waters is eventually denitrified in the sediments via coupled nitrification/denitrification, assuming a C:N ratio of 6.625:1 for phytoplankton. The model calculated denitrification rates compare favorably with rates reported for several shelf regions in the North Atlantic.

The model-predicted average denitrification rate for continental shelf sediments in the North Atlantic Basin is 0.69 mmol N m^{-2} d^{-1}. Denitrification rates (per unit area) predicted by the model are highest for the continental shelf region in the western North Atlantic between Cape Hatteras and South Florida and lowest for Hudson Bay, the Baffin Island region, and Greenland. Within latitudinal belts, average denitrification rates were lowest in the high latitudes, intermediate in the tropics and highest in the mid-latitudes. Although denitrification rates per unit area are lowest in the high latitudes, the total N removal by denitrification (53×10^{10} mol N y^{-1}) is similar to that in the mid-latitudes (60×10^{10} mol N y^{-1}) due to the large area of continental shelf in the high latitudes. The Gulf of St. Lawrence/Grand Banks area and the North Sea are responsible for seventy-five percent of the denitrification in the high latitude region. N removal by denitrification in the western North Atlantic (96×10^{10} mol N y^{-1}) is two times greater than in the eastern North Atlantic (47×10^{10} mol N y^{-1}). This is primarily due to differences in the area of continental shelf in the two regions, as the average denitrification rate per unit area is similar in the western and eastern North Atlantic.

We calculate that a total of 143×10^{10} mol N y^{-1} is removed via coupled nitrification/denitrification on the North Atlantic continental shelf. This estimate is expected to underestimate total sediment denitrification because it does not include direct denitrification of nitrate from the overlying water. The rate of coupled nitrification/denitrification calculated is greater than the nitrogen inputs from atmospheric deposition and river sources combined, and suggests that onwelling of nutrient rich slope water is a major source of N for denitrification in shelf regions. For the two regions where N inputs to a shelf region from onwelling have been measured, onwelling appears to be able to balance the denitrification loss.

Introduction

A number of studies have suggested that denitrification is a major sink for fixed nitrogen in the oceans (e.g., Bender et al. 1977; Christensen et al. 1987a, b; Christensen 1994; Codispoti & Christensen 1989; Devol & Christensen 1993; Lui & Kaplan 1984). Recent studies suggest that denitrification losses in the oceans are on the order of 9.2×10^{12} mol N y^{-1} which exceed known oceanic nitrogen inputs (Christensen 1994). More than half of this denitrification (4.8×10^{12} mol N y^{-1}) takes place in sediments, with the remainder in pelagic oxygen minimum zones (Christensen 1994).

Although sediments are commonly thought of as sources of inorganic nitrogen to the overlying water through regeneration, they also serve as a sink for both externally supplied and regenerated nitrogen via denitrification (Seitzinger 1988). Denitrification is the reductive respiration of nitrate or nitrite to N_2 or N_2O, and is carried out by a diverse group of bacteria under anaerobic conditions. Sediments are ideal sites for denitrification because they generally are anaerobic a few mm below the surface (Revsbech et al. 1980; Archer & Devol 1992). Nitrate can be supplied to the sediments either directly from the overlying water (direct denitrification) or through nitrification in the sediments of ammonia released during the decomposition of organic matter (coupled nitrification/denitrification) (Seitzinger 1988; Devol & Christensen 1993).

Studies have indicated that most of the denitrification taking place in marine sediments occurs on the continental shelf (Christensen et al. 1987a; Christensen 1994). Estimates of the amount of N removed on the continental shelf are on the order of 3.6×10^{12} mol N y^{-1}, which accounts for more than 50% of the total nitrogen inputs to the oceans (Christensen et al. 1987a). In contrast, estuarine areas remove about 0.9×10^{12} mol N y^{-1} (Seitzinger 1988) and deep sea sediments remove about 0.5×10^{12} mol N y^{-1} (Christensen 1994). However, in spite of the apparent importance of denitrification on continental shelves to the oceanic N budget there have been very few measurements of denitrification in shelf sediments. Direct measurements of denitrification (N_2 production) in continental shelf sediments are limited to the northwest coast of the U.S. (Devol 1991; Devol & Christensen 1993). Estimates of denitrification have been made in other shelf regions using indirect methods including models of pore water solute profiles (Billen 1978; Christensen 1989; Raaphorst et al. 1992), acetylene block measurements (Haines et al. 1981; Raaphorst et al. 1992; Law & Owens 1990; Lohse et al. 1993), $^{15}NO_3^-$ additions to sediment slurries (Koike & Hattori 1979), and stoichiometric calculations (Florek & Rowe 1983; Gardner et al. 1993).

Because of the paucity of direct measurements of denitrification, estimates of global rates of denitrification in shelf sediments generally have been based on the application of a single rate to all shelf areas. For example, Christensen et al. (1987 a) estimated that 3.6×10^{12} mol N y^{-1} are removed by denitrification in continental shelf sediments by applying an average nitrate consumption rate that he calculated for continental shelf plus some slope sediments to the global shelf area.

In this paper we estimate the denitrification rate of continental shelf sediments in the North Atlantic. We have focused on the importance of denitrification from coupled nitrification/denitrification because newer studies on shelf sediments have shown that this is an important (Devol & Christensen 1993) and in many cases, the major (Raaphorst et al. 1990; Kelly & Nowicki 1993; Lohse et al. 1993) mechanism by which denitrification occurs in shelf sediments. Like other studies we have had to extrapolate from a very few rate measurements to large areas. However, in this study we have attempted to examine how denitrification may change spatially throughout the North Atlantic shelf by examining the link between denitrification in the sediments and primary production in the overlying water.

Approach and data sources

To estimate the large scale spatial distribution of the denitrification N sink in various continental shelf regions throughout the North Atlantic basin, we assumed that coupled nitrification/denitrification is directly related to carbon oxidation, as measured by oxygen uptake, in shelf sediments. The relationship between denitrification and sediment oxygen consumption has been demonstrated in a wide variety of estuarine and freshwater sediments (Seitzinger 1987 & 1990). Denitrification and sediment oxygen consumption also are related in Washington shelf sediments (Devol & Christensen 1993). Using data from a wide range of continental shelf regions, we developed a relationship between denitrification rates and sediment oxygen consumption rates. We used denitrification rates that were either measured directly (N_2 production) or that were calculated stoichiometrically, as described below. Then, because there still were numerous continental shelf regions with no measured oxygen uptake data, we developed a relationship between sediment oxygen uptake and primary production in the overlying waters. Finally, denitrification rates for shelf areas throughout the North Atlantic were calculated based on: 1) the primary production rate for each region, 2) the derived relationship between denitrification and primary production, and 3) the shelf area for each region.

Denitrification versus O_2 uptake

For continental shelf sediments, a relationship between sediment oxygen consumption and denitrification coupled to sediment nitrification was developed using two types of data. The first type was data from continental shelf studies in which denitrification was measured directly (N_2 flux) in conjunction with sediment oxygen consumption. These included *in situ* measurements made during summer at twelve locations on the Washington continental shelf (Devol & Christensen 1993) and laboratory measurements using cores from five locations in Massachusetts Bay (Kelly & Nowicki 1993; Giblin et al. 1994). For the above studies, we subtracted direct denitrification (which was equated to measured nitrate influx from the overlying water) from the total N_2 flux, to obtain denitrification rates coupled to sediment nitrification. The second type of data included continental shelf studies in which sediment O_2 consumption and DIN (ammonia, nitrate, and nitrite) fluxes were measured simultaneously. In those studies we estimated denitrification rates based on stoichiometric calculations of the DIN deficit (Nixon et al. 1976; Giblin et al. 1992; Gardner et al. 1993). The DIN deficit was calculated as the difference between the predicted and measured DIN effluxes. Predicted DIN effluxes were estimated based on measurements of oxygen consumption rates and an assumed Redfield O:N ratio of 6.625 (O_2:N). Denitrification was equated to the DIN deficit, based on the assumption that any mineralized N not released as DIN from the sediments was denitrified. The data include sandy and finer grained sediments with sites ranging in depth from 16 to 106 m. Data for the North Sea include summer measurements at eleven locations (Raaphorst et al. 1990) and measurements over an annual cycle at two other locations (Raaphorst et al. 1992). Data from four stations in the Kattegatt were means of seven sampling events over an annual cycle (Enoksson et al. 1990). Gulf of Mexico data were from two studies: one summer study at four locations (Gardner et al. 1993) and the other a single location on four dates between July and January (Flint & Kamykowski 1984).

The validity of the stoichiometric approach for continental shelf sediments is indicated in studies which have both types of data available. In Washington shelf sediments there was a strong linear relationship between direct measurements of denitrification (N_2 production) and denitrification rates estimated from stoichiometry (slope = 0.98; $r = 0.82$) (data from Devol and Christensen 1993; N_2 flux adjusted for contribution of nitrate from the overlying water). In Massachusetts Bay, the annual denitrification rate measured directly (N_2 flux) compared well with the rate calculated from stoichiometry at various stations, although for individual dates there was sometimes considerable scatter between the measured and calculated rates (Giblin et al. 1993, 1994).

A linear relationship between denitrification and sediment oxygen consumption was found for continental shelf sediments (r = 0.80; significant at 0.001 level) (Figure 1):

$$DNF_c = 0.116^{*}SOD \quad (1)$$

where DNF_c equals denitrification coupled to sediment nitrification (mmol N as N_2 m^{-2} d^{-1}) and SOD equals sediment oxygen consumption (mmol O_2 m^{-2} d^{-1}). The regression line was fit through the origin, based on the assumption that no coupled denitrification/nitrification will occur if there is no oxygen consumption. The slope of the line for the relationship between denitrification and sediment oxygen consumption for the combined continental shelf data (slope = 0.116) is similar to, although slightly lower than, that found for the combined data from seven estuarine and freshwater ecosystems (slope = 0.142; Seitzinger 1990). The latter relationship was developed from direct N_2 flux measurements in conjunction with simultaneous SOD measurements. Those denitrification rates included both coupled nitrification/denitrification and direct denitrification of nitrate from the overlying water. The inclusion of both sources of nitrate for denitrification in the estuarine and freshwater relationship may be responsible for the higher slope relative to our continental shelf relationship which just considers coupled nitrification/denitrification.

We did not use denitrification rates that were estimated from acetylene block measurements. In addition to blocking N_2O reduction to N_2, acetylene also blocks nitrification (Hynes & Knowles 1978, 1984; Bremner & Blackmer 1979). The acetylene block method has been shown to markedly underestimate denitrification rates in estuarine and freshwater sediments, particularly when nitrification and denitrification are closely coupled (Kemp et al. 1990; Seitzinger et al. 1993). In this paper we are attempting to estimate denitrification coupled to nitrification.

Oxygen uptake vs. primary production

In order to apply the relationship developed above between denitrification and sediment oxygen consumption (eq. 1) to a wide range of geographical locations, we investigated the relationship between primary production and sediment oxygen consumption. The assumption was that because the source of organic matter decomposed in the sediments, resulting in oxygen consumption, largely originates from phytoplankton production, the large scale spatial distribution of sediment oxygen consumption should reflect the large scale spatial distribution of primary production.

Data were compiled for sixteen continental shelf regions in which both primary production and sediment oxygen consumption rates were available

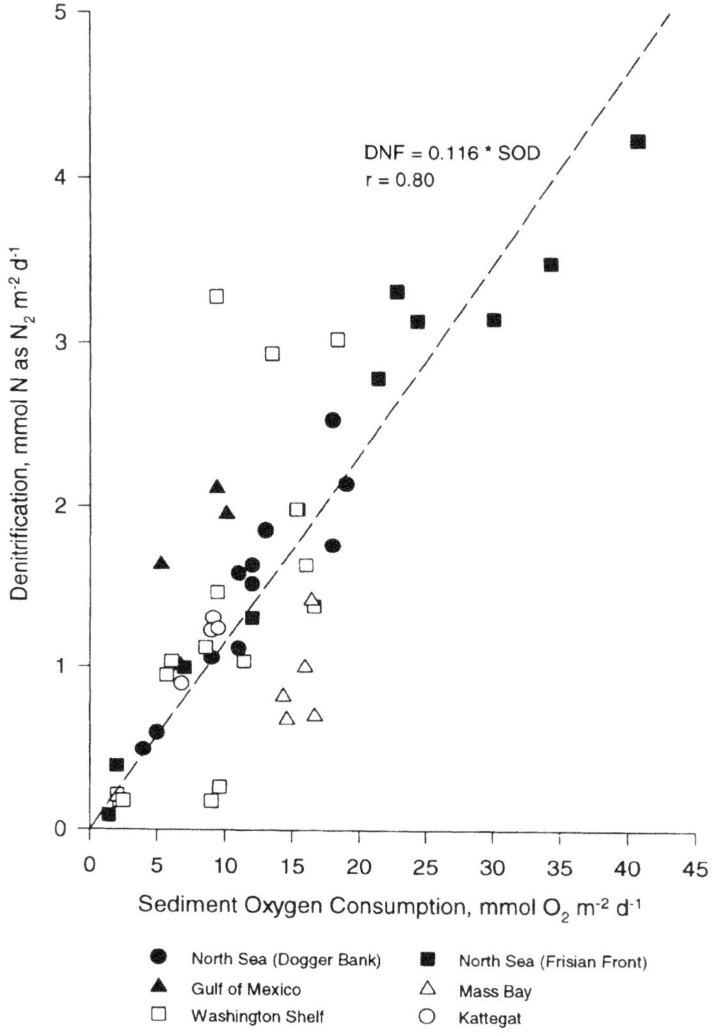

Figure 1. Relationship between denitrification rates coupled to sediment nitrification and sediment oxygen consumption rates in various continental shelf regions. Dashed line indicates regression line. Washington shelf (Devol & Christensen 1993, adjusted for N_2 production due to nitrate from overlying water) and Massachusetts Bay (Kelly & Nowicki 1993; Giblin et al. 1994) data are from direct N_2 flux measurements; North Sea (Raaphorst et al. 1990, 1992), Gulf of Mexico (Gardner et al. 1993), and Kattegat region of the Baltic (Enoksson et al. 1990) data are from stoichiometric calculations (see text for details). Data from Flint & Kamykowski (1984) are not shown, although they follow the same general pattern; oxygen uptake and denitrification rates were very high relative to the data shown which resulted in compression of the graphed data.

(Figure 2). In most cases the two processes were not measured concurrently; whenever possible we used data for similar time frames (e.g., summer or annual) (Figure 2). The data include sites from depths ranging from 14–200 meters. For five regions, (Gulf of Maine, Georges Bank, and Scotian, Newfoundland and Labrador shelves), the authors used data from a variety of stations to determine the relationship between O_2 uptake and temperature; this temperature response then was used to calculate an annual average O_2 uptake to compare to annual primary production data (Grant et al. 1991). In that study, measurements in both sandy and muddy sediments were made and an average O_2 uptake for these shelf regions was calculated based on the amount of each sediment type in the region (Grant et al. 1991). Data from the other twelve studies represent the average of a number of measurements at specific locations and were not corrected for sediment type on a regional basis. However, the measurements come from a variety of sediment types ranging from coarse sands to fine muds. Over half of the data sets are annual averages; the rest generally are concentrated during the summer season.

Based on the above data, a linear relationship was found between sediment oxygen consumption and phytoplankton production (Figure 2):

$$\text{SOD} = 0.164^{*}\text{PhytoProd} \tag{2}$$

where SOD is in mmol O_2 m^{-2} d^{-1} and PhytoProd equals phytoplankton production as mmol C m^{-2} d^{-1} (regression line fit through origin). Although the regression is significant (at 0.001 level), the scatter in the data is fairly large and the number of areas is small. Nixon (1983) examined a number of estuaries and calculated that 23.8% of the primary production and allochthonous carbon inputs were mineralized in the sediments. The correlation coefficient of the regression of the data collected by Nixon (1981) was much higher than we obtained from the regression of the shelf data. One important source of variation in our data set may be the portion of carbon produced in various shelf areas that is advected off the shelf and not mineralized *in situ* (Lee et al. 1991; Walsh 1983; Christensen 1989).

This relationship (eq. 2) suggests that approximately 16% of the carbon produced in surface waters is mineralized on the bottom, assuming a respiratory coefficient (organic carbon:O_2 by moles) of 1. This is consistent, if somewhat conservative, relative to other shelf studies where more complete carbon budgets have been constructed or where carbon delivery to the bottom has been measured using sediment traps. For example, about 25% of the shelf primary production is used by the benthos off southern New England (Rowe et al. 1988) and off the Amazon (Aller et al. 1991). In the Canadian Arctic, 31–53% of the primary production in the overlying water reached the bottom at 33 m, based on sediment trap data (Atkinson & Wacasey 1987).

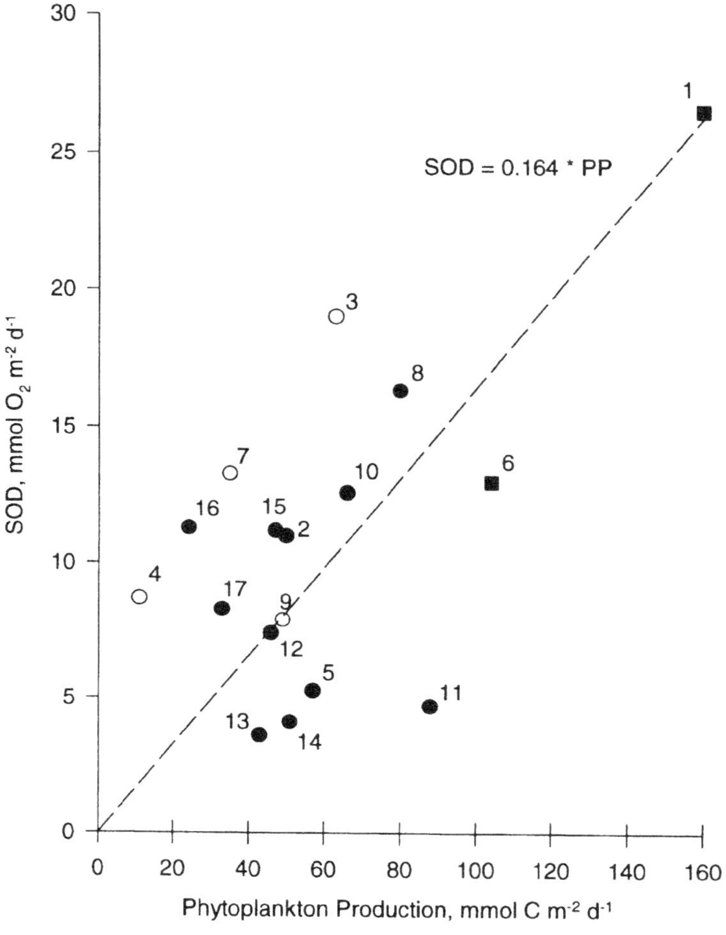

Figure 2. Relationship between sediment oxygen consumption rates (SOD) and phytoplankton production rates in a variety of continental shelf regions. Dashed line indicates regression line. Data sources: 1) South Atlantic Bight (Hanson et al. 1981; Yoder et al. 1985), 2) New York Bight Apex (Rowe et al. 1988), 3) Bering shelf (Grebmeier & McRoy 1989), 4) Alaska Coastal (Grebmeier & McRoy 1989), 5) North Sea, Dogger Bank (deWilde et al. 1984), 6) North Sea, Dogger Bank (Raaphorst et al. 1990), 7) Belt Sea and Kattegat (Jorgensen & Revsbech 1989), 8) Massachusett Bay, 30–33 m (Giblin et al. 1994; Kelly & Nowicki 1993), 9) Gulf of Mexico (Gardner et al. 1993; Sklar & Turner 1981; Flint & Rabalais 1981), 10) Gulf of Maine, 11) Georges Bank, 12) Scotian Shelf, 13) Newfoundland Shelf, and 14) Labrador Shelf (10–14 from Grant et al. 1991), 15) North Sea, 25–31 m and 16) North Sea, 32–81 m (Upton et al. 1993), 17) Kattegat (Enoksson et al. 1990; Rydberg et al. 1990). summer data ■; annual data ●; other ○.

This is similar to sediment trap data from the North Sea where 20–35% of the primary production reached the bottom at 100 m (Davies & Payne 1984) and in the Baltic where at 40 m, 25% of the production reached the bottom (Forsskahl et al. 1982). Vertical carbon fluxes below the photic zone in the Mississippi River plume and inner Gulf of Mexico shelf region (Redalje et al. 1992) are on the order of 20% to 30% of the primary production values reported (see below). The somewhat lower percent of organic carbon mineralization in sediments found in our relationship (16%) relative to the above studies may be due to a number of factors. For example, reduced end products of anaerobic metabolism (e.g. sulfides) may be stored in the sediments, and/or some of the organic matter deposited to the sediment surface may be decomposed in the water column following resuspension.

In this data set the variation in the percentage of the primary production that reaches the bottom cannot be explained by depth. Even when data within regions are compared there is no strong correlation between depth and primary production mineralized on the bottom. For example, in the North Sea, sites at 31–81 m had a greater percentage of the primary production reaching the bottom than sites at 25–31 m (Upton et al. 1993) (Figure 2). At sites off Eastern Canada, sediment type rather than depth explained most of the variation in oxygen uptake (Grant et al. 1991).

Our finding that depth is not a major factor controlling the percentage of primary production which reaches the bottom may seem surprising because several other studies have shown a strong correlation of oxygen uptake with water depth (Devol & Christensen 1993; Christensen 1989; Rowe et al. 1988; Martin et al. 1987) or depth to the mixed layer (Hargrave 1973). However, those studies included data from stations extending to depths far greater than 200 m. When only the data from 200 m and less are examined, the correlation becomes much weaker or disappears. Our findings are similar to those of Nixon (1981) who found no relationship between sediment oxygen uptake and carbon inputs to the bottom in coastal sediments with water depths from 3 m to 70 m. Based on the information presented above, we did not use depth as a variable in our calculations.

Extrapolation to the North Atlantic Shelf

The relationship between sediment oxygen consumption and denitrification (eq. 1) was combined with the relationship between sediment oxygen consumption and phytoplankton production (eq. 2) to arrive at a model of denitrification as a function of phytoplankton production:

$$DNF_c = 0.019^{*}PhytoProd \qquad (3)$$

where DNF_c equals denitrification coupled to sediment nitrification (mmol N as N_2 m^{-2} d^{-1}) and PhytoProd equals phytoplankton production (mmol C m^{-2} d^{-1}) in the overlying water. This relationship suggests that approximately 13% of the N incorporated into phytoplankton in shelf waters is eventually denitrified in the sediments via coupled nitrification/denitrification, assuming a C:N ratio of 6.625:1 for phytoplankton.

Phytoplankton production rates for each region within the North Atlantic Basin were compiled and used to calculate regional denitrification rates (eq.3) (Table 1). The primary production data were taken from a wide variety of sources (see footnotes, Table 1). In a few cases, such as for the Mid-Atlantic Bight and part of the North Sea, the estimates are based upon data with good spatial and temporal coverage. However, for most regions we had to extrapolate from short term studies made in a restricted area. For areas where only limited data sets were available, we used lower estimates so that our calculated rates of denitrification would also be conservative.

The total amount of N removed by coupled nitrification/denitrification in each region was calculated from the estimated denitrification rate per unit area for that region (eq. 3) and the area of the continental shelf in that region (Pilson & Seitzinger, this vol.). The continental shelf boundary was chosen operationally as the 200 m bathymetric contour.

Results and discussion

Denitrification (coupled nitrification/denitrification) rates per unit area predicted by the model are highest for the continental shelf region between Cape Hatteras and the southern extent of Florida (2.34 mmol N as N_2 m^{-2} d^{-1}) and lowest for Hudson Bay, the Baffin Island region, and Greenland (0.11 to 0.15 mmol N as N_2 m^{-2} d^{-1}) (Table 1; Figure 3a). Within latitudinal belts, average denitrification rates were lowest (0.45 mmol N as N_2 m^{-2} d^{-1}) in the high latitudes (>45° N), intermediate (0.71 mmol N as N_2 m^{-2} d^{-1}) in the tropics (0–20° N), and highest (1.3 mmol N as N_2 m^{-2} d^{-1}) in the mid-latitudes (20° N–45° N) (Table 1). The area weighted average denitrification rate for continental shelf sediments in the North Atlantic Basin is 0.69 mmol N as N_2 m^{-2} d^{-1}.

Although denitrification rates per unit area are lowest in the high latitudes, due to the large area of continental shelf in the high latitudes (3,252,268 km^2; Pilson and Seitzinger, this vol.), the total N removal by denitrification (53 $\times$ 10^{10} mol N y^{-1}) is similar to that in the mid-latitudes (60 $\times$ 10^{10} mol N y^{-1}) (Table 1; Figure 4). The Gulf of St. Lawrence/Grand Banks area and the North Sea are responsible for seventy-five percent of the denitrification in the high latitude region (Figure 3b). N removal by denitrification in the western

Table 1. Estimated N removal by sediment denitrification coupled to nitrification in various continental shelf regions. See text for explanation of calculations.

Region	Area[1] km^2	Prim. prod.[2] mmol C m^{-2} d^{-1}	Denitrification mmol N as N_2 m^{-2} d^{-1}	Denitrification for region mol N y^{-1}
>45° N latitude				
Greenland	159,897	8	0.15	0.89×10^{10}
Iceland	107,272	18	0.34	1.34×10^{10}
Hudson Bay	780,178	6	0.11	3.25×10^{10}
Baffin Is. region	348,350	6	0.11	1.45×10^{10}
Labrador	147,955	51	0.97	5.23×10^{10}
G. St. Law. & Gr. Bank	518,998	43	0.82	1.55×10^{11}
North Sea	1,095,734	32	0.61	2.43×10^{11}
Norway Shelf	93,884	23	0.44	1.50×10^{10}
Total	3,252,268			5.35×10^{11}
Average			0.45	
20° N–45° N				
N Scotia/Hatteras	361,042	71	1.35	1.78×10^{11}
Hatteras/S. Florida	137,364	123	2.34	1.17×10^{11}
Gulf of Mexico	573,565	49	0.93	1.95×10^{11}
African self	144,476	83	1.58	8.32×10^{10}
Spain and Portugal	51,435	80	1.52	2.85×10^{10}
Total	1,267,882			6.02×10^{11}
Average			1.30	
0–20° N				
Central and S. America	279,799	49	0.93	9.51×10^{10}
Amazon Shelf	389,562	31	0.59	8.37×10^{10}
Caribbean Islands	194,956	20	0.38	2.70×10^{10}
African Shelf	273,848	46	0.87	8.74×10^{10}
Total	1,138,165			2.93×10^{11}
Average			0.71	
Overall totals	5.66×10^6			143×10^{10}
Average			0.69	

[1] Continental shelf areas from Pilson and Seitzinger (this vol.) calculated from the NOAA ETOPO5 data base.

[2] *Greenland, Iceland, Hudson Bay and Canadian Arctic* – For Hudson's Bay and most other Arctic shelf regions off Canada, including the area around Baffin Island, we used a primary production value of 27 gC m^{-2} y^{-1} (Harrison et al. 1982). We used a value of 80 gC m^{-2} y^{-1} for the western and southern tip of Greenland (Atkinson & Wacasey 1987; Steemann-Nielson

Table 1. Continued.

& Hansen 1960) for an average of 35 gC m^{-2} y^{-1} for the entire Greenland shelf. We used 80 gC m^{-2} y^{-1} for the Iceland shelf by extrapolating Steemann-Nielson & Hansen's (1960) data to a 120 day growing season.

Labrador Shelf – We used the estimate of Grant et al. (1991) who calculated a yearly average production of 225 gC m^{-2} y^{-1} based upon July and Sept production data at three sites.

Newfoundland Shelf including Grand Banks – We used Grant et al.'s (1991) value of 190 gC m^{-2} y^{-1}. Grant et al. (1991) used published ^{14}C production data and assumed a 10 month growing season. Their estimate compares well with that of Prasad & Haedrich's (1994) who used satellite detectable Chl a and a model to estimate production as 193 gC m^{-2} y^{-1}.

North Sea and surrounding areas – We used the primary production of Joint & Pomroy (1993) for the southern and central North Sea. Their estimates for the southern areas are roughly consistent with estimates that Reid et al. (1990) made from a compilation of the literature, but Reid et al.'s (1990) estimated values for primary production in the central North Sea (250 gC m^{-2} y^{-1}) are nearly twice that measured by Joint and Pomroy. We used Joint and Pomroy's lower estimate which was less heavily biased toward the more productive coastal areas. For the northern North Sea we used a value of 150 gC m^{-2} y^{-1} which is the low end of the range given in Reid et al. (1990). We used a value of 102 gC m^{-2} y^{-1} for the primary production of the Celtic Sea (Joint et al. 1986) and 144 gC m^{-2} y^{-1} for the Kattegat. The area weighted average for the entire North Sea region was 140 gC m^{-2} y^{-1}.

Norway – Harrison et al. (1982) cite a study where rates of 1.12 gC m^{-2} d^{-1} were measured. Assuming 120 day growing season this would yield a production of 134 gC m^{-2} y^{-1}. Off shore in the Norwegian Sea, Bodungen et al. (1990) has measured somewhat lower values of about 90 gC m^{-2} y^{-1}. For our calculations we assumed a value of 100 gC m^{-2} y^{-1}.

Northwestern Atlantic Shelf: Gulf of Maine to Cape Hatteras – We used a value of 310 gC m^{-2} y^{-1} from the data of O'Reilly and Bush (1984) for shelf primary production between Nova Scotia and Cape Hatteras. Their primary production data was based upon the results of 23 surveys (628 stations occupied) made along the Northwestern Atlantic shelf between 1977–1980.

South Atlantic Bight – Cape Hatteras to the tip of Florida – The complicated hydrographic controls on productivity in the South Atlantic Bight make it difficult to determine the primary production of the entire region from the limited data available (Lee et al. 1991; Yoder et al. 1985). We have used a value of 630 gC m^{-2} y^{-1} for the production of the inner shelf areas. This value is equal to the lower value found by Verity et al. (1993) for two stations off Skidaway Island but somewhat higher than the production of 550 gC m^{-2} y^{-1} found by Thomas (1966) to the south. We used a value of 360 gC m^{-2} y^{-1} for production on the outer shelf (Yoder 1985). We assumed that the outer shelf is approximately a third the area of the entire shelf (Lee et al. 1991).

Gulf of Mexico – We used an average rate of 290 gC m^{-2} y^{-1} for the Northern Gulf of Mexico which is what has been found on the shelf offshore of Barataria Bay (Skalar & Turner 1981). Flint and Rababais (1981) report a primary production value of 177 gC m^{-2} y^{-1} for the shelf off of Texas. We used this value for the southwestern Gulf. Primary production off Western Florida is lower than that found in the Northern Gulf (Walsh 1983). In the absence of good data for this region we also used a primary production rate of 177 gC m^{-2} y^{-1}. When adjusted for area this gives an average of 215 gC m^{-2} y^{-1} for the entire Gulf of Mexico.

African Shelf 20° N to 35° N From 20° N to the Straits of Gibraltar at 35° N, we used an average annual value of 365 gC m^{-2} y^{-1}. Most of the studies in this region suggest that the primary productivity averages over 1gC m^{-2} d^{-1} over the annual cycle (Gabric et al. 1993; Huntsman & Barber 1977 and references therein; Berguis et al. 1993).

Table 1. Continued.

Spain and Portugal – We used a value of 350 gC m^{-2} y^{-1} for the entire region. Average daily production values for the Bay of Biscay and Galician coast off of Spain are on the order of 1 gC m^{-2} d^{-1} (Fernandez et al. 1993; Varela 1992).

Central and South America – Older data suggest that on average the production over much of this area is about the same as the southwestern Gulf of Mexico with higher rates off Venezuela (Koblentz-Mishke et al. 1970 as redrawn in Berger 1989). The absolute values given for primary productivity given in Koblentz-Mishke et al. (1970) appear very low (less than 60 gC m^{-2} y^{-1}) throughout the entire region. However, the relative distribution of production is consistent with satellite maps of chlorophyll a and more recent production data. Based upon the distribution of production in Koblentz-Mishke et al. (1970) we assigned the same value of 215 gC m^{-2} y^{-1} to the Central and South American shelves as to the southwestern Gulf of Mexico.

Amazon Shelf – DeMaster et al. (1991) measured ^{14}C primary production from 0–4° N off the Amazon River out to the 100 m contour during falling and high river discharge periods (August and May, respectively). Their average productivity for the entire region was 2.5 gC m^{-2} d^{-1} which would extrapolate to 900 gC m^{-2} y^{-1}. This rate is not typical of the production further offshore where nutrient concentrations are low (Ryther et al. 1967). We used an average of 450 gC m^{-2} y^{-1} for the Amazon shelf region between 0–4° N. For the Amazon shelf north of 4° N (4° N–12° N) we assigned a value of 87 gC m^{-2} y^{-1} based on the broad scale distribution of production and nutrients in Koblentz-Mishke et al. (1970) and Ryther et al. (1967). The area weighted average for the entire Amazon Shelf region between 0 and 12° N was135 gC m^{-2} y^{-1}.

Caribbean Islands – Very few studies have been carried out in the shelf regions of Caribbean Islands. Hopcroft et al. (1990) reported that the phytoplankton productivity in shelf waters off Kingstown, Jamaica were 240 gC m^{-2} y^{-1} but this region appears to be influenced by nutrient inputs from Kingston. Offshore, values of less than 100 gC m^{-2} y^{-1} are more typical of this area (Beers et al. 1968). We used a value of 88 gC m^{-2} y^{-1} for this region.

Africa Shelf 0 to 20° N – Between 0–20° N we used a value of 200 gC m^{-2} y^{-1} based upon the spring and summer rates measured offshore by Mahnken (1969) (which extrapolated over the year average 200–300 gC m^{-2} y^{-1}) and Longhurst's (1983) lower numbers (100–160 gC m^{-2} y^{-1}) from a annual primary productivity model of the inshore region at 8° N.

North Atlantic (96×10^{10} mol N y^{-1}) is two times greater than in the eastern North Atlantic (47×10^{10} mol N y^{-1}) (Figure 4). This is primarily due to differences in the area of continental shelf in the two regions, as the average denitrification rate per unit area is similar in the western and eastern North Atlantic (0.67 and 0.73 mmol N as N_2 m^{-2} d^{-1}, respectively) (Table 1).

We are attempting to predict average denitrification rates for relatively large regions of the continental shelf. Rates of denitrification for any particular area within a region may be higher or lower than the predicted rates, due to smaller scale variability in such factors as sediment type or locally high primary production rates in upwelling regions. Still, it is instructive to compare the rates of denitrification estimated from our model to rates of denitrification measured or predicted by other methods within various regions. The only direct measurements of denitrification for North Atlantic shelf sediments are in Massachusetts Bay. Model predicted rates of 1.35 mmol

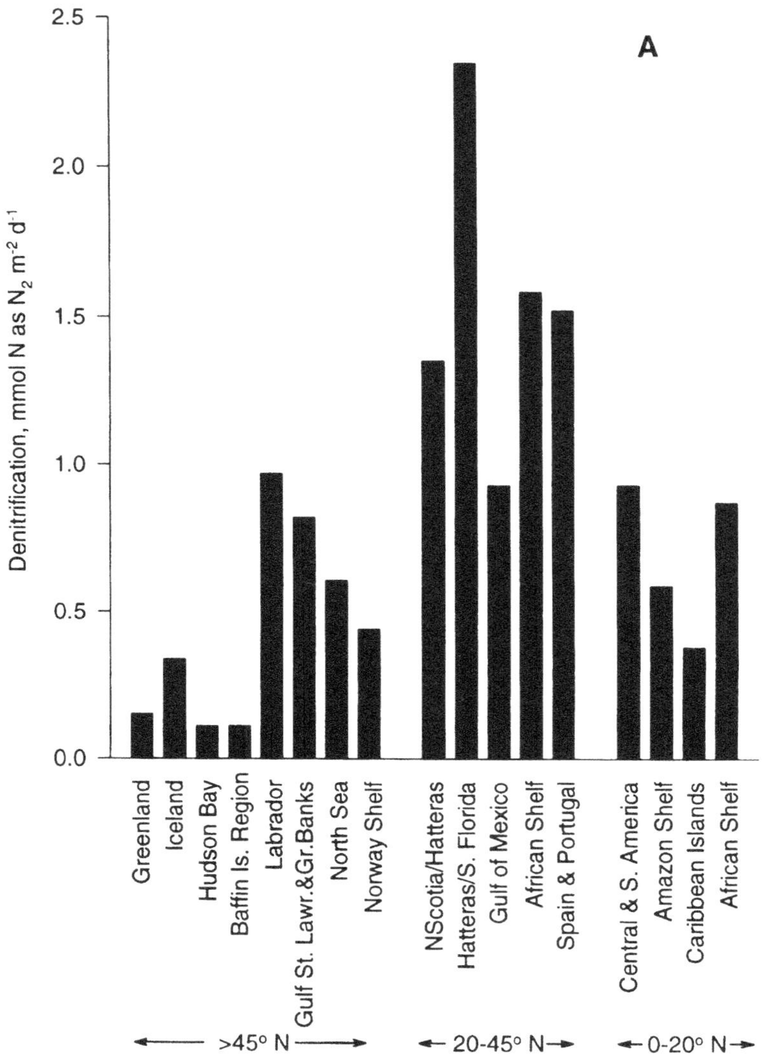

Figure 3. Model predicted rates of denitrification coupled to sediment nitrification for various continental shelf regions in the North Atlantic Basin: a) denitrification rates per unit area (mmol N as N_2 m^{-2} d^{-1}), and b) N removal by region (mol N $\times$ 10^{10} y^{-1}; e.g. Nova Scotia/Hatteras = 17.8 $\times$ 10^{10} mol N y^{-1}).

N as N_2 m^{-2} d^{-1} (Nova Scotia to Cape Hatteras region) (Table 1) are within the range of the directly measured rates of $\leq$0.24 to 2.56 mmol N as N_2 m^{-2} d^{-1} in Massachusetts Bay (Giblin et al. 1994). Denitrification rates in the Gulf of Mexico estimated from our model calculations were 0.93 mmol

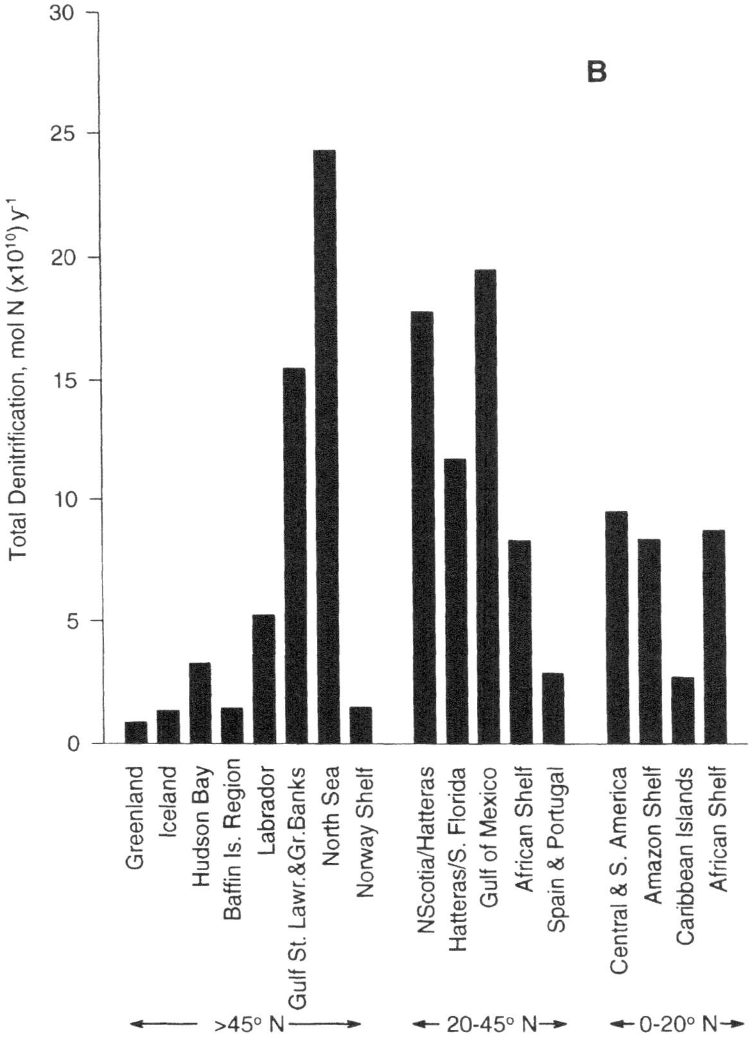

Figure 3. Continued.

N as N_2 m^{-2} d^{-1}, which is within the range of rates reported by Gardner et al. (1993) for the northern Gulf of Mexico (0.5 to 1.68 mmol N as N_2 m^{-2} d^{-1}) based on stoichiometric calculations. In the South Atlantic Bight (located within our Cape Hatteras to S. Florida region), denitrification rates in excess of 3.2 mmol N as N_2 m^{-2} d^{-1} were calculated for a hard bottom community based on the deficit of the measured benthic DIN effluxes relative

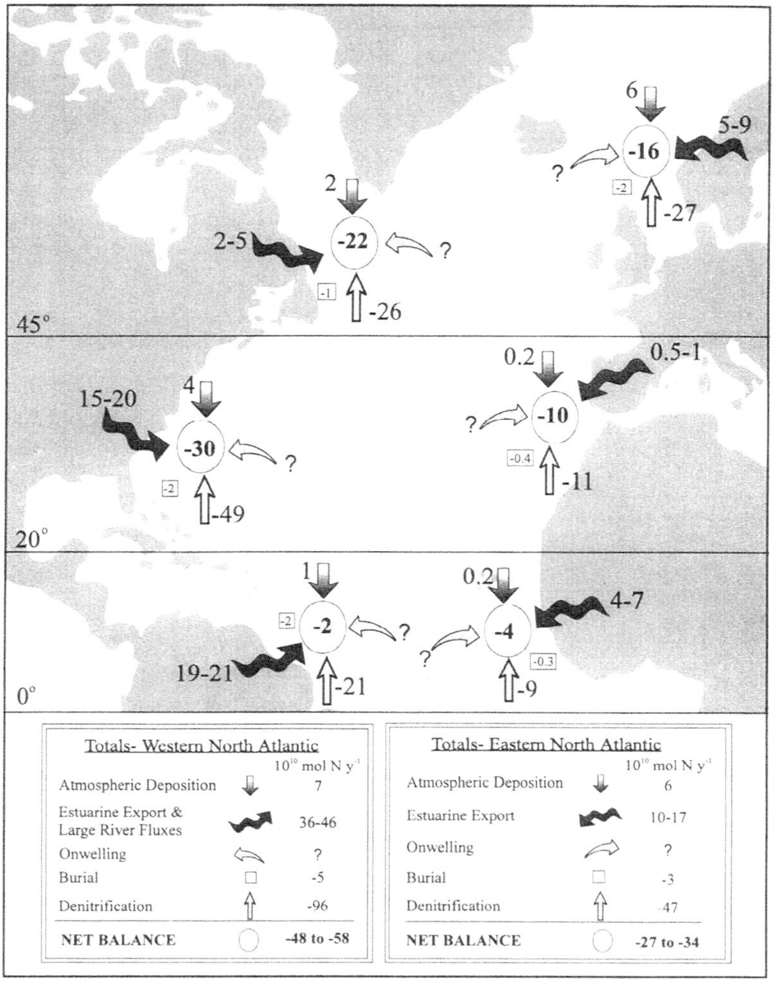
6
-16
5-9
?
-2
-27
2
-22
2-5
?
-1
-26
45°
0.2
0.5-1
4
15-20
-10
?
-30
?
-0.4
-11
-2
-49
20°
1
0.2
4-7
-2
-2
?
-4
?
19-21
-0.3
-21
-9
0°
Totals- Western North Atlantic
10^10 mol N y^-1
Atmospheric Deposition 7
Estuarine Export & Large River Fluxes 36-46
Onwelling ?
Burial -5
Denitrification -96
NET BALANCE -48 to -58
Totals- Eastern North Atlantic
10^10 mol N y^-1
Atmospheric Deposition 6
Estuarine Export 10-17
Onwelling ?
Burial -3
Denitrification -47
NET BALANCE -27 to -34

to the phosphate effluxes and assuming a Redfield stoichiometry of 16:1 (N:P by moles) (Hopkinson et al. 1991); those rates are similar to our predictions (2.34 mmol N as N_2 m^{-2} d^{-1}). Model-predicted rates for the North Sea are 0.61 mmol N as N_2 m^{-2} d^{-1} which is consistent with denitrification rates reported for that region based on models of pore water nutrient profiles (0.34 to 2.74 mmol N as N_2 m^{-2} d^{-1}, Billen (1978); 0.06 to 0.54 mmol N as N_2 m^{-2} d^{-1}, Raaphorst et al. (1990)). However, rates of denitrification for the North Sea estimated from acetylene block measurements (0.0012 to 0.2 mmol N as N_2 m^{-2} d^{-1}, Law & Owens (1990); 0–0.3 mmol N as N_2 m^{-2} d^{-1}, Lohse et al. 1993) are considerably lower than our model calculated rates. One explanation is that our model estimates coupled nitrification/denitrification, while the acetylene block method has been found to markedly underestimate coupled nitrification/denitrification in some aquatic sediments (Seitzinger et al. 1993). In summary, our model estimates of denitrification for continental shelf regions appear conservative based on comparisons of model estimates for specific regions relative to rates reported in the literature.

We calculate that a total of 143×10^{10} mol N y^{-1} is removed via coupled nitrification/denitrification on the North Atlantic continental shelf (Table 1). Our model is expected to underestimate total sediment denitrification because it does not include direct denitrification of nitrate from the overlying water. The stoichiometric approach that we used to calculate denitrification rates only provides an estimate of denitrification coupled to sediment nitrification. Nitrate diffusing, or transported by macrofauna irrigation, into the sediments is another source for denitrification and was not included. There are some shelf sites where nitrate uptake from the overlying water is relatively unimportant in supporting denitrification when compared to coupled nitrification/denitrification (Raaphorst et al. 1990; Hopkinson et al. 1991; Kelly & Nowicki 1993; Giblin et al. 1994). However, in Washington shelf and Gulf of Maine sediments, denitrification of nitrate from the overlying water was estimated to be similar to, or greater than, denitrification coupled to

⟶

Figure 4. Known sources and sinks of N for continental shelf regions in the western and eastern North Atlantic Basin by latitudinal zone. The input of N to the continental shelf from estuarine export plus fluxes from rivers that discharge directly on the shelf was calculated from Nixon et al. (this vol.). Atmospheric deposition of $NO_y + NH_x$ (wet plus dry deposition) was calculated from Nixon et al.'s (this vol.) summary of Prospero et al. (this vol.). Burial of N in shelf sediments exclusive of major river deltas was calculated based on the primary production data in Table 1 and Nixon et al.'s (this vol.) assumption that 0.25% to 1.25% of the primary production with a C/N (molar) ratio of 10 is buried. The balance of the indicated sources and sinks is indicated within the closed circles; the mid-point of the range in estuarine export/large river fluxes for a region was used for that calculation. The contribution of N from onwelling of nutrient rich slope water is not known, but is hypothesized to account for the additional N input needed to balance the N budget.

sediment nitrification (Christensen et al. 1987a; Devol 1991; Devol & Christensen 1993). Similarly, over an annual cycle in the Aarhus Bight, nitrate from the overlying water and nitrification within the sediment were equally important in supporting denitrification but the relative importance of each changed seasonally (Jensen et al. 1990). Sediments took up nitrate in about half of the benthic flux measurements made off Cap Blanc, Africa (Rowe et al. 1977) and in the New York Bight (Florek & Rowe 1983). The rates of nitrate disappearance in these studies could support direct denitrification rates of 0.1 to 1.6 mmol N as N_2 m^{-2} d^{-1}. This range is comparable to the range of coupled nitrification/denitrification rates predicted by our model (Table 1), and suggests that N removal by direct denitrification may be of a similar magnitude to coupled nitrification/denitrification. While coupled nitrification/denitrification appears to be related to organic matter deposition, the magnitude of direct denitrification is likely to vary spatially as a function of bottom water nitrate concentration. Clearly, a complete analysis of the contribution of the continental shelf to the nitrogen budget should include denitrification of nitrate from both direct and coupled sources. One approach that could be taken to estimate direct denitrification of nitrate from the overlying water is to model nitrate input into sediments due to diffusion and to transport by macrofauna based on bottom water nitrate and macrofaunal distributions in shelf regions. However, such an analysis awaits future efforts.

Sources of nitrogen to the continental shelf region to support the estimated denitrification include nitrogen transport by rivers, atmospheric deposition and onwelling of deeper slope water. The rate of coupled nitrification/denitrification that we estimate for North Atlantic shelf regions (143×10^{10} mol N y^{-1}) is approximately one and a half times as great as the total nitrogen transport by rivers in the North Atlantic Basin (94×10^{10} mol N y^{-1}; Howarth et al., this vol.). Given that a significant amount of N is removed by denitrification within estuaries before being exported to the shelf (Seitzinger 1988), the actual input of river N to continental shelf areas is considerably less. Nixon et al. (this vol.) estimated that of the 94×10^{10} mol N y^{-1} transported by rivers in the North Atlantic Basin, only 46 to 63×10^{10} mol N y^{-1} are transported to the shelf, with the remainder removed within estuaries by denitrification, burial or fish harvesting (Table 2). Thus, inputs of N to the shelf from river transport can account for only a third to less than half of the N required to support the estimated denitrification on the shelf. Additional sources of N to the shelf from atmospheric deposition (Nixon et al., this vol. based on Prospero et al., this vol.) or biological N_2-fixation (Nixon et al., this vol.) are considerably smaller than needed to balance the shelf N budget (Table 2). Total known inputs amount to 59 to 76×10^{10} moles N y^{-1}, which accounts for only fifty percent or less of the N needed to supply the

Table 2. Known sources and sinks of nitrogen for the North Atlantic continental shelf (see text).

	10^{10} moles N y^{-1}
Known inputs	
Estuarine export + large river fluxes[1]	46–63
Biological N_2-fixation[3]	0.2
Atmospheric deposition[2]	13
Total	59–76
Known sinks	
Denitrification (coupled nitrification/denitrification)	143
Burial in shelf sediments[3]	2–9
Fish landings[3]	3–4
Export from shelf	?
Total	148–156
Net balance (onwelling?)	−72 to −97

[1] Estuarine export plus fluxes from large rivers that discharge directly on the shelf from Nixon et al. (this vol.)
[2] Atmospheric deposition of NO_y + NH_x (wet plus dry deposition) from Nixon et al.'s (this vol.) summary of Prospero et al. (this vol.)
[3] Nixon et al. (this vol.)

estimated denitrification (143×10^{10} mol N y^{-1}). A comparison of known inputs and known sinks (including denitrification, burial in shelf sediments, and fish landings) suggests that at least 72 to 97 $\times$ 10^{10} mol N y^{-1} would be needed from other sources to balance the shelf N budget (Table 2). This is a minimum estimate as direct denitrification of nitrate from the overlying water is not included in the denitrification removal term.

A view of the known sources and sinks of N in the western and eastern North Atlantic and by latitudinal zone provides an additional perspective (Figure 4). Overall, in the western North Atlantic known inputs can account for only approximately half of N required to supply the estimated denitrification; in the eastern North Atlantic known inputs are only about a third to half of the N required for denitrification. (The burial term is a small number in all regions, amounting to a N sink less than ten percent of that removed by denitrification.) A more detailed comparison can be made by latitudinal zone. In some regions the known N inputs are close to sufficient to supply the estimated N removal by denitrification. For example, in the tropical regions (0 to 20° N) of both the western and eastern North Atlantic, inputs of N from estuarine export, fluxes from large rivers directly onto the shelf, and atmospheric deposition are approximately equal to the loss through denitrification and burial (Figure 4). In other regions, the known N inputs can account for only approximately half of the estimated N removal by denitrification; this is the case in the mid-latitudes (20° to 45° N) of the western North Atlantic and the high latitudes (>45° N) of the eastern North Atlantic. In the high latitudes of the western North Atlantic and mid-latitudes of the eastern North Atlantic, the known inputs are very out of balance with the N removal terms. In those regions, known inputs can account for only about 10% of the N needed to supply the estimated removal by denitrification.

Onwelling of nitrate rich slope water onto continental shelves is known to be a major source of nitrogen in various shelf regions. Unfortunately, slope transport of nitrogen onto the shelf has not been adequately quantified for the North Atlantic Basin as a whole. However, in the two regions for which we have data, it appears that slope transport is of the correct magnitude to balance the shelf N budget. A minimum estimate of slope water nitrate transport to the South Atlantic Bight (our Cape Hatteras to S. Florida region) from Lee et al. (1991) (9.3×10^{10} mol N y^{-1}) combined with N export from estuaries (0.6 to 1.2×10^{10} mol N y^{-1}; Nixon et al., this vol.) is similar to our estimated N removal by denitrification (11.7×10^{10} mol N y^{-1}; Table 1). In the Mid-Atlantic Bight (our Nova Scotia to Cape Hatteras region), Csanady (1990) estimated that 11.4×10^{10} mol N y^{-1} are transported by slope water onto the shelf along a 900-km stretch, which when combined with N export from estuaries in this region (1.3 to 2.6×10^{10} mol N y^{-1}; Nixon et al., this vol.), is within approximately 30% of our estimated N removal by denitrification (17.8×10^{10} mol N y^{-1}; Table 1). The above studies suggest that onwelling of slope water is of the magnitude needed to balance the shelf N budgets in the two regions where data are available. However, further studies of N transported by slope water onto various shelf regions throughout the North Atlantic Basin are required to determine if this holds true for other regions.

The magnitude of all nitrogen sources to the total North Atlantic is uncertain. It is presently not known if there are sufficient sources of new nitrogen entering the North Atlantic gyres to support the onwelling of N to continental shelf areas needed to support the denitrification rates calculated here. However, recent estimates of nitrate creation in the upper 1000 m of the Sargasso Sea (3.7 to 6.4×10^{12} moles NO_3^- y^{-1}) by Michaels et al. (this vol.) could potentially supply the required denitrification in the shelf regions. Michaels et al. (this vol.) have postulated N_2-fixation as a likely source for the created nitrate, or dissolved organic N mineralization.

The North Atlantic shelf makes up only 20% of the total continental shelf area in the ocean. If the coupled nitrification/denitrification rates we calculate for the North Atlantic can be extrapolated to the globe, the continental shelf sediments are an important oceanic sink of nitrogen removing more than 7.2 $\times$ 10^{12} mol N y^{-1} (100 Tg N y^{-1}). This is a minimum estimate of sediment denitrification since it only includes coupled nitrification/denitrification and does not include direct denitrification of nitrate from the overlying water. The current estimate is two times higher than that made by Christensen et al. (1987a) (3.6 $\times$ 10^{12} mol N y^{-1}) for continental shelves, whose estimate in turn was considerably higher than the earlier estimates of between 0.4 $\times$ 10^{12} to 1.8 $\times$ 10^{12} mol N y^{-1} (Hattori 1983; Liu & Kaplan 1984). Although this global loss is large (7.2 $\times$ 10^{12} mol N y^{-1}), it appears that it could easily be supported by the onwelling of nitrate from the deep sea to the global shelves (40 $\times$ 10^{12} mol N y^{-1}) as estimated by Walsh (1991).

The analyses presented here reinforce and flesh out the idea of Christensen et al. (1989), that denitrification in shelf sediments is removing a significant amount of N inputs to continental shelf regions. Furthermore, denitrification in Atlantic continental shelf regions appears to be an important sink for N in the Atlantic Basin as a whole (Michaels et al., this vol.). This large N sink in continental shelf regions likely contributes to N limitation in those regions. The areal rates of denitrification that we calculate are consistent with recent measurements and do not appear to be extremely high. However, the number of areas where measurements have been made is small and has been biased towards temperate regions. Clearly more measurements of denitrification in a variety of regions and sediment types are needed.

Acknowledgments

This work was sponsored, in part, by NOAA, Office of Sea Grant, Department of Commerce, under Grants No. NA85AA-D-SG084 and NA89AA-D-SG057 to SPS (NJSG 96-344). Salary support for AEG came from NSF-OCE-9214461, NOAA Sea Grant and Massachusetts Water Resources Authority. Funding to support the overall efforts of the North Atlantic N Project was also provided by UNEP, the Mellon Foundation and WMO. We would like to thank Michael Pilson, Sue Watts and anonymous reviewers for their comments on earlier versions of the manuscript, and Renee Styles for preparing the graphics.

References

Aller RC, Aller JY, Blair NE, Makin JE & Rude PD (1991) Biogeochemical processes in Amazon shelf sediments. Oceanography 4: 27–32

Archer D & Devol A (1992) Benthic oxygen fluxes on the Washington shelf and slope: A comparison of in situ microelectrode and chamber flux measurements. Limnol. Oceanogr. 37(3): 614–629

Atkinson EG & Wacasey JW (1987) Sedimentation in Arctic Canada: Particulate organic carbon flux to a shallow marine benthic community in Frobisher Bay. Polar Biol. 8: 3–7
Beers JR, Steven DM & Lewis JB (1968) Primary productivity in the Caribbean. Bull. Mar. Sci. 18: 86–104
Bender ML, Fanning KA, Froelich PN, Heath GH & Maynard V (1977) Interstitial nitrate profiles and oxidation of sedimentary organic matter in the eastern equatorial Atlantic. Science 198: 605–609
Berger WH (1989) Apendix: Global maps of ocean productivity. In: Berger WH, Smetacek VS & Wefer G (Eds) Productivity of the Ocean: Present and Past (pp 429–455). John Wiley and Sons
Berghuis EM, Duineveld GCA & Hegeman J (1993) Primary production and distribution of phytopigments in the water column and sediments on the upwelling shelf off the Mauritanian coast (Northwest Africa). Hydrobiologia 258: 81–93
Billen G (1978) A budget of nitrogen recycling in North Sea sediments off the Belgian coast. Est. Coast. Shelf Sci. 7: 127–146
Bodungen B, Bathmann U, Voss M & Wunsch M (1990) Vertical particle flux in the Norwegian Sea-resuspension and interannual variation. In: Wassmann P, Heiskanen A & Lindahl O (Eds) Sediment Trap Studies in the Nordic Countries (pp 116–136). Kristineberg Marine Biological Station, Sweden
Bremner JM & Blackmer AM (1979) Effects of acetylene and soil water content on emission of nitrous oxide from soils. Nature 280: 380–381
Christensen JP (1994) Carbon export from continental shelves, denitrification and atmospheric carbon dioxide. Cont. Shelf Res. 14: 547–576
Christensen JP (1989) Sulfate reduction and carbon oxidation rates in continental shelf sediments, an examination of offshelf carbon transport. Cont. Shelf Res. 9(3): 223–246
Christensen JP, Murray JW, Devol AH & Codispoti LA (1987a) Denitrification in continental shelf sediments has major impact on the oceanic nitrogen budget. Global Biogeochem. Cycles 1: 97–116
Christensen JP, Smethie WM Jr & Devol AH (1987b) Benthic nutrient regeneration and denitrification on the Washington continental shelf. Deep-Sea Res. 34(5/6): 1027–1047
Codispoti LA & Christensen JP (1985) Nitrification, denitrification, and nitrous oxide cycling in the eastern tropical South Pacific Ocean. Marine Chemistry 16: 277–300
Csanady GT (1990) Physical basis of coastal productivity, The SEEP and MASAR experiments. Eos. Trans. AGU 71(36): 1060–1065
Davies JM & Payne R (1984) Supply of organic matter to the sediment in the northern North Sea during a spring phytoplankton bloom. Mar. Biol. 78: 315–324
DeMaster DJ, McKee BA, Moore WS, Nelson DM, Showers WJ & Smith WO Jr. (1991) Geochemical processes occurring in the waters at the Amazon River/ocean boundary. Oceanography 4: 15–20
De Wilde PAWJ, Berghuis EM & Kok A (1984) Structure and energy demand of the benthic community of the oyster ground, Central North Sea. Neth. J. Sea Res. 18(1/2): 143–159
Devol AH & Christensen JP (1993) Benthic fluxes and nitrogen cycling in sediments of the continental margin of the eastern North Pacific. J. Mar. Res. 51: 345–372
Devol AH (1991) Direct measurement of nitrogen gas fluxes from continental shelf sediments. Nature 349: 319–321
Enoksson V, Sorensson F & Graneli W (1990) Nitrogen transformations in the Kattegat. Ambio. 19: 159–166
Falkowski PG, Flagg CN, Rowe GT, Smith SL, Whitledge TE & Wirick CD (1988) The fate of a spring phytoplankton bloom: export or oxidation? Cont. Shelf Res.8(5–7): 457–484
Falkowski PG, Vidal J, Hopkins TS, Rowe GT, Whitledge TE & Harrison WG (1983) Summer nutrient dynamics in the Middle Atlantic Bight: Primary production and utilization of phytoplankton carbon. J. Plankton Res. 5: 515–537

Ferandez E, Cabal J, Acuna JL, Bode A, Botas A & Garcia-Soto C (1993) Plankton distribution across a slope curent-induced front in the southern Bay of Biscay. J. Plankton. Res. 15: 619–641
Flint RW & Kamykowski D (1984) Benthic nutrient regeneration in South Texas coastal waters. Est. Coast. and Shelf Sci. 18: 221–230
Flint RW & Rabalais NN (1981) Gulf of Mexico shrimp production: a food web hypothesis. Fisheries Bulletin 79: 737–748
Florek RJ & Rowe GT (1983) Oxygen consumption and dissolved inorganic nutrient production in marine coastal and shelf sediments of the Middle Atlantic Bight. Int. Rev. Ges. Hydrobiol. 68(1): 73–112
Forsskahl M, Laakkone A & Leppanen JM (1982) Seasonal cycle of production and sedimentation of organic matter at the entrance of the Gulf of Finland. Neth. J. Sea Res. 16: 290–299
Gabric AJ, Garcia L, Van Camp L, Nykjaer L, Eifler W & Schrimpf W (1993) Offshore export of shelf production in the Cape Blanc (Mauritania) giant filament as derived from coastal zone color scanner imagery. J. Geophys. Res. 98: 4697–4712
Gardner WS, Briones EE, Kaegi EC & Rowe GT (1993) Ammonium excretion by benthic invertebrates and sediment-water nitrogen flux in the Gulf of Mexico near the Mississippi River outflow. Estuaries 16(4): 799–808
Giblin AE, Hopkinson CS & Tucker J (1992) Metabolism and nutrient cycling in Boston Harbor sediments. MWRA Enviro. Quality Dept. Tech. Rept. Series No. 92–1, Massachusetts Water Resources Authority, Boston, MA. 42 pp
Giblin AE, Hopkinson CS & Tucker J (1993) Metabolism, nutrient cycling and denitrification in Boston Harbor and Massachusetts Bay sediments. MWRA Enviro. Quality Dept. Tech. Rept. Series No. 93–2, Massachusetts Water Resources Authority, Boston, MA. 42 pp
Giblin AE, Hopkinson CS, Tucker J, Nowicki B & Kelly JR (1994) Metabolism, nutrient cycling and denitrification in Boston Harbor and Massachusetts Bay sediments in 1993. MWRA Enviro. Quality Dept. Tech. Rept. Series No. 94–, Massachusetts Water Resources Authority, Boston, MA. 54 pp
Grant J, Emerson CW, Hargrave BT & Shortle JL (1991) Benthic oxygen consumption on continental shelves off Eastern Canada. Cont. Shelf Res. 11: 1083–1097
Grebmeier JM & McRoy CP (1989) Pelagic-benthic coupling on the shelf of the northern Bering and Chukchi Seas. III. Benthic food supply and carbon cycling. Mar. Ecol. Prog. Ser. 53: 79–91
Haines JR, Atlas RM, Griffiths RP & Morita RY (1981) Denitrification and nitrogen fixation in Alaskan continental shelf sediments. Appl. and Environ. Microbiol. 41(2): 412–421
Hanson RB, Tenore KR, Bishop S, Chamberlain C, Pamatmat MM & Tietjen J (1981) Benthic enrichment in the Georgia Bight related to Gulf Stream intrusions and estuarine outwelling. J Mar. Res. 39(3): 417–441
Hargrave BT (1973) Coupling carbon flow through some pelagic and benthic communities. J. Fish. Res. Bd. Can. 30: 1317–1326.
Harrison WG, Platt T & Irwin B (1982) Primary production and nutrient assimilation by natural phytoplankton populations of the Eastern Canadian Arctic. Can. J. Fish. Aquat. Sci. 39: 335–345
Hattori A (1983) Denitrification and dissimilatory nitrate reduction. In: Carpenter EJ & Capone DG (Eds) Nitrogen in the Marine Environment (pp 191–232). Academic Press, NY
Hopcroft RR, Roff JC & Berges JA (1990) Size-fractionated sedimentation in a tropical neritic ecosystem near Kingston, Jamaica. Cont. Shelf Res. 10: 795–806
Hopkinson Jr CS, Fallon RD, Jansson B-O & Schubauer JP (1991) Community metabolism and nutrient cycling at Gray's Reef, a hard bottom habitat in the Georgia Bight. Mar. Ecol. Prog. Ser. 73: 105–120
Howarth RW, Billen G, Swaney D, Townsend A, Jaworski N, Lajtha K, Downing JA, Elmgren R, Caraco N, Jordan T, Berendse F, Freney J, Kudeyarov V, Murdoch P & Zhao-liang Z

(1996) Regional nitrogen budgets and riverine N and P fluxes for the drainages to the North Atlantic Ocean: Natural and human influences. Biogeochem. (this vol.)

Huntsman SA & Barber RT (1977) Primary production off northwest Africa: The relationship to wind and nutrient conditions. Deep-Sea Research 24: 25–33

Hynes RK & Knowles R (1978) Inhibition by acetylene of ammonia oxidation in *Nitrosomonas europaea*. FEMS (Fed. Eur. Microbiol. Soc.) Microbiol. Lett. 4: 319–321

Hynes RK & Knowles R (1984) Production of nitrous oxide by *Nitrosomonas europaea*: Effects of acetylene, pH, and oxygen. Can. J. Microbiol. 30: 1397–1404

Jensen MH, Lomstein E & Sorensen J (1990) Benthic NH_4^+ and NO_3^- flux following sedimentation of a spring phytoplankton bloom in Aarhus Bigh, Denmark. Mar. Ecol. Prog. Ser. 61: 87–96

Joint I & Pomroy A (1993) Phytoplankton biomass and production in the southern North Sea. Mar. Ecol. Prog. Ser. 99: 169–182

Joint IR, Owens NJP & Pomroy AJ (1986) Seasonal production of photosynthetic picoplankton and nanoplankton in the Celtic Sea. Mar. Ecol. Prog. Ser. 28: 251–258

Jorgensen BB & Revsbech NP (1989) Oxygen uptake, bacterial distribution, and carbon-nitrogen-sulfur cycling in sediments from the Baltic Sea–North Sea Transition. Ophelia 31(1): 29–49

Kelly JR & Nowicki BL (1993) Direct measurements of denitrification in Boston Harbor and Massachusetts Bay sediments. Massachusetts Water Resources Authority Environ. Quality Technical Report. Series No. 93–3. 21 pp

Kemp WM, Sampou P, Caffrey J & Mayer M (1990) Ammonium recycling versus denitrification in Chesapeake Bay sediments. Limnol. Oceanogr. 35: 1545–1563

Koblentz-Mishke OJ, Volkvinsky VV & Kabanova JG (1970) Scientific Exploration of the South Pacific. Nat. Acad. Sciences, Washington DC: 183–193

Koike I & Hattori A (1979) Estimates of denitrification in sediments of the Bering Sea shelf. Deep-Sea Res. 26A: 409–415

Law CS & Owens NJP (1990) Denitrification and nitrous oxide in the North Sea. Neth. J. Sea Res. 25(1/2): 65–74

Lee TN, Yoder JA & Atkinson LP (1991) Gulf Stream frontal eddy influence on productivity of the southeast U.S. continental shelf. J. Geophys. Res. 96(C12): 22,191–22,205

Liu K-K & Kaplan IR (1984) Denitrification rates and availability of organic matter in marine environments. Earth. Plan. Sci. Letters. 68: 88–100

Lohse L, Malschaert JFP, Slomp CP, Helder W & Raaphorst W van (1993) Nitrogen cycling in North Sea sediments: interaction of denitrification and nitrification in offshore and coastal areas. Mar. Ecol. Prog. Ser. 101: 283–296.

Longhurst A (1983) Benthic-pelagic coupling and export of organic carbon from a tropical Atlantic continental shelf – Sierra Leone. Est. Coast. and Shelf Sci. 17: 261–285

Mahnken CVW (1969) Primary organic production and standing stock of zooplankton in the tropical Atlantic Ocean-Equalant I and II. Bull. Mar. Sci. 19: 550–567

Martin JH, Knauer GA, Karl DM & Broenkow WW (1987) VERTEX: carbon cycling in the northeast Pacific. Deep-Sea Res. 34: 267–285

Michaels AF, Olson D, Sarmiento J, Ammerman J, Fanning K, Jahnke R, Knap AH, Lipschultz F & Prospero J (1996) Inputs, losses and transformations of nitrogen and phosphorus in the deep North Atlantic Ocean. Biogeochem. (this vol.)

Nixon SW (1981) Remineralization and nutrient cycling in coastal communities. In: Neilson BJ & Cronin LE (Eds) Estuaries and Nutrients (pp 111–138) Humana Press, NJ

Nixon SW, Ammerman J, Atkinson L, Berounsky V, Billen G, Boicourt W, Boynton W, Church T, DiToro D, Elmgren R, Garber J, Giblin A, Jahnke R, Owens N, Pilson MEQ & Seitzinger S (1996) The fate of nitrogen and phosphorus at the land-sea margin of the North Atlantic Ocean. Biogeochem. (this vol.)

Nixon SW, Oviatt CA & Hale S (1976) Nitrogen regeneration and the metabolism of coastal marine bottom communities. In: Anderson JM & MacFadyen A (Eds) The Role of Terrestrial and Aquatic Organisms in Decomposition Processes (pp 269–283). Blackwell

O'Reilly JE & Busch DA (1984) Phytoplankton primary production on the northwestern Atlantic shelf. Rapp P-v Reun. Cons. Int. Explor. Mer. 183: 255–268
Pilson MEQ & Seitzinger SP (1996) Areas of shallow water in the North Atlantic. Biogeochem. (this vol.)
Prasad KS & Haedrick RL (1994) Satellite-derived primary production estimates from the Grand Banks: comparison to other oceanic regimes. Cont. Shelf Res. 14: 1677–1687
Prospero J et al. (1996) Atmospheric deposition of nutrients to the North Atlantic Basin. Biogeochem. (this vol.)
Raaphorst W van, Kloosterhuis HT, Cramer A & Bakker KJM (1990) Nutrient early diagenesis in the sandy sediments of the Dogger Bank area, North Sea: pore water results. Neth. J. Sea Res. 26(1): 25–52
Raaphorst W van, Kloosterhuis HT, Berghuis EM, Gieles AJM, Malschaert JFP & van Noort GJ (1992) Nitrogen cycling in two types of sediments of the southern North Sea (Frisian Front, Broad Fourteens): Field data and mesocosm results. Neth. J Sea Res. 28(4): 293–316
Redalje DG, Lohrenz SE & Fahnenstiel GL (1992) Phytoplankton dynamics and the vertical flux of organic carbon in the Mississippi river plume and inner Gulf of Mexico shelf region. In: Falkowski PG and Woodhead AD (Eds) Primary Production and Biogeochemcial Cycling in the Sea (p 526). Plenum Press
Reid PC, Lancelot C, Gieskes WWC, Hagmeier E & Weichart G (1990) Phytoplankton of the North Sea and its dynamics: A review. Neth. J. Sea Res. 26: 295–331
Revsbech PN, Sorensen J, Blackburn TH & Lomholt JP (1980) Distribution of oxygen in marine sediments measured with microelectrodes. Limnol. Oceanogr. 25: 403–411
Rowe GT, Theroux R, Phoel W, Quinby H, Wilke R, Koschoreck D, Whitledge TE, Falkowski PG & Fray C (1988) Benthic carbon budgets for the continental shelf south of New England. Cont. Shelf Res. 8(5–7): 511–528
Rowe GT, Clifford CH & Smith KL Jr (1977) Nutrient regeneration in sediments off Cap Blanc, Spanish Sahara. Deep Sea Res. 24: 57–63
Rydberg L, Edler L, Floderus S & Graneli W (1990) Interaction between supply of nutrients, primary production, sedimentation and oxygen consumption in SE Kattegat. Ambio 19: 134–141
Ryther JH, Menzel DW & Corwin N (1967) Influence of the Amazon river outflow on the ecology of the Western Tropical Atlantic. I. Hydrography and nutrient chemistry. J. Mar. Res. 25: 69–83
Seitzinger SP (1987) Nitrogen biogeochemistry in an unpolluted estuary: The importance of benthic denitrification. Mar. Ecol. Prog. Ser. 41: 177–186
Seitzinger SP (1988) Denitrification in freshwater and coastal marine ecosystems: ecological and geochemical importance. Limnol. Oceanogr. 33: 702–724
Seitzinger SP, Nielsen LP, Caffrey J & Christensen PB (1993) Denitrification measurements in aquatic sediments: A comparison of three methods. Biogeochem. 23: 147–167
Sklar RH & Turner RE (1981) Characteristics of phytoplankton production off Barataria Bay in an area influenced by the Mississippi River. Contributions in Marine Science 24: 93–106
Steemann-Nielsen E & Hansen VK (1960) The primary production in the waters west of Greenland during July 1958. Rapp P-v Reun. Cons. Int. Explor. Mer. 149: 150–151
Thomas JP (1966) Influence of the Altamaha River on primary production beyond the mouth of the river. M.S. thesis, University of Georgia, Athens. 88 pp
Upton AC, Nedwell DB, Parkes RJ & Harvey SM (1993) Seasonal benthic microbial activity in the southern North Sea; Oxygen uptake and sulfate reduction. Mar. Ecol. Prog. Ser. 101: 273–281
Varela M (1992) Upwelling and phytoplankton ecology in Galician (NW Spain) rias and shelf waters. Bol. Inst. Esp. Oceanogr. 8: 57–74
Verity PG, Yoder JA, Bishop SS, Nelson JR, Craven DB, Blanton JO, Robertson CY & Tronzo CR (1993) Compostion, productivity and nutrient chemistry of a coastal ocean plankton food web. Cont. Shelf Res. 13: 741–776

Walsh JJ (1991) Importance of continental margins in the marine biogeochemical cycling of carbon and nitrogen. Nature 350: 53–55

Walsh JJ (1983) Death in the sea: Enigmatic phytoplankton losses. Prog. Oceanog. 12: 1–86

Yoder JA (1985) Environmental control of phytoplankton production on the southeastern U.S. continental shelf. In: Atkinson LP, Menzel DW & Bush KA (Eds) Oceanography of the Southeastern United States Continental Shelf Chpt. 7. AGU, Washington

Yoder JA, Atkinson LP, Bishop SS, Blanton JO, Lee TN & Pietrafesa LJ (1985) Phytoplankton dynamics within Gulf Stream intrusions on the southeastern United States continental shelf during summer 1981. Cont. Shelf Res. 4: 611–635

Biogeochemistry **35**: 261–274, 1996.

An assessment of nitrogen fixation as a source of nitrogen to the North Atlantic Ocean

FREDRIC LIPSCHULTZ[1] & NICHOLAS J. P. OWENS[2]
[1]*Bermuda Biological Station for Research, Inc., Ferry Reach, GE01, Bermuda;* [2]*Department of Marine Sciences and Coastal Management, The Ridley Building, University of Newcastle, Newcastle-upon-Tyne, NE1 7RU, United Kingdom*

Received 23 February 1995; accepted 11 August 1995

Key words: budget, new production, nitrogen, nitrogen fixation, North Atlantic, *Trichodesmium*

Abstract. The role of nitrogen fixation in the nitrogen cycle of the North Atlantic basin was re-evaluated because recent estimates had indicated a far higher rate than previous reports. Examination of the available data on nitrogen fixation rates and abundance of *Trichodesmium*, the major nitrogen fixing organism, leads to the conclusion that rates might be as high as 1.09×10^{12} mol N yr^{-1}. Several geochemical arguments are reviewed that each require a large nitrogen source that is consistent with nitrogen fixation, but the current data, although limited, do not support a sufficiently high rate. However, recent measurements of the fixation rates per colony are higher than the historical average, suggesting that improved methodology may require a re-evaluation through further measurements. The paucity of temporally resolved data on both rates and abundance for the major areal extent of the tropical Atlantic, where aeolian inputs of iron may foster high fixation rates, represents another major gap.

Introduction

The North Atlantic is arguably the most studied region of the world's ocean, including not solely the marine component but also the adjacent coastal, estuarine and atmospheric ecosystems. In addition, the highly populated and industrialized watersheds surrounding the North Atlantic and its relatively small volume compared to other ocean basins suggests that man's impact might be discernible. As part of a basin scale evaluation of the various terms of the nitrogen budget for the North Atlantic (Galloway et al. this volume), a review of the nitrogen cycle of the oceanic component was undertaken to identify the important terms, especially imports and exports (Michaels et al. this volume).

Nitrogen fixation, the conversion of atmospheric dinitrogen to biologically usable ammonium, is one source of nitrogen to the ocean. Studies in the 1970's focused on measuring this input of nitrogen by the most obvious organism, *Trichodesmium* sp., a filamentous cyanobacteria that forms colonial associations approximately 1–1000 mm in size and can aggregate at the surface in extensive blooms (Carpenter & Capone 1992). These studies were extensively reviewed by Carpenter (1983), with the conclusion that fixation in the entire Atlantic contributed 0.94×10^{11} mol N yr^{-1} to a global rate of 3.4×10^{11} mol N yr^{-1}. On a global basis, this value is a small input of nitrogen to the photic zone compared to other sources such as convective transport and upwelling.

A recent reassessment of the magnitude of nitrogen fixation, however, concluded that nitrogen fixation rates in the Caribbean region were considerably higher than earlier reports (Carpenter & Romans 1991). Using a calculated rate of 10–50 mg N m^{-2} d^{-1}, and applying it over the area of the Caribbean and Atlantic Ocean from 0–30° N of 24×10^{12} m^2 implies a source of 6.3 to 31.3×10^{11} mol N yr^{-1}, a dramatic upward revision of the nitrogen fixation rate from 1983 to 1991. We therefore examined the literature on nitrogen fixation rates in this region to identify the cause of the increase.

Results and discussion

Methodological considerations

The high fixation rate calculated by Carpenter & Romans (1991) is largely based on data collected as part of tests of the acetylene reduction assay (Carpenter et al. 1987). Martinez et al. (1983) had noted that diver-collected mats of *Rhizosolenia* sp. had much higher rates of nitrogen fixation than mats collected by net tows. In addition, Carpenter & Price (1976) had noted a negative relationship between fixation rates by *Trichodesmium* and turbulence, again suggesting that net tows would underestimate the estimation of nitrogen fixation. However, a comparison of rates between hand and net collected colonies of *Trichodesmium* found no significant difference in several trials in the Caribbean and the Sargasso Sea (Carpenter et al. 1987). Subsequent investigations of nitrogen fixation by *Rhizosolenia* have failed to detect any activity (Villareal & Carpenter 1989).

Realization by the oceanographic community that trace metal contamination could significantly reduce measurements of primary productivity in oligotrophic waters (Fitzwater et al. 1982) had also raised the question whether measurements of nitrogen fixation were too low. *Trichodesmium* is sensitive to metals such as copper (Rueter et al. 1979) and the high requirement of the nitrogen fixation enzyme (nitrogenase) for metals such as Fe (Rueter et al. 1992) and possibly Mo (Howarth & Cole 1985) suggests that contamination could act to either enhance or inhibit fixation. Again, Carpenter et al. (1987) found no significant difference between rates using colonies collected using extremely strict precautions and less stringent, "standard" protocols.

Although both of these methodological issues revealed no problem with earlier measurements, the doubling times cited by Carpenter & Romans (1991), from all geographic locales, were considerably higher than previously published values. For example, Carpenter & Price (1977) found values of 45 days for surface colonies from the Caribbean and 105 days for the Sargasso Sea, compared to 1–2 days for the same locations in the 1991 estimate. Since the later doubling time is the basis for the recent re-assessment of the fixation rate (Carpenter & Romans 1991), it is important to understand its provenance. Calculation of the doubling time from rates requires evaluating the biomass. Carpenter et al. (1987) used a biomass of 0.986 ng N trichome^{-1} and a rate of 0.6 ng N trichome^{-1} d^{-1} to calculate a doubling time of 1.13 d (ln 2/0.6/0.986). However, although the biomass of individual colonies does vary somewhat, 2 ug N colony^{-1} or about 10 ng trichome^{-1} (based on 200 trichomes colony^{-1}) seems to be the most commonly reported value (Carpenter & Price 1977; McCarthy & Carpenter 1979; Carpenter et al. 1993). Since the low biomass estimate was utilized by Carpenter et al. (1987), calculated doubling times were tenfold faster than if the higher biomass was utilized. If the doubling times from 1987 were in fact 10–12 days, they are more in line with earlier reports of 10–150 days and suggest an average doubling time longer than the 10 days used by Carpenter & Romans (1991) would be appropriate.

Another factor contributing to the high areal fixation rate is the assumption of depth dependence of abundance and rate as well as the time period of fixation used in the calculation compared to those used in earlier calculations. The rate of nitrogen fixation diminishes rapidly with depth (see for example, Goering et al. 1966; Carpenter & McCarthy 1975; Carpenter & Price 1977), due to a decline in both colony abundance and fixation rate per colony. Carpenter (1983) accounted for this by estimating a mean rate per colony over the photic zone and then halving that rate to account for reduced fixation at the beginning and end of the daylight period (which was assumed to be 12 hrs). Since the majority of estimates for colony abundance are from surface tows, the depth distribution of colony abundance was estimated by assuming a constant abundance from the surface to 15 m and then a linear decrease to 50 m where the abundance was assumed to be 10% of that at the surface (Carpenter 1983). In contrast, the 1991 rate estimate assumed a constant depth distribution of both abundance and rate for colonies present at the surface and a 12 hour period of active fixation. The net effect of these different assumptions are summarized in Table 1 and account for a nearly 10 fold lower areal rate based on the different assumptions. The assumptions employed by Carpenter (1983) still seem realistic and we employ them for an estimate of the basin scale rate of fixation.

Table 1. Effect of various assumptions on calculated areal nitrogen fixation rates.

Parameter	Calculations based on Carpenter (1983) Rate from Carpenter & Romans (1991)	Measured rate from Carpenter et al. (1987)	Mean Photic zone rate from Carpenter (1983)	Calculation and rate from Carpenter & Romans (1991)
Surface rate (ng N colony^{-1} day^{-1})	200	120	17	200
Daylength correction[1] (ng N colony^{-1} day^{-1})	100	60	8.5	200
Mean Photic zone[2] (ng N colony^{-1} day^{-1})	68	41	8.5	200
Colony abundance[2] (Colonies m^{-2})	34,000	34,000	34,000	50,000
Areal Rate (mg N m^{-2} d^{-1})	2.3	1.4	0.3	10

[1] Assuming half the maximal rate over the entire daylight period (Carpenter 1983).
[2] Assuming a constant value from 0–15 m and a linear decrease to 10% of the surface value at 50 m (Carpenter 1983).

Vertical, horizontal and basin-scale distributions

A major requirement for a basin scale estimate of nitrogen fixation is to account for the temporal and horizontal variation in the rates and distribution of the organisms. There is considerable variability in the individual colony rates of nitrogen fixation, ranging from values of 0.4 ng N colony hr^{-1} to values in blooms of 17.4 to 50.4 ng N colony hr^{-1} (Carpenter & Capone 1992). Some of this variation is due to different colony sizes, but even on a per cell basis, the rates are quite variable for reasons not yet understood. A compilation of the extant measurements of fixation rates over the past 20 years reveals the paucity of measurements over the North Atlantic Ocean, arguably the most studied region (Figure 1). The only seasonal series of measurements of nitrogen fixation to define the temporal scale of the variability is not for *Trichodesmium* but for cyanobacteria associated with *Sargassum sp.* off Florida (Phlips et al. 1986). Choosing a representative rate for the entire North Atlantic is therefore admittedly difficult but the data summarized above suggest that the use of just the higher rates of Carpenter & Romans (1991) for the entire basin is currently unwarranted.

In addition to variability in per colony rates, the number of colonies can vary tremendously. Recasting the data summarized previously (Carpenter 1983) to emphasize stations in the North Atlantic (Figure 2) again reveals the lack of information for vast areas of the ocean, especially in view of the

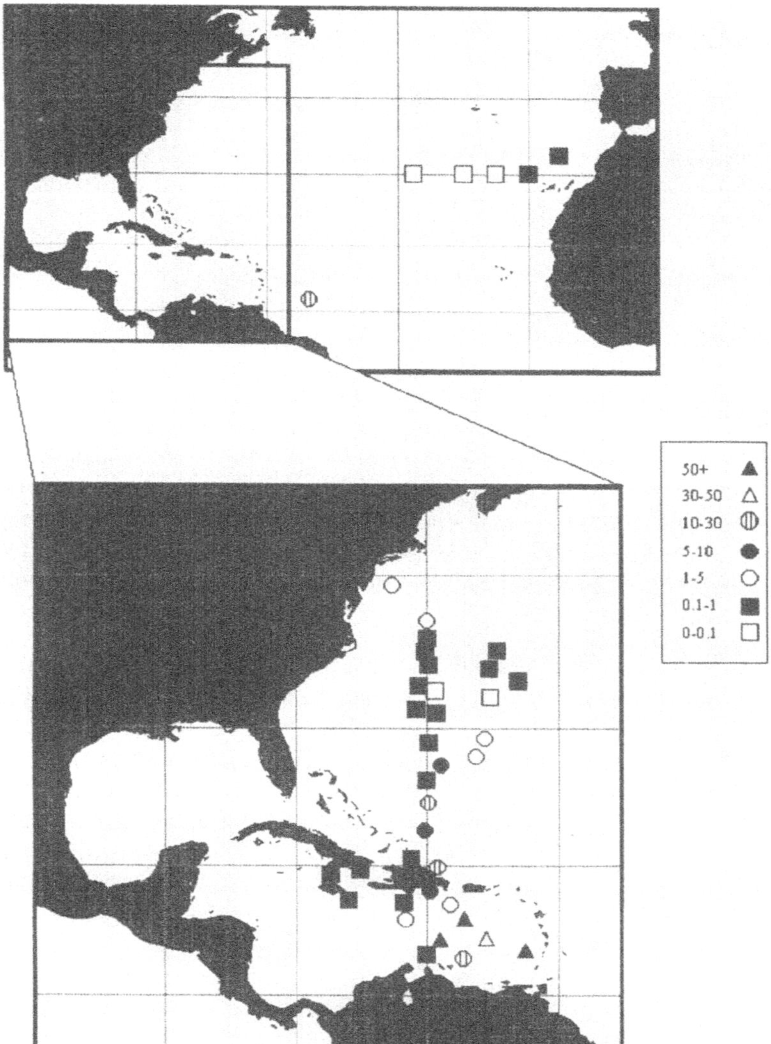

Figure 1. Areal rates of nitrogen fixation (μmol N m^{-2} d^{-1}) measured in the North Atlantic Ocean. Rates based on references cited in Carpenter (1983).

fact that each station represents a single realization of a patchy distribution. For example, there is almost no information in the region from 0° to 30° N and from 20° W to the Caribbean. A basin scale estimate currently requires extrapolation of both rates and abundances from the Caribbean over the entire area.

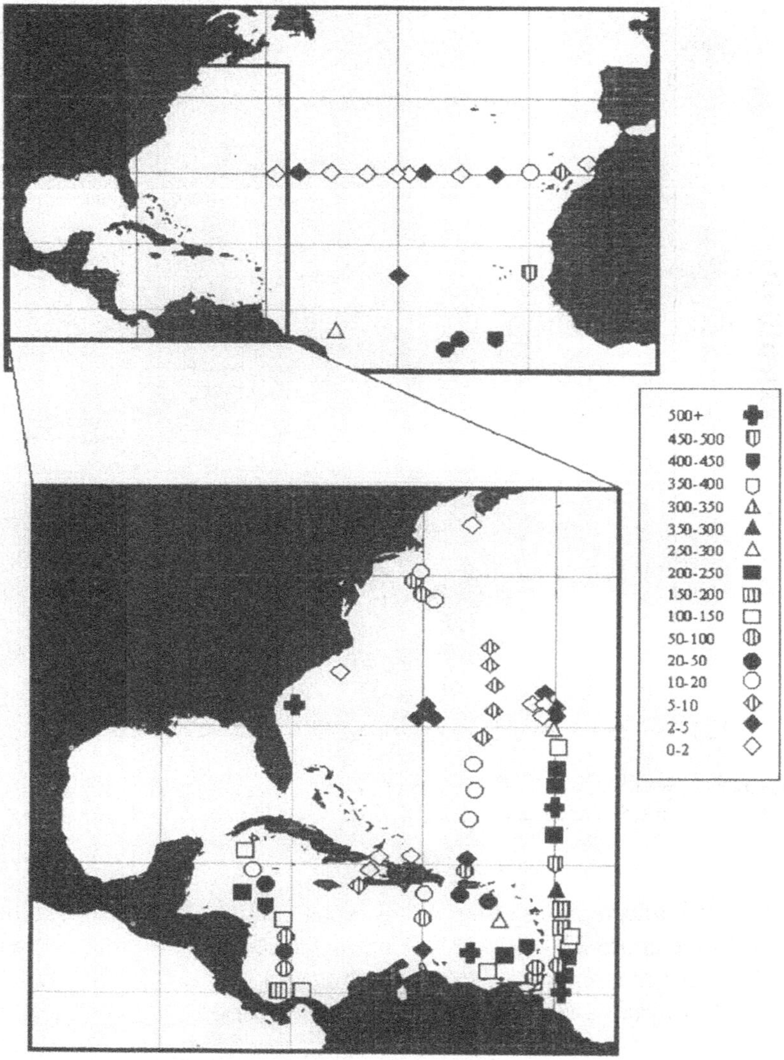

Figure 2. Surface distribution of *Trichodesmium* trichomes (trichomes $m^{-3} \times 10^{-3}$) in the North Atlantic Ocean. Data based on references cited in Carpenter (1983) with the addition of Carpenter & Romans (1991).

In addition to re-examining specific studies on nitrogen fixation, we examined the extensive Continuous Plankton Recorder (CPR) survey data for the northern region of the basin (Edinburgh, Oceanic Laboratory 1967). The CPR survey is a plankton distribution study using instruments towed at a

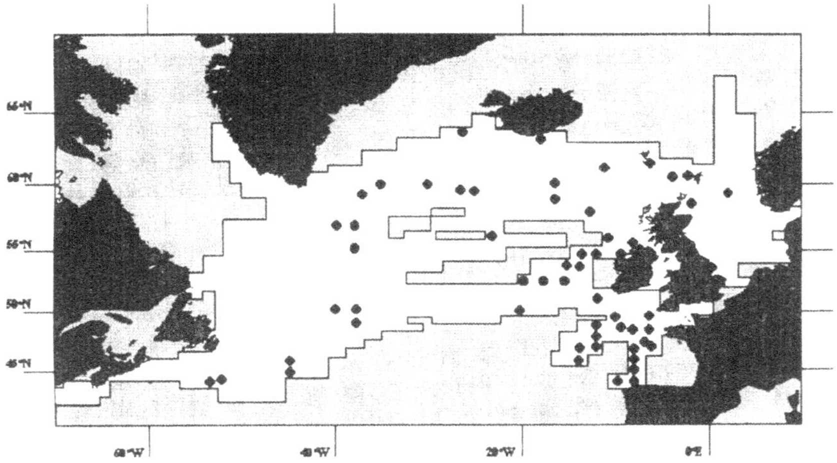

Figure 3. Occurrence of *Trichodesmium* spp in the northern North Atlantic. Dots represent presence in 1° latitude 2° longitude squares pooled for all months for the years 1958–1968 inclusive. Hatched oceanic areas represent no data. Data source: Continuous Plankton Survey, UK (The Oceanic Laboratory, Edinburgh, 1973).

fixed depth of 10 m, and carried out by ships of opportunity. Although the extent of the survey has varied since its introduction in the 1930's, there has been reasonable coverage over the northern North Atlantic basin. Figure 3 summarizes the locations where *Trichodesmium* has been recorded. Given the general view that *Trichodesmium* is a tropical or sub-tropical species, it is notable that the genus has been found in many parts of the central and eastern regions of the basin, clearly indicating that it is not confined to the lower latitudes. However, the general pattern of the distribution suggests that the *Trichodesmium* may be passively transported by the Gulf Stream/North Atlantic Drift, rather than growing *in situ*. Unfortunately, because of its nature the CPR survey data merely indicate presence or absence, thus no abundance information can be inferred. Even assuming the (unlikely) case of similar cell densities as in the tropics in all the regions indicated by the CPR data, the question remains whether the individual colonies were active and capable of nitrogen fixation. Carpenter (1983) concluded that 20 °C was the minimum temperature for fixation activity by *Trichodesmium* and thus, given the low average temperatures of the northern North Atlantic basin, we assume that rates of fixation would be relatively low. A low rate combined with the small areal extent of the northern region compared to the southern region, leads us to conclude that nitrogen fixation in this area of the basin makes a small contribution to the nitrogen cycle. Nevertheless, the widespread distribution and potential activity of the genus warrants more attention than hitherto displayed.

Given the paucity of data on the distribution and activity of *Trichodesmium* over much of the Atlantic basin, is it possible that under-sampling could result in a major underestimate of its importance, and thus account, in part at least, for the apparent imbalance in the nitrogen budget? As pointed out by others, the presence of relatively rare blooms can dramatically increase the input of nitrogen to the surface ocean, yet our ability to quantify their occurrence is limited to chance encounters on scattered cruises (Carpenter & Capone 1983; Carpenter & Capone 1992; Karl et al. 1992). One of the major gaps in our understanding of *Trichodesmium* physiology is knowledge of the requirements for bloom formation.

Formation of a bloom *de novo* rather than merely a surface aggregation of a vertically dispersed population would seem to require a faster growth rate than the average values measured for nitrogen. The highest turnover rates of about 10 d for nitrogen (Carpenter et al. 1987) are still slow compared to what one might expect to overcome physical dispersion. However turnover times of 1–8 d for carbon (Lewis et al. 1988; Carpenter et al. 1993; Carpenter & Roenneberg 1995) and for cellular chlorophyll (Prufert-Bebout et al. 1993) are certainly sufficient for bloom formation but are inconsistent with balanced growth of the cells.

There are few contemporaneous measurements of carbon and nitrogen turnover times. Carpenter & Price (1977) measured both rates for Caribbean populations, from which turnover times of 44.8 d for nitrogen and 14.5 - 29 d for carbon can be calculated. More recently, Carpenter et al. (1993) reported carbon doubling times of 3–3.8 d. For comparison, nitrogen doubling times of 17–33 d can be calculated based on acetylene reduction rates of 0.5 nmol ethylene colony^{-1} hr^{-1} for *T. erythraeum* and 1.0 nmol colony^{-1} hr^{-1} for *T. thiebautii* and measured N content per colony of 2.1–2.4 ug. Applying a 6.3:1 conversion factor from acetylene to dinitrogen as measured by Carpenter & Price (1977), rather than a 3:1 ratio, would proportionally increase the nitrogen doubling time and exacerbate the difference. Use of alternate nitrogen sources such as inorganic nitrogen would provide nitrogen to support high carbon-based growth rates but assimilation of both NH_4^+ and NO_3^- has been shown to be negligible compared to fixation (Carpenter & McCarthy 1975). Resolving the nutrient demands of balanced growth clearly requires more contemporaneous measurements both nitrogen and carbon utilization but should also include use of phosphorus (see below).

Other nitrogen fixing organisms

Traditionally, *Trichodesmium* has been considered the primary nitrogen fixing organism in the pelagic ocean. There are other organisms that can also contribute but our knowledge of their impact is limited. For instance, *Hemiaulus* sp., a marine diatom, frequently contains a cyanobacterial symbiont that fixes nitrogen (Villareal 1991). Villareal (1991) estimated that *Hemiaulus*, assuming a cell density of 100 l^{-1} ,would contribute 4.8 ug-at N m^2 d^{-1} but noted that densities of 10^5 cells l^{-1} have been reported off Florida. Although not all cells contained symbionts, the majority (91–100%) in the Caribbean did (Villareal 1994), suggesting that these organisms could also be important if blooms were frequent. Unfortunately, our knowledge of their distribution or activity is extremely limited so that a basin scale assessment of the contribution of *Hemiaulus* sp. to the nitrogen cycle of the North Atlantic is premature.

Basin scale speculation

The motivation for these deliberations sprang from a desire to independently evaluate the value of 6–31 $\times$ 10^{11} mol N yr^{-1} as compared to the earlier value of 1 $\times$ 10^{11} mol N yr^{-1} for nitrogen fixation in the North Atlantic. Using an historically high rate (rather than a turnover time) of 0.060 ng N $trichome^{-1}$ hr^{-1} (Carpenter et al. 1987) for the entire region and trichome densities from Carpenter (1983), we roughly calculate an input of about 10.7 $\times$ 10^{11} mol N yr^{-1} to the region (Table 2). Based on the available evidence summarized above (other than the additional data in 1987, nearly as limited today as in 1983) we are forced to conclude that Carpenter & Romans (1991) somewhat overstated the role of nitrogen fixation in the North Atlantic.

Interestingly, several lines of geochemical evidence imply an even higher fixation rate than we calculate can be supported by available evidence. There is a puzzling mismatch in the central Sargasso Sea between the depth of appearance of nitrate (nitracline) and the deeper depth of detectable levels of phosphate (Michaels et al. 1994b). Combining the residence time of different depth layers in the volume from 15° N to 45° N with the amount of "excess" nitrate compared to phosphate leads to a calculated production rate of 34 to 61 $\times$ 10^{11} mol N yr^{-1} required to sustain the anomaly (Michaels et al. this volume) and they argue that nitrogen fixation is required to sustain the observed NO_3^-:$PO_4^=$ anomaly. A second puzzle is the net depletion of dissolved inorganic carbon from the surface waters near Bermuda during a period when the water column is stratified and fully oligotrophic (Michaels et al. 1994a). The amount of new nitrogen required to balance the observed net community fixation of carbon is comparable to that required to sustain the NO_3^-:$PO_4^=$ anomaly and again, high rates of nitrogen fixation would resolve the puzzle.

Other potential sources of new nitrogen such as atmospheric inputs of ammonium and nitrate in wet or dry deposition have been evaluated for the waters around Bermuda and found to supply small amounts of nitrogen compared to internal processes such as convective transport of nitrate from

Table 2. Basin scale assessment of the nitrogen fixation rate.

Region	Fixation rate[1] (pmoles N col^{-1} hr^{-1})	Days[2] (d yr^{-1})	Depth interval (m)	Colony density (colonies m^{-3})	Area (10^{12} m^2)	Area rate[3] (10^{12} mol N yr^{-1})
Southern (0–30° N)	800	365	50	1000	18	1.05
mid-Atlantic (31–50° N)	400	250	50	200	14	0.036
Northern (>50° N)	80	100	25	20	6	7×10^{-5}

[1] The southern rate is based on the Carpenter at al. (1987) rate of 6 pmole C_2H_4 $trichome^{-1}$ h^{-1}, 200 trichomes $colony^{-1}$ and a 3:1 ratio. The mid-Atlantic and northern rates are estimates.

[2] The number of days of active fixation are diminished northward to account for the winter period of high turbulence and reduced light intensity.

[3] Includes 6 hrs of daylight to account for diminished fixation at the beginning and end of the daylight period (Carpenter 1983), a factor of 0.7 to reduce the number density of surface colonies with depth (Carpenter 1983) and another factor of 0.7 to replicate the depth dependence of fixation activity

deeper water to the euphotic zone (Knap et al. 1986; Michaels et al. 1993). A basin scale assessment of atmospheric inputs from North America and Europe reached a similar conclusion for the pelagic regime, although in certain coastal waters an anthropogenic influence can be documented (Prospero et al. this volume).

These geochemical arguments for high fixation rates in the North Atlantic also imply a difference from other basins where similar evidence is not present. Nitrogen fixation rates in the tropical Atlantic might actually be stimulated by large inputs of Fe from atmospheric dust (Prospero et al. this volume). Limited numbers of modern evaluations of Fe concentrations in the Atlantic find higher concentrations (1 nM Fe) in the Northern Atlantic (T. Church, pers. comm.), which receives aeolian inputs from Africa, than in the Pacific (0.01 nM Fe) where dust storms from Asia are infrequent (Martin et al. 1991). Release from iron limitation might therefore foster higher fixation in parts of the Atlantic (0°–30° N) where (conveniently for such speculation) there are no measurements! This difference might also explain the lack of a $NO_3^-:PO_4^=$ anomaly in the Pacific Ocean.

In addition to iron, phosphorus is also required for balanced growth. Phosphorus is in limited supply in the oligotrophic waters of the Atlantic, raising the interesting question of the source of phosphorus to balance nitrogen fixation. Carpenter & Price (1977) suggested that increased phosphorus concentrations as oceanic water entered the Caribbean fostered higher numbers and greater rates of fixation by *Trichodesmium*. Phosphorus limitation might therefore reduce fixation rates over the tropical Atlantic region compared to extrapolation from the Caribbean rates. However, abundances along a transect from the Caribbean to Bermuda can be quite constant (Carpenter & Romans 1991) and fixation rates between the Caribbean population and Sargasso Sea population varied only from 92 to 120 ng N colony day^{-1} (Carpenter et al. 1987) and hence do not lend support to such a limitation.

Karl (1992) suggested that the ability of *Trichodesmium* to vertically migrate (Villareal & Carpenter 1990) permitted acquisition and storage of required phosphorus at the nutricline to support carbon and nitrogen fixation at the surface. Such a mechanism would help explain the anomalous distribution of nitrate and phosphorus in the Sargasso Sea (Michaels et al. this volume) by providing a mechanism to remove phosphorus and leave nitrate below the euphotic zone. In addition to migration, Trichodesmium also has the capacity for high rates of alkaline phosphatase activity (McCarthy and Carpenter 1979; Elardo et al. 1994) which would permit utilization of organic phosphorus in the surface waters and alleviate the phosphorus constraint but would not provide a sink in the nitracline. In this scenario, remineralization of high N:P (125) *Trichodesmium* sp. (Karl et al. 1992) at depth would provide "excess" nitrate compared to phosphate. Either mechanism has major implications for the phosphorus cycle in the same way that high rates of nitrogen fixation have for the nitrogen cycle and ultimately the carbon cycle of the North Atlantic.

Conclusions

The primary conclusion from this investigation is that although the most recent basin estimate of nitrogen fixation (Carpenter & Romans 1991) is of the order required by a basin scale budget of nitrogen, such high rates cannot be supported on current evidence. However, both our assessment and the 1991 estimate are based on very limited information. The new basin-scale nitrogen budget (Michaels et al. this volume) dramatically illustrates the inadequacy of our knowledge of ocean scale nitrogen fixation and clearly points to the need for a major re-assessment. The consequences for biogeochemical cycling of a nitrogen fixation rate of the suggested magnitude are profound, and would lead to a major re-adjustment of our concepts of global biogeochemical cycles.

To provide a better constraint on the fixation rate over the basin, future research directions must begin to address not only spatial and temporal variability, but also the physiology of the organisms. For instance Karl et al. (1995) attribute major ecosystem changes to changes in abundance of *Trichodesmium* trichomes rather than colonies. The JGOFS time series station at Bermuda and the newly established German-Spanish station in the Canary Islands represent two important platforms from which to launch studies of the temporal variation and more importantly, the mechanisms behind the variability such as the causal factor(s) permitting bloom formation. Spatial variation would also be partially addressed by such a fundamental understanding but some form of remote sensing would seem to be required to deal with the scales of variability (Borstad et al. 1992; Subramaniam & Carpenter 1994).

Acknowledgements

We wish to acknowledge and express our appreciation for the funding of this project from the UNEP and the Mellon Foundation to support the efforts of the SCOPE Nitrogen Project, and from the World Meteorological Organization to investigate the interaction between the N cycles of the atmosphere and the NAO&W. We wish to acknowledge Chris Young who drew the figures.

References

Borstad GA, Carpenter EJ & Gower JF (1992) Development of algorithms for remote sensing of *Trichodesmium* blooms. In: Carpenter EJ, Capone DG & Rueter JG (Eds) Marine Pelagic

Cyanobacteria: *Trichodesmium* and Other Diazotrophs (pp 193–210). Kluwer Academic Publishers, Dordrecht

Carpenter EJ (1983) Nitrogen fixation by marine *Oscillatoria* (*Trichodesmium*) in the world's oceans. In: Carpenter EJ, Capone DG (Eds) Nitrogen in the Marine Environment (pp 65–103). Academic Press, Inc., New York

Carpenter EJ & Capone DG (1983) Nitrogen in the Marine Environment. Academic Press, Inc., New York

Carpenter EJ & Capone DG (1992) Nitrogen fixation in *Trichodesmium* blooms. In: Carpenter EJ, Capone DG & Rueter J (Eds) Marine Pelagic Cyanobacteria: *Trichodesmium* and Other Diazotrophs (pp 211–217). Kluwer Academic Publishers, Dordrecht

Carpenter EJ & McCarthy JJ (1975) Nitrogen fixation and uptake of combined nitrogenous nutrients by *Oscillatoria (Trichodesmium) thiebautii* in the western Sargasso Sea. Limnol. Oceanogr. 20: 389–401

Carpenter EJ, O'Neil J, Dawson R, Capone DG, Siddiqui JA, Roenneberg T & Bergmann B (1993) The tropical diazotrophic phytoplankton *Trichodesmium*: Biological characteristics of two common species. Mar. Ecol. Prog. Ser. 95: 295–304

Carpenter EJ & Price CC (1976) Marine *Oscillatoria* (*Trichodesmium*): Explanation for aerobic nitrogen fixation without heterocysts. Sci. 191: 1278–1281

Carpenter EJ & Price CC (1977) Nitrogen fixation, distribution, and production of *Oscillatoria* (*Trichodesmium*) spp. in the western Sargasso and Caribbean Seas. Limnol. Oceanogr. 22: 60–72

Carpenter EJ & Roenneberg T (1995) The marine planktonic cyanobacteria *Trichodesmium* spp.: Photosynthetic rate measurements in the SW Atlantic Ocean. Mar. Ecol. Prog. Ser. 118, 267–273

Carpenter EJ & Romans K (1991) Major role of the cyanobacteria *Trichodesmium* in nutrient cycling in the North Atlantic Ocean. Sci. 254: 1356–1358

Carpenter EJ, Scranton MI, Novelli PC & Michaels A (1987) Validity of N_2 fixation rate measurements in marine *Oscillatoria (Trichodesmium)*. J. Plankton Res. 9: 1047–1056

Edinburgh, Oceanic Laboratory (1973) Continuous plankton records: A plankton atlas of the North Atlantic and the North Sea. Bull. Mar. Ecol. 7:1–174

Elardo K, Ammerman JW & Gundersen K (1994) Exoenzyme activity associated with *Trichodesmium* colonies. EOS. 75: 111

Fitzwater SW, Knauer GA & Martin JH (1982) Metal contamination and its effect on primary production measurements. Limnol. Oceanogr. 27: 544–551

Goering JJ, Dugdale RC & Menzel DW (1966) Estimates of in situ rates of nitrogen uptake by *Trichodesmium* sp. in the tropical Atlantic Ocean. Limnol. Oceanogr. 11: 614–620

Howarth RW & Cole JJ (1985) Molybdenum availability, nutrient limitation, and phytoplankton growth in natural waters. Science 229: 653–655

Karl DM, Letelier R, Hebel DV, Bird DF & Winn CD (1992) *Trichodesmium* blooms and new nitrogen in the North Pacific gyre. In: Carpenter EJ, Capone DG & Rueter JG (Eds) Marine Pelagic Cyanobacteria: *Trichodesmium* and Other Diazotrophs. Kluwer Academic Publishers, Dordrecht

Karl DM, Letelier R, Hebel D, Tupas L, Dore J, Christian J & Winn C (1995) Ecosystem changes in the North Pacific subtropical gyre attributed to the 1991–1992 El Nino. Nature 373: 230–234

Knap A, Jickells T, Pszenny A & Galloway J (1986) Significance of atmospheric-derived fixed nitrogen on productivity of the Sargasso Sea. Nature 320: 158–160

Lewis MR, Ulloa O & Platt T (1988) Photosynthetic action, absorption, and quantum yield spectra for a natural population of *Oscillatoria* in the North Atlantic. Limnol. Oceanogr. 33: 92–98

Martin JH, Gordon RM & Fitzwater SE (1991) The case for iron. Limnol. Oceanogr. 6: 1793–1802

Martinez L, Silver MW, King JM & Alldredge AL (1983) Nitrogen fixation by floating diatom mats: A source of new nitrogen to oligotrophic ocean waters. Sci. 221: 152–154

McCarthy JJ & Carpenter EJ (1979) *Oscillatoria (Trichodesmium) thiebautii* (Cyanophyta) in the central North Atlantic Ocean. J. Phycol. 15: 75–82

Michaels AF, Bates NR, Buesseler KO, Carlson CA & Knap AH (1994a). Carbon system imbalances in the Sargasso Sea. Nature 372: 505–508

Michaels AF, Knap AH, Dow RL, Gundersen K, Johnson RJ, Sorensen J, Close A, Knauer GA, Lohrenz SE, Asper VA, Tuel M & Bidigare RR (1994b) Seasonal patterns of biogeochemistry at the US JGOFS Bermuda Atlantic Time-Series Study site. Deep-Sea Res. 41: 1013–1038

Michaels AF, Siegel DA, Johnson RJ, Knap AH & Galloway JN (1993) Episodic inputs of atmospheric nitrogen to the Sargasso Sea: Contributions to new production and phytoplankton blooms. Global Biogeochem. Cycles 7: 339–352

Phlips EJ, Willis M & Verchick A (1986) Aspects of nitrogen fixation in *Sargassum* communities off the coast of Florida. J. Exp. Mar. Biol. Ecol. 102: 99–119

Prufert-Bebout L, Paerl HW & Lassen C (1993) Growth, nitrogen fixation, and spectral attentuation in cultivated *Trichodesmium* species. Appl. Environ. Microbiol. 59: 1367–1375

Rueter JG, McCarthy JJ & Carpenter EJ (1979) The toxic effect of copper on *Oscillatoria (Trichodesmium) theibautii*. Limnol. Oceanogr. 24: 558–562

Rueter JG, Hutchins DA, Smith RW & Unsworth NL (1992) Iron nutrition of *Trichodesmium*. In: Carpenter EJ, Capone DG & Rueter J (Eds) Marine Pelagic Cyanobacteria: *Trichodesmium* and Other Diazotrophs (pp 289–306). Kluwer, Netherlands

Subramaniam A & Carpenter EJ (1994) An empirically derived protocol for the detection of blooms of the marine cyanobacterium, *Trichodesmium*, using CZCS imagery. Intl. J. Remote Sensing 15: 1559–1569

Villareal TA (1991) Nitrogen-fixation by the cyanobacterial symbiont of the diatom genus *Hemiaulus*. Mar. Ecol. Prog. Ser. 76: 201–204

Villareal TA (1994) Widespread occurrence of the *Hemialus*-cyanobacterial symbiosis in the southwest North Atlantic ocean. Bull. Mar. Sci. 54: 1–7

Villareal TA & Carpenter EJ (1989) Nitrogen fixation, suspension characteristics, and chemical composition of *Rhizosolenia* mats in the Central North Pacific gyre. Biol. Oceanogr. 6: 327–345

Villareal TA & Carpenter EJ (1990) Diel buoyancy regulation in the marine diazotrophic cyanobacterium *Trichodesmium thiebautii*. Limnol. Oceanogr. 35: 1832–1837

Biogeochemistry **35:** 275–304, 1996.

Estimation of the air/sea exchange of ammonia for the North Atlantic Basin

P.K. QUINN[1], K.J. BARRETT[2], F.J. DENTENER[3], F. LIPSCHULTZ[4] & K.D. SIX[5]
[1]*NOAA Pacific Marine Environmental Laboratory, Seattle, WA, USA;* [2]*Norwegian Meteorological Institute, Oslo, Norway;* [3]*Wageningen Agricultural University, Wageningen, Netherlands;* [4]*Bermuda Biological Station for Research, Bermuda;* [5]*Max-Planck-Institute for Meteorology, Hamburg, Germany*

Received 19 December 1994; accepted 6 March 1996

Abstract. As gas phase atmospheric ammonia reacts with acidic aerosol particles it affects the chemical, physical, and optical properties of the particles. A knowledge of the source strengths of NH_3 is useful in determining the effect of NH_3 on aerosol properties on a regional basis. Here, an attempt is made to determine the direction and magnitude of the air/sea flux of ammonia for the North Atlantic Basin from both measured and modeled seawater and atmospheric ammonia concentrations. Previously reported measured seawater concentrations range from less than 30 to 4600 nM with the highest concentrations reported for the Caribbean Sea, the North Sea, and the Belgium coast. Measured atmospheric ammonia concentrations range from 2 to 500 nmol m^{-3} with the largest values occurring over the Sargasso Sea, the Caribbean Sea, and the North Sea. For comparison to the measurements, seawater ammonia concentrations were calculated by the Hamburg Model of the Ocean Carbon Cycle (HAMOCC3). HAMOCC3 open ocean values agree well with the limited number of reported measured concentrations. Calculated coastal values are lower than those measured, however, due to the coarse resolution of the model. Atmospheric ammonia concentrations were calculated by the Acid Deposition Model of the Meteorological Synthesizing Center (MSC-W) and by the global 3-dimensional model Moguntia. The two models predict similar annually averaged values but are about an order of magnitude lower than the measured concentrations. Over the North Sea and the NE Atlantic, the direction and magnitude of the air/sea ammonia flux calculated from MSC-W and Moguntia agree within the uncertainty of the calculations. Flux estimates derived from measured data are larger in both the positive and negative direction than the model derived values. The discrepancies between the measured and modeled concentrations and fluxes may be a result of sampling artifacts, inadequate chemistry and transport schemes in the models, or the difficulty in comparing point measurements to time-averaged model values. Sensitivity tests were performed which indicate that, over the range of values expected for the North Atlantic, the accuracy of the calculated flux depends strongly on seawater and atmospheric ammonia concentrations. Clearly, simultaneous and accurate measurements of seawater and atmospheric ammonia concentrations are needed to reduce the uncertainty of the flux calculations, validate the model results, and characterize the role of oceanic ammonia emissions in aerosol processing and nitrogen cycling for the North Atlantic.

1. Introduction

In the marine atmosphere, gas phase acidic species that can act as aerosol particle precursors include H_2SO_4, HNO_3 and organic acids. Because gas phase ammonia, NH_3 (g), is the dominant basic species available to interact with these acidic compounds, it has a unique role in determining the chemical, physical, and optical properties of both marine and continental aerosol. It can have a significant impact on the formation of new particles by lowering the vapor concentration of other gases required for the initiation of nucleation. It also can result in particle growth by reacting with acidic particles and lowering the surface acid vapor pressure thereby enhancing the condensation of gas phase acidic species. In addition, the degree of neutralization of non-seasalt sulfate aerosol by NH_3 (g) will influence the aerosol hygroscopicity and, hence, the growth response of particles to changes in atmospheric relative humidity. As light scattering by aerosol particles is determined, in part, by the size dependent aerosol mass concentration, NH_3 (g) will influence the scattering efficiency of the aerosol by adding particulate mass and by changing its hygroscopic properties.

A knowledge of the sources of NH_3 (g) as well as their magnitude is needed to determine the spatial and temporal variability of the effect of NH_3 (g) on aerosol chemical, physical, and optical properties. Much work has been reported on quantifying the sources of NH_3 (g) in continental regions (e.g., Buijsman et al. 1987; Asman et al. 1988). Comparatively little research has been done, however, in characterizing the direction and magnitude of the ammonia flux in oceanic regions. Hence, the importance of the oceanic source of NH_3 on global and regional scales is unknown as is the impact of continental sources on oceanic regions. The goal of the present work is to determine the direction and magnitude of the air/sea ammonia flux for different regions of the North Atlantic and for continental regions bordering the Atlantic. The deposition of ammonium to the sea surface and its importance in ocean processes is described in a separate article (Prospero et al., this issue).

NH_3 (g) reacts primarily with submicron non-seasalt sulfate aerosol, nss $SO_4^=$ (p), to form ammoniated sulfate salts ranging in composition from NH_4HSO_4 to $(NH_4)_2SO_4$ (Quinn et al. 1987). In remote ocean regions, the major source of atmospheric NH_3 will be the ocean itself (Quinn et al. 1988, 1990). In coastal regions, NH_3 (g) also may be derived from continental sources such as biomass burning, animal wastes, emissions from soil and vegetation, and fertilizer losses (Dentener & Crutzen 1994; Warneck 1988). Both the ocean- and the continentally-derived NH_3 (g) will react with unneutralized sulfate aerosol. The resulting particulate phase ammonium, NH_4^+ (p), may be transported long distances (100's to 1000's of miles depending on prevailing wind speeds and precipitation frequency) before it is removed from the atmosphere through wet and dry deposition.

Ammonia exists in seawater as both ionized ammonium, NH_4^+ (s), and dissolved ammonia, NH_3 (s). NH_3 (s) contributes about 10% to the total seawater ammonium concentration, NH_x (s), at a pH of 8.2 and temperature of 25 °C. NH_x (s) is produced in the upper ocean from the degradation of organic nitrogen containing compounds and excretion from zooplankton. It also is released from bottom sediments to overlying waters. Loss processes for NH_x (s) include bacterial nitrification, uptake by phytoplankton and bacteria, and emission across the air/sea interface. It is the unionized dissolved form, NH_3 (s), that is exchanged across the ocean surface and serves as a source of atmospheric ammonia. At steady state, an equilibrium is reached between air and seawater concentrations of ammonia so that both NH_3 (s) and NH_3 (g) determine the direction and magnitude of the air/sea flux.

Attempts to estimate the air/sea flux of ammonia for the North Atlantic Basin are hindered by a lack of simultaneous measurements of oceanic and atmospheric concentrations. This lack of measurements can be attributed to the difficulty in determining very low concentrations of NH_3 (g) using currently available techniques and in avoiding sample contamination (Williams et al. 1992). Similarly, NH_x (s) concentrations frequently are below the detection limits (50 to 100 nM) of conventional techniques (D'Elia 1983). The result is that little is known about the magnitude and direction of the air/sea ammonia flux. Only one study from the North Atlantic Basin which took place in the North Sea reports results of simultaneously determined NH_3 (g) and NH_x (s) concentrations (Asman et al. 1994). From these measurements, it was determined that there is a variable net flux of ammonia ranging from -75 to 43 μmol m^{-2} d^{-1} where negative values indicate a flux from the atmosphere to the ocean. The direction of the flux appears to depend on the highly variable atmospheric NH_3 (g) concentrations. This variability is, in turn, a result of the proximity of the measurements to continental source regions.

Simultaneous measurements of NH_3 (g) and NH_3 (s) from the central and NE Pacific Ocean indicate that there is a net flux of ammonia from the ocean to the atmosphere ranging between 1.8 and 16 μmol m^{-2} d^{-1} (Quinn et al. 1988, 1990). Based on observations of an increasing NH_4^+ (p) to nss $SO_4^=$ (p) molar ratio, Clarke & Porter (1993) have inferred a similar efflux of ammonia from the ocean to the atmosphere of about 10 μmol m^{-2} d^{-1} over the equatorial Pacific. These few measurements from the North Sea and the Pacific Ocean indicate that the ocean may serve as a local source of NH_3 in remote regions but that there will be a net deposition to the surface in regions impacted by continental sources.

In this paper, we estimate regional air/sea fluxes of ammonia from reported measurements of NH_3 (g) and NH_x (s). We then compare these measurement-derived fluxes with those estimated from two sets of model calculations. The two models used are the Acid Deposition Model of the Meteorological

Table 1. Notation used for ammonia species in the marine environment.

Species	Notation	Concentration notation
Atmospheric gas phase NH_3	NH_3 (g)	$(NH_3)_g$, nmol m^{-3}
Atmospheric particulate phase NH_4^+	NH_4^+ (p)	$(NH_4^+)_p$, nmol m^{-3}
Seawater total NH_3 and NH_4^+	NH_x (s)	$[NH_x]_s$, nM
Seawater dissolved, unionized NH_3	NH_3 (s)	$[NH_3]_s$, nM
NH_3 in equilibrium with dissolved NH_3 in the sea surface	NH_3 (s,eq)	$(NH_3)_{s,eq}$, nmol m^{-3}

Synthesizing Center-West of EMEP (MSC-W) (Tuovinen et al. 1994) and the global atmospheric transport/chemistry model Moguntia (Dentener & Crutzen 1994; and references therein) together with the Hamburg Model of the Ocean Carbon Cycle (HAMOCC3) (Kurz 1993). The sensitivity of the flux calculation to changes in surface seawater temperature, pH, salinity, and NH_x (s) and NH_3 (g) concentrations also is determined.

2. Model descriptions

2a. *Moguntia and HAMOCC3 models*

Moguntia, a global transport model of the troposphere, has a horizontal resolution of 10 by 10 degrees and 10 vertical layers which are spaced every 100 hPa. It uses monthly averaged winds and an eddy-diffusion parameterization based on the standard deviations of the winds together with a parameterization for deep cumulus convection. A complete description of the use of Moguntia to simulate the global atmospheric NH_3 cycle is given by Dentener & Crutzen (1994).

In the model, NH_3 (g) is removed from the atmosphere through wet and dry deposition and reaction with acidic sulfate aerosols and cloud droplets. Moguntia-calculated values of ammonium deposition in the North Atlantic basin are presented by Prospero et al. (this issue). NH_3 (g) is allowed to react with sulfuric acid aerosol until an NH_4^+ (p) to non-seasalt $SO_4^=$ (p) molar ratio of 1.5 is reached. This neutralization limit is motivated by the possibility that the equilibrium vapor pressure over nearly-neutralized sulfate aerosols may be substantial. The formation of particulate phase NH_4NO_3 occurs primarily in continental regions where there are high concentrations of HNO_3 (g) and NH_3 (g). Due to the large spatial scales of Moguntia, the formation of NH_4NO_3 is not included in this work. Omission of NH_4NO_3 may add uncertainty to the calculated flux of NH_3 to the ocean surface. This uncertainty is assumed to be negligible, however, as calculations with the more detailed MSC-W of EMEP model (see Section 2b) indicate that only a small fraction of the aerosol exists as NH_4NO_3. The air/sea flux of NH_3 is calculated using NH_x (s) concentrations estimated by HAMOCC3 and the method described in Section 5a. In the previous work of Dentener & Crutzen (1994), NH_3 fluxes were scaled to those of DMS. In this work, however, calculated oceanic NH_3 concentrations and piston velocities are used.

The global ocean circulation model HAMOCC3 has a horizontal resolution of 3.5 by 3.5 degrees and 15 vertical layers with a finer resolution in the upper 350 m (Maier-Reimer et al. 1993). The model operates on a monthly time step. It originally was designed to simulate the global marine carbon cycle and has been modified to include a simplified plankton model (Kurz 1993). Processes taken into account include the uptake of NH_x (s) by phytoplankton, regeneration of NH_x (s) by grazers, and bacterial utilization of NH_x (s). Monthly averaged exchange velocities are calculated from 6 hour ECMWF windfields following the method of Duce et al. (1991). A complete description of the use of HAMOCC3 to model the air/sea exchange of NH_3 is given by Kurz & Dentener (1994).

Both global models, Moguntia and HAMOCC3, are not designed to resolve processes within the subgrid coastal regions. Therefore, comparison with measurements is most appropriate in and above the open ocean.

2b. *MSC-W of EMEP model*

The MSC-W model provides estimates of NH_3 (g) and NH_4^+ (p) concentrations and depositions arising from emissions in each European country as well as fluxes between each country and to the sea (Tuovinen et al. 1994). The region covered by the model includes the northeast Atlantic, the European Arctic Ocean, and the North, Norwegian, Baltic, Black, and Mediterranean Seas. The model has a single layer Lagrangian formulation representing transport in the European boundary layer. It has a resolution of 150 km at 60° N and calculates mass-balance changes in air parcels traveling on 4-day trajectories to defined receptor points. The model uses actual meteorology which is updated every 6 hours. Wet and dry deposition rates are determined at this frequency. With a resolution of 150 km, the model can not resolve processes in the coastal transition zone but can describe regional sea sub-basins as well as wider oceanic regions.

Nitrogen and sulfur chemistry routines are combined such that ammonia exists as NH_3 (g) and NH_4^+ (p) associated with nitrate or sulfate aerosol. Three NH_3 molecules combine irreversibly with two of $SO_4^=$ to form $(NH_4)_{1.5}SO_4$ (assuming that $(NH_4)_2SO_4$ and NH_4HSO_4 occur equally) until either no ammonia or no sulfate remains. Remaining ammonia then enters into equilibrium with nitric acid. The nitrate equilibrium point is defined according to K = $NH_{3(eq)}HNO_{3(eq)}$ with (eq) indicating equilibrium concentrations. If the product of non-equilibrium concentrations is larger than K, NH_4NO_3 is formed. If the product is smaller, NH_4NO_3 dissociates to form its precursor gases. In remote oceanic regions with abundant sulfur present, NH_4NO_3 may be of little consequence. Where NH_3 (g) emissions exceed the sulfur capacity of the aerosol, as with the southern North Sea, 20% of NH_4^+ (p) may exist as NH_4NO_3. This compares to 4.5% for the entire model domain for this time period.

The model has been modified for this study to estimate the upper limit of the sea-to-land flux of ammonia for European sea areas that border the North Atlantic Ocean by incorporating emissions from the ocean. An air/sea equilibrium between NH_x (s) and NH_3 (g) has been added to the model and is described by Henry's law as given in Section 5a.

3. Measured and modeled seawater NH_x (s) concentrations

3a. *Measured values*

A representative selection of seawater NH_x (s) concentrations is listed in Table 2. The colorimetric Berthelot method was used to determine the NH_x (s) concentrations in all cases except those reported by Brzezinski (1988) and Lipschultz (this work). Brzezinski (1988) used the Berthelot method followed by solid phase extraction to concentrate the indophenol dye and lower the detection limit. The method used by Lipschultz involves the derivatization of NH_x (s) to form a highly fluorescent compound (Jones 1991).

The measurement of NH_x (s) is difficult due to contamination during sampling, storage, and analysis (D'Elia 1983). These analytical difficulties become critical at the nanomolar levels of NH_x (s) present in most of the world's oceans. The development of new techniques (Brzezinski 1988; Jones 1991) has made a more precise analysis (± 2 nM) with a lower detection limit (<10 nM) possible but an intercomparison of these methods is needed to determine their accuracy. The wide range of values reported for the Sargasso Sea near Bermuda illustrates these analytical difficulties. Concentrations near the detection limit of the colorimetric method, 30 to 50 nM, have been reported for this region (Glibert & McCarthy 1984; Glibert et al. 1988) as well as values as high as 2720 nM (Carpenter 1973). Measurements made with methods which have lower detection limits reveal surface concentrations in the 10 nM range (Brzezinski 1988; Lipschultz, this work). As shown in Figure 1, the only seasonal measurements for this region suggest that values do not exceed 200 nM (Lipschultz, this work) indicating that an average concentration near 50 nM or less is appropriate.

Table 2. Measured $[NH_x]_s$ for the North Atlantic.

Site	Date	$[NH_x]_s$, nM	Method	Reference
Caribbean 12N, 59W to 17N, 69W	Aug. to Sept., 1977	70–960	Berthelot reaction	Glibert & McCarthy 1984
Caribbean to Bermuda, 23N to 34N, 71W	May	1920–2720	Berthelot reaction	Carpenter 1973
Sargasso Sea	June to July, 1979	<30	Berthelot reaction	Glibert & McCarthy 1984
Sargasso Sea 34N, 73W	May to June, 1982	<30	Berthelot reaction	Glibert et al. 1988
Sargasso Sea 38N, 65W	Sept., 1988	90 ± 65	Berthelot reaction	Longhurst et al. 1989
Sargasso Sea 32N, 62W	June, 1987	10 to 160	Berthelot reaction with solid phase extraction	Brzezinski 1988
Near Bermuda	Sept. to Oct.	60–1070 mean = 460	Berthelot reaction	Carpenter & McCarthy 1975
Bermuda 32N, 64W	Sept. 1992 to April, 1994	10 to 200	OPA derivatization	Lipschultz, this work
SE US Shelf, 32N, 80W	April	300 to 400	Berthelot reaction	Hanson & Robertson 1988
Equator, 0 to 2N, 4W	Various	<50	Berthelot reaction	Le Bouteiller 1986
NW African Coast 31N, 10W to 15N, 18W	March to April, 1971	500 to 1000	Berthelot reaction	Slawyk & MacIsaac 1972
Belgium Coast, 51N, 3E	Spring	600 to 4600	Berthelot reaction	Lancelot et al. 1986

Table 2. Continued.

Site	Date	$[NH_x]_s$, nM	Method	Reference
Scotian Shelf, 43.5N, 62.5W	Various	500 to 1000	Berthelot reaction	Cochlan 1986
W. Canadian Arctic, 67N, 56W	July	240	Berthelot reaction	Harrison 1983
North Atlantic 40N, 47W 45N, 41W	April to May, 1989	80–300	Berthelot reaction	Harrison et al. 1993
North Atlantic 47N, 20W	May	50	Berthelot reaction	Garside & Garside 1993
North Sea	Feb. to Oct., 1989	300 to 4300	Berthelot reaction	Asman et al. 1994 Brockman et al. 1990

Values reported for the equatorial Atlantic near 4° W also are near the detection limit of the colorimetric technique (<50 nM) (Le Bouteiller 1986). In this southern region, NH_x (s) concentrations remain low throughout the year as the water column is permanently stratified and the euphotic zone is isolated from the deep water nitrate reservoir. The northern Atlantic Basin, however, experiences a bloom of phytoplankton in the springtime when the water column restratifies after wintertime convection of deep water brings nitrate to the surface. As the bloom ends, decomposition of the increased biomass can lead to higher levels of NH_x (s) for a short period. This is reflected in the higher concentrations reported for the North Atlantic (up to 300 nM) (Harrison et al. 1993) and the North Sea (up to 4300 nM) (Asman et al. 1994). Reported values also are high in coastal regions including the northwest African coast (500 to 1000 nM) (Slawyk & MacIsaac 1972), the Belgium coast (600 to 4600 nM) (Lancelot et al. 1986), the Scotian Shelf (500 to 1000 nM) (Cochlan 1986), and along the southeastern U.S. (300 to 400 nM) (Hanson & Robertson 1988).

3b. *Modeled values*

Surface NH_x (s) concentrations calculated from the HAMOCC3 model for the North Atlantic are shown in Figures 2a through 2d for the months of October, January, April, and July, respectively. Annual mean concentrations

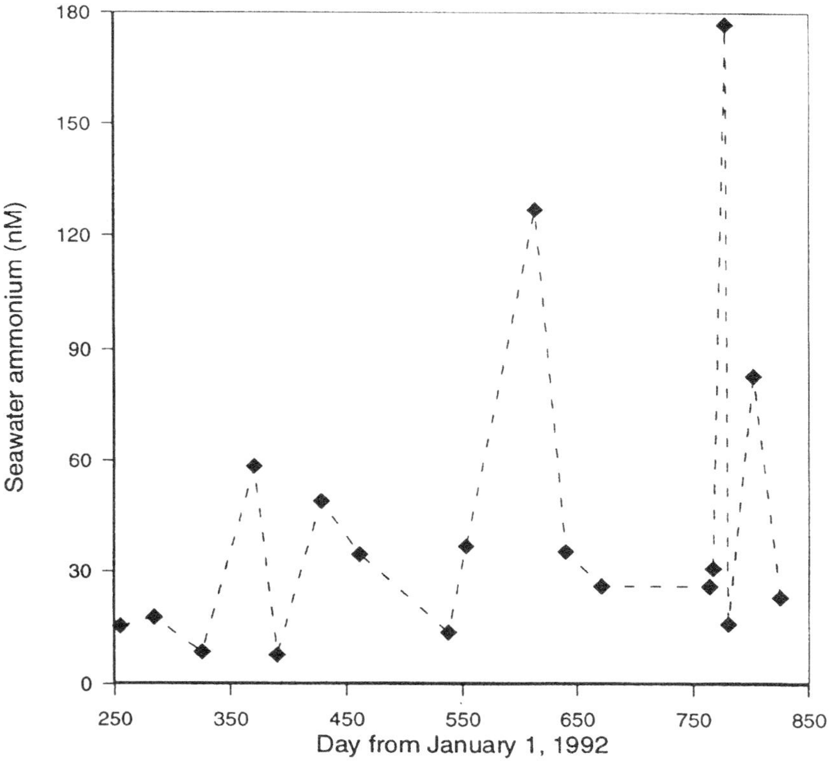

Figure 1. Results of time series measurements of NH_x (s) concentrations made near Bermuda at 31.8° N and 64.1° W with the fluorescence technique of Jones (1991). Concentrations are in units of nM.

range from 10 to 50 nM (Figure 2e) with lower values in the subtropical gyres and higher values at latitudes between 45° and 65° N. Seasonally, the lowest concentrations occur in fall and winter. Highest concentrations are found during and after plankton blooms; values up to 200 nM occur in the mid-Atlantic at 30° N in April and at 45° to 65° N in July. Concentrations higher than 200 nM are not predicted for the North Atlantic at latitudes below 65° N.

The modeled and measured open ocean NH_x (s) concentrations fall within the same range; both are <30 nM for the Sargasso Sea and 10 to 100 nM for Bermuda. The model underestimates coastal values by as much as an order of magnitude, however. For example, off the northwestern African coast, values in the range of 500 to 1000 nM have been observed during March and April (Slawyk & McIsaac 1972) but the model values remain between

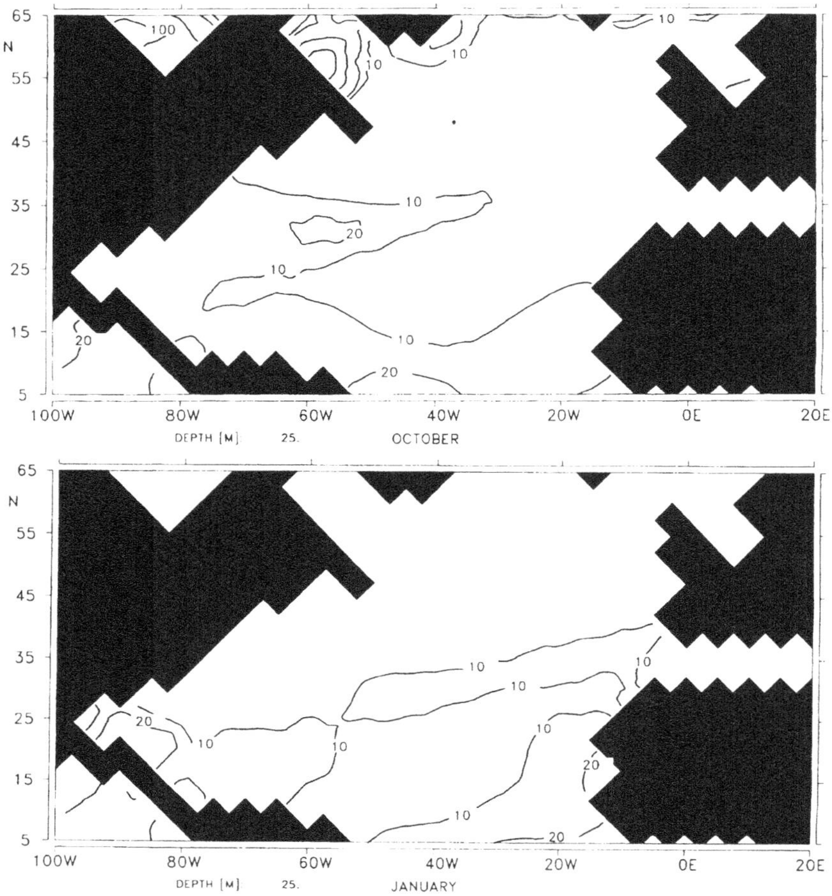

Figure 2. Global distribution of NH_x (s) concentrations calculated by the HAMOCC3 model for a) October, b) January, c) April, and d) July. Annually averaged values are shown in e). Concentrations are in units of nM. Isolines are given in logarithmic scaling: 10, 20, 50, 100, and 200 nM.

10 and 20 nM throughout the year. This discrepancy between measured and modeled results in coastal regions can be explained by the coarse resolution of the model. The complex small scale structure of dynamical and biological processes that exist in coastal areas is not well represented by a global ocean model. In addition, year-to-year variability can not be reproduced accurately by HAMOCC3 as it is based on climatological forcing. Hence, the model results will not reflect the change in the measured concentration from 5 to 40 nM near Bermuda in February, 1993 to 180 nM in February, 1994 (Figure 1). The comparison of point measurements with HAMOCC3 values is most

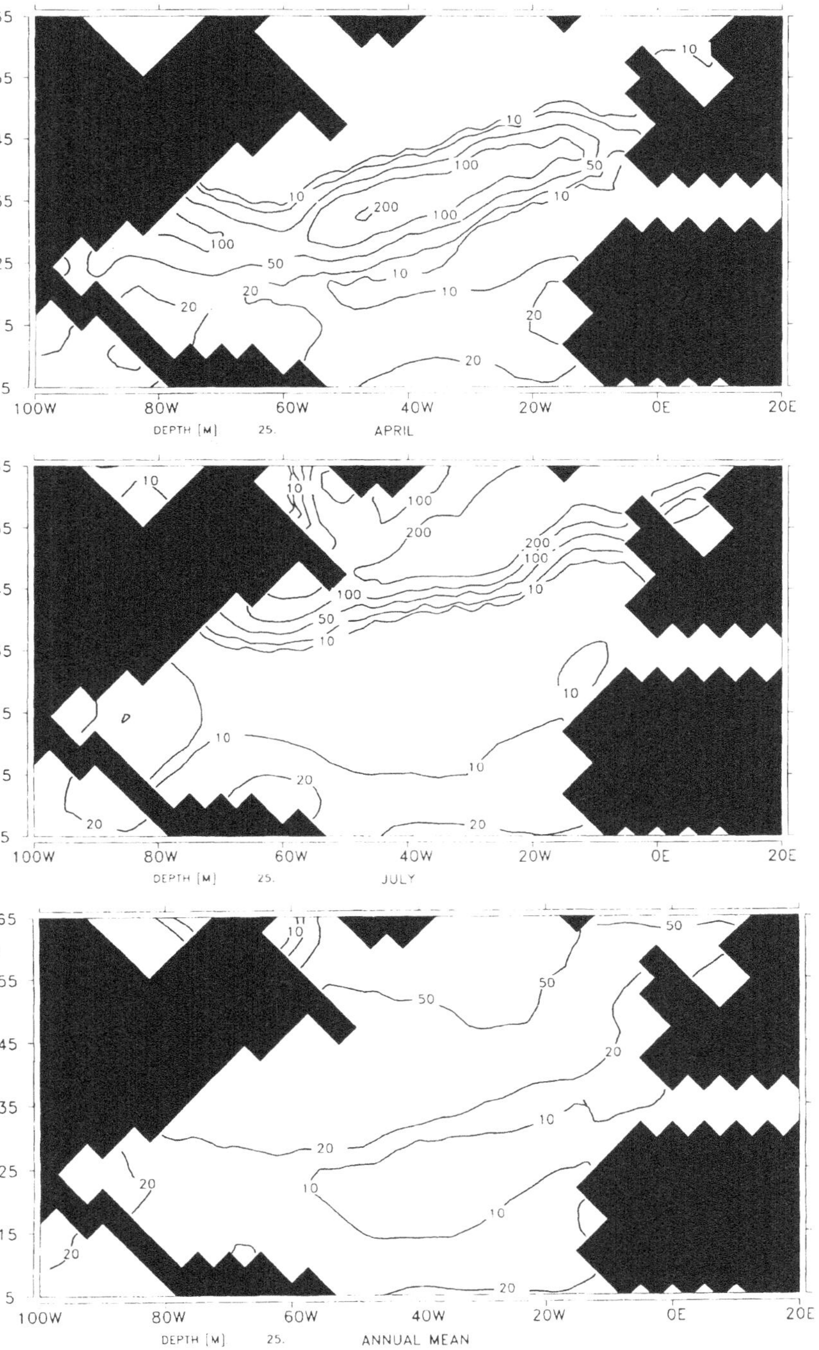

Figure 2. Continued.

appropriate in open ocean regions where spatial and temporal variability is minimized. The agreement within a factor of 2 in large scale structures between observations and HAMOCC3 predictions indicates that the simulated fields can serve as lower ocean boundary conditions for Moguntia.

4. Measured and modeled atmospheric NH_3 (g) concentrations

4a. *Measured values*

Reported values for atmospheric NH_3 (g) concentrations are listed in Table 3. The data are a result of four field experiments in which different measurement methods and, presumably, different protocols to prevent sample contamination were used. Georgii & Gravenhorst (1977) report results from a cruise taken in October of 1973 that traveled between Hamburg, Germany and the Caribbean Sea. No indication is given of the measurement method or precautions taken to avoid sample contamination. High NH_3 (g) concentrations were found near the Azores (50 to 170 nmol m^{-3}) and over the Caribbean (220 nmol m^{-3}) and Sargasso Seas (110 to 500 nmol m^{-3}). Lower concentrations of 11 to 60 nmol m^{-3} were measured over the tropical Atlantic along a transect from 40W to 10W. The high concentrations over the Sargasso Sea prompted the authors to suggest that the ocean was acting as a source of NH_3 in this region. This can not be confirmed, however, without simultaneously measured NH_x (s) concentrations. The high concentrations near the Azores and over the Sargasso Sea contrast with more recent measurements. Zhuang & Huebert (1994) measured concentrations at Santa Maria island in the Azores with annular denuders and found values between 1 and 30 nmol m^{-3} with an average and standard deviation of 9 ± 9.4 nmol m^{-3}. LeBel et al. (1985) measured concentrations at Bermuda of 20 nmol m^{-3} using the WO_x denuder technique. It is unclear if these large discrepancies are a result of sampling artifacts or natural variability.

More recently, Asman et al. (1994) reported concentrations of 2 to 140 nmol m^{-3} for the North Sea using H_3PO_4 impregnated filters. This is the only experiment of the four in which simultaneously measured NH_x (s) concentrations were reported.

4b. *Modeled values*

NH_3 (g) concentrations for the months of October, January, April, and July calculated with the Moguntia model are shown in Figures 3a through 3d, respectively. Annually averaged values are shown in Figure 3e. In the fall and winter, concentrations are low throughout the open ocean of the North

Table 3. Measured atmospheric NH_3 (g) concentrations, $(NH_3)_g$, for the North Atlantic Basin.

Site	Date	$(NH_3)_g$, nmol m^{-3}	Method	Reference
North Sea South of 56N	March to Sept., 1989	2–140	Impregnated H_3PO_4 filter	Asman et al. 1994
North Sea, English Channel	October, 1973	42–180	Not given	Georgii & Gravenhorst 1977
Azores 30N to 40N	October, 1973	50–170	Not given	Georgii & Gravenhorst 1977
Azores 37N, 25W	June, 1992	1–30	Denuder	Zhuang & Huebert 1994
Tropical Atlantic 40W to 10W	October, 1973	11–60	Not given	Georgii & Gravenhorst 1977
Caribbean 70W	October, 1973	220	Not given	Georgii & Gravenhorst 1977
Sargasso Sea 20N to 30N	October, 1973	110–500	Not given	Georgii & Gravenhorst 1977
Bermuda	1982	20	WO_3 Denuder	LeBel et al. 1985

Atlantic; values are around 1.0 nmol m^{-3} at 30° N and increase to 5 to 10 nmol m^{-3} over the Sargasso Sea and between 5° and 15° N. Concentrations are higher over the Gulf of Mexico (5 to 10 nmol m^{-3}), Caribbean Sea (10 to 20 nmol m^{-3}), and North Sea (2 to 10 nmol m^{-3}). Along the coasts of Ireland, France, and Portugal, concentrations range from 2 to 10 nmol m^{-3}. During the summer, concentrations increase to 5 nmol m^{-3} over the North Atlantic but decrease to 0.5 nmol m^{-3} further south near 25° N. Summer concentrations are similar to those calculated for the winter months over the Gulf of Mexico, the Sargasso Sea and the Caribbean Sea. They are higher, however, over the North Sea and the coasts of Ireland, France, and Portugal reaching values of 20 nmol m^{-3}

Annually averaged NH_3 (g) concentrations calculated with the MSC-W model for the year 1990 are shown in Figure 4. These calculations use a polynomial interpolation between grid squares. The values are based on NH_3 emissions resulting from agricultural and industrial activity and, hence, are of purely anthropogenic origin. Over the open ocean, NH_3 (g) concentrations

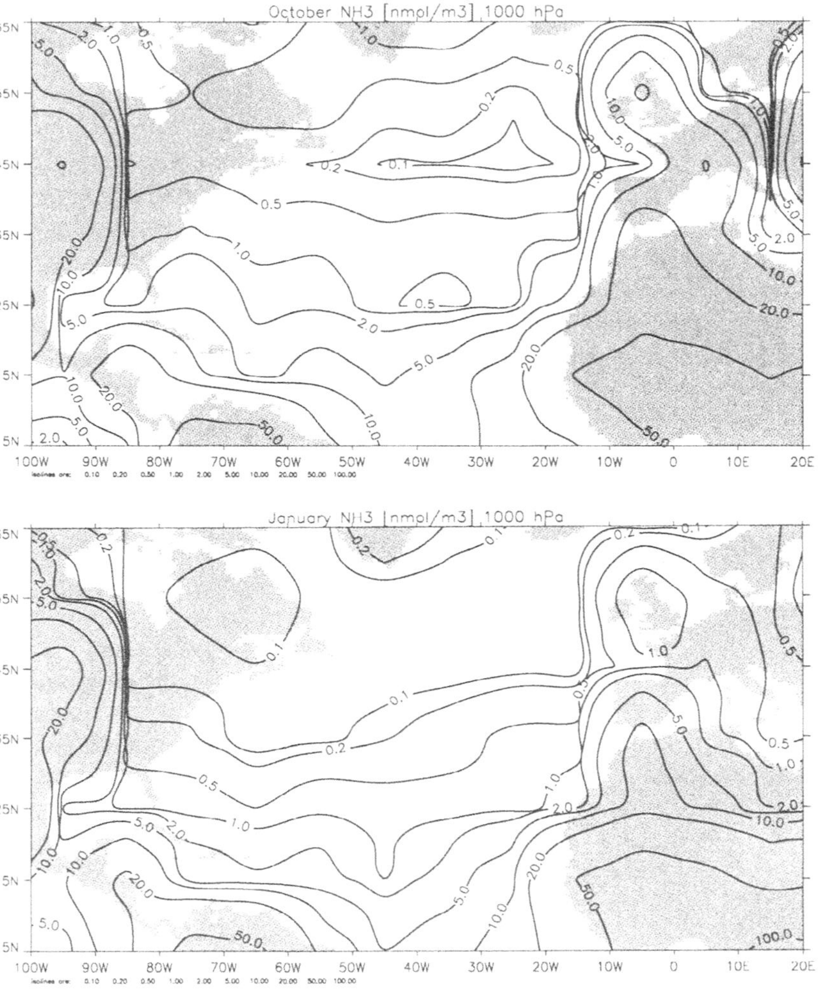

Figure 3. Global distribution of NH_3 (g) calculated by Moguntia for a) October, b) January, c) April, and d) July. Annually averaged values are shown in e). Concentrations are in units of nmol m^{-3}.

are below 2 nmol m^{-3}. Over the North Sea, concentrations range from 2 to 100 nmol m^{-3} with values decreasing northwards. Along the coast of Ireland, gradients are steep with concentrations on land being over 50 nmol m^{-3} and those 150 km from the coast being below 2 nmol m^{-3}. Adjacent to the coasts of France and Portugal they are over 50 nmol m^{-3}. Throughout this region, concentrations decrease with distance from the shore.

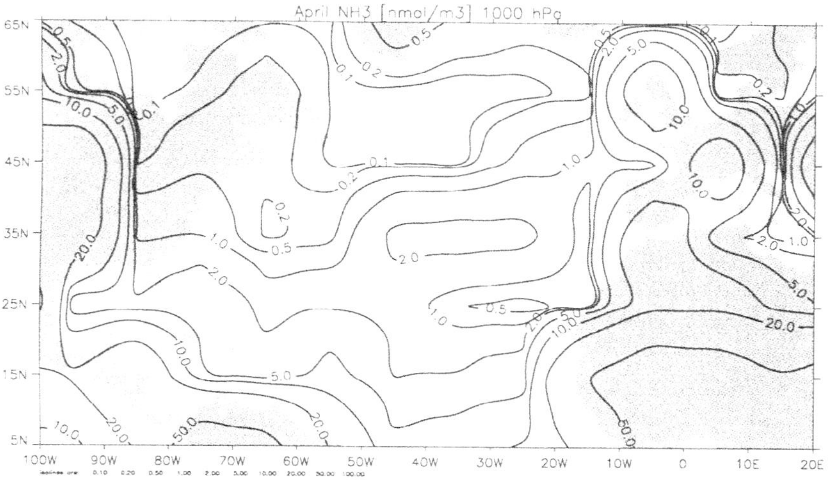
April NH3 [nmol/m3] 1000 hPa
65N
55N
45N
35N
25N
15N
5N
100W 90W 80W 70W 60W 50W 40W 30W 20W 10W 0 10E 20E

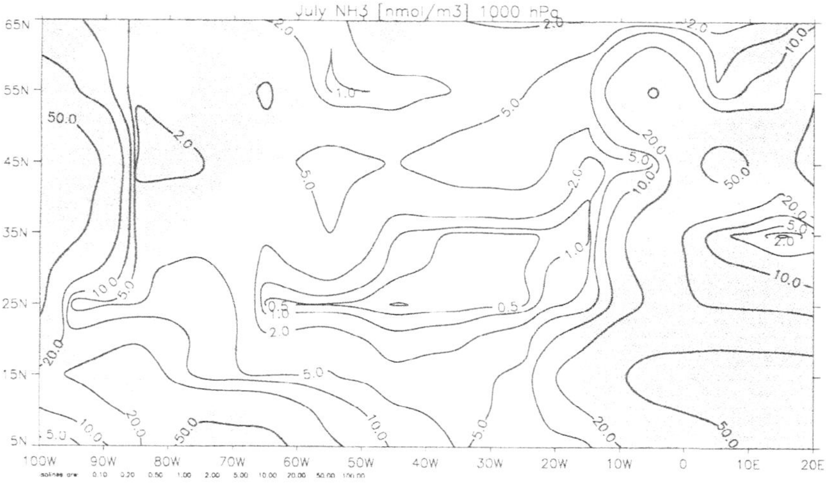
July NH3 [nmol/m3] 1000 hPa
65N
55N
45N
35N
25N
15N
5N
100W 90W 80W 70W 60W 50W 40W 30W 20W 10W 0 10E 20E

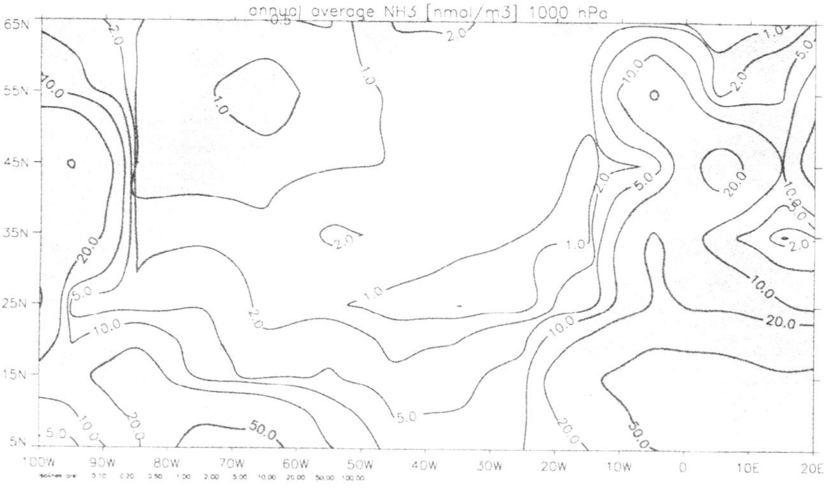
annual average NH3 [nmol/m3] 1000 hPa
65N
55N
45N
35N
25N
15N
5N
100W 90W 80W 70W 60W 50W 40W 30W 20W 10W 0 10E 20E

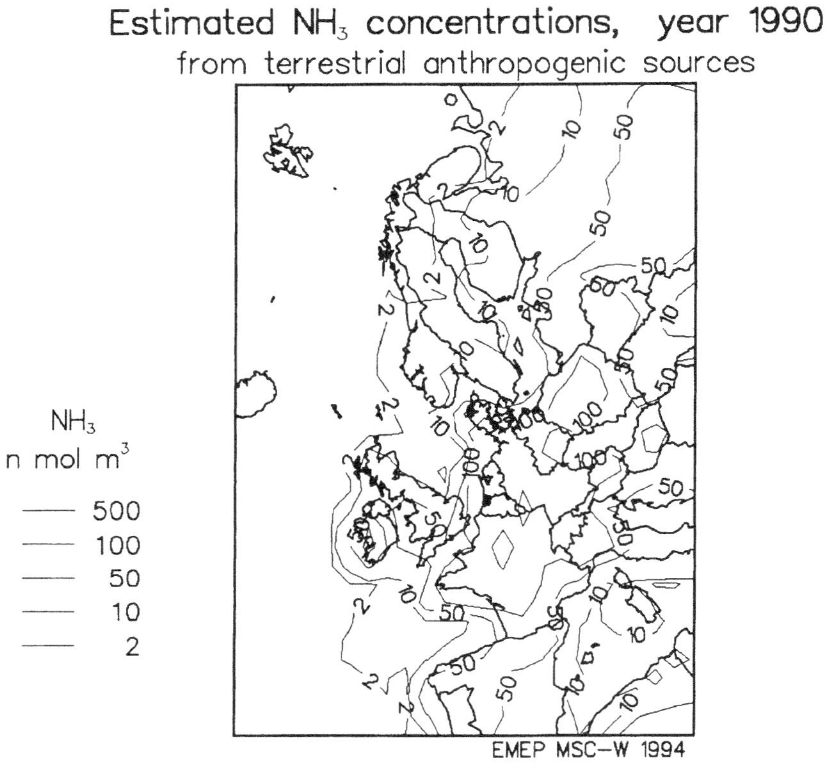

Figure 4. Annually averaged distribution of atmospheric NH_3 (g) for the northeast Atlantic Basin calculated from the MSC-W model for the year 1990. Concentrations are in nmol m^{-3}.

The two models predict similar annually averaged values for the North Sea (Moguntia: 2 to 10 and MSC-W: 2 to over 100 nmol m^{-3}) and the NE Atlantic (Moguntia: 1 to 5 and MSC-W: below 2 nmol m^{-3}). As expected, however, this level of agreement does not extend to regions over land where large point sources can exist. This lack of agreement is due to differences in the model characteristics. MSC-W has a resolution of 150 km. The resolution of Moguntia is 1000 km which leads to a dilution of sources and a lowering of the calculated NH_3 (g) concentrations. Transport schemes also are different as MSC-W uses actual meteorology and Moguntia uses monthly averaged winds. In addition, emissions from MSC-W are more specific in location than those from Moguntia.

In all regions where both measured and modeled concentrations of NH_3 are available, the measured values are about an order of magnitude larger. One exception is over the North Sea where modeled values are within the range of those that were measured. The systematically higher measured values may be a result of contamination during sample collection and analysis in areas with low concentrations. Additionally, the models could be underestimating the emissions of NH_3 (g) or overestimating the fraction of NH_3 (g) that is converted to NH_4^+ (p). The latter seems unlikely considering the fast reaction rate between NH_3 (g) and sulfate aerosol (Quinn et al. 1992). A final difficulty is that continental modeling at scales in excess of 100 km resolution can not resolve sharp concentration gradients. Hence, for regions with large NH_3 sources, values derived from point measurements are expected to differ from those derived from wider-scale models.

5. Air/sea NH_3 fluxes

5a. *Description of flux calculation*

The following method (Asman et al. 1994) was used to calculate the air/sea NH_3 flux from the measured concentrations of NH_3 (g) and NH_x (s). The same method is used in the MSC-W model and the Moguntia/HAMOCC3 models. The flux of ammonia across the air/sea interface is parameterized as the product of the exchange velocity, V_e, and the difference between the concentration of ammonia in equilibrium with dissolved ammonia in the sea surface, NH_3 (s,eq), and the concentration of atmospheric ammonia, NH_3 (g), such that

$$F = V_e[(NH_3)_{s,eq} - (NH_3)_g] \qquad (1)$$

For flux estimates derived from measured concentrations, a constant exchange velocity of 0.8 cm s^{-1} was used. This exchange velocity is based on experimentally determined transfer rates of water at air/water interfaces (Liss & Galloway 1993) recognizing that NH_3 and H_2O have similar molecular structures and behaviors. Values of V_e used in the MSC-W model are determined at 6 hour intervals and vary with wind speed and stability assuming no explicit surface resistance to transfer (Tuovinen et al. 1994). Values of V_e used in the Moguntia/HAMOCC3 models vary with observed wind speeds and range from 0.8 to 2 cm s^{-1}.

Values of $(NH_3)_{s,eq}$ are derived from

$$(NH_3)_{s,eq} = \frac{M_{NH_3}[NH_x]_s}{RTH_{NH_3}\left(\frac{1}{\alpha} + \frac{10^{-pH}}{\beta K_{NH_4}}\right)} \qquad (2)$$

which takes into account the equilibrium in seawater between NH_3 (s) and NH_4^+ (s) and the Henry's law equilibrium between NH_3 (s) and NH_3 (s,eq).

Here, M_{NH_3} is the molecular weight of NH_3 (17 g mol^{-1}), $[NH_x]_s$ is the seawater NH_x concentration in umol l^{-1}, R is the gas constant (8.2075×10^{-5} atm m^3 mol^{-1} K^{-1}), and T is temperature in K. H_{NH_3} is the Henry's law constant for NH_3 in mol l^{-1} atm^{-1} and is given by

$$H_{NH_3} = 56\exp\left(4092\left(\frac{1}{T} - \frac{1}{298.15}\right)\right) \qquad (3)$$

(Dasgupta & Dong 1986). α is the activity coefficient of NH_3 (s) and is given by

$$\alpha = 1 + 0.085I \qquad (4)$$

(Garrels & Christ 1965) where I is the seawater ionic strength such that

$$I = 0.00147 + 0.01988S + 2.08357 \times 10^{-}5S^2 \qquad (5)$$

and S is salinity (Lyman & Fleming 1940). The activity coefficient of NH_4^+ is β where

$$\beta = 10^{\gamma} \qquad (6)$$

(Seinfeld 1986) and

$$\gamma = -A_i z(\sqrt{I}/(1 + \sqrt{I}) - 0.2I) \qquad (7)$$

Here, $A_i = A + BT + CT^2$, $A = 0.4896844$, $B = -0.0007477$, $c = 2.729 \times 10^{-6}$, and z is the charge on the ion. The dissociation constant of NH_4^+ is in units of mol l^{-1} and is given by

$$K_{NH_4} = 5.67x10^{-10}\exp\left(-6286\left(\frac{1}{T} - \frac{1}{298.15}\right)\right) \qquad (8)$$

(Bates & Pinching 1950).

5b. *Flux estimates derived from measured concentrations*

Reported concentrations of atmospheric NH_3 (g) and seawater NH_x (s) were used to calculate fluxes for those regions of the North Atlantic where both atmospheric and seawater data are available. These regions include the Caribbean and Sargasso Seas, Bermuda, and the North Sea. The measurements from the North Sea are the only ones in which the atmospheric and oceanic data were collected simultaneously. For the other regions, the atmospheric and seawater data were collected at different times by different researchers. Samples for atmospheric NH_3 (g) are collected for longer periods of time than are samples for NH_x (s) and will be representative of a larger region. The seawater samples may be collected at a depth that does not represent the surface layer concentration that determines the flux. As a result, the flux estimates reported here are intended to be a first approximation for the North Atlantic Ocean. As these estimates are derived from discrete measurements, they are "point-in-time" fluxes and are not time averaged net balances of deposition and emission.

Estimated fluxes as well as the input parameters for the calculation are shown in Table 4. A uniform surface seawater pH of 8.1 was assumed and annually averaged sea surface temperatures (Pickard & Emery 1982) and salinities (Taylor & Stevens 1980) were used. For the Caribbean Sea, a fixed atmospheric NH_3 (g) concentration but a varying seawater concentration was used which resulted in flux values ranging from -150 to -110 μmol m^{-2} d^{-1}. The low concentrations of NH_x (s) resulted in a flux from the atmosphere to the ocean.

Two sets of flux calculations were performed for the region of the Sargasso Sea. In the first, measured NH_x (s) and NH_3 (g) concentrations from the wider region of the Sargasso Sea were used so that NH_x (s) ranged between 10 and 160 nM (Brzezinski 1988) and the NH_3 (g) ranged from 110 to 500 nmol m^{-3} (Georgii & Gravenhorst 1977). The large measured concentrations of atmospheric NH_3 led to a consistent flux from the atmosphere to the ocean ranging between -340 and -6 μmol m^{-2} d^{-1}. In the second estimation, data taken near and on Bermuda were used. A range of NH_x (s) concentrations from 10 to 200 nM (Lipschultz, this work) and an average NH_3 (g) concentration of 20 nmol m^{-3} (LeBel 1985) resulted in an estimated flux of -13 to -5.8 μmol m^{-2} d^{-1}. The discrepancy between these two calculated fluxes for the Sargasso Sea results from a large difference in the measured NH_3 (g) concentrations. It is likely that the early measurements of Georgii & Gravenhorst (1977) suffered from artifacts or contamination during sampling that led to artificially high concentrations. This is supported by the factor of ten difference in the concentrations measured in the Azores by Zhuang & Huebert (1994) (1 to 30 nmol m^{-3}) and Georgii & Gravenhorst (1977) (50 to 170 nmol m^{-3}). This assumption of an artifact in the Georgii & Gravenhorst (1977) measurements can not be confirmed, however, as no indication of the technique used or precautions that were taken to prevent contamination is given.

For the North Sea, where seawater and atmospheric concentrations were measured simultaneously (Asman et al. 1994), flux estimates vary from -75 to 43 μmol m^{-2} d^{-1}. The net direction of the flux appears to be determined

Table 4. NH_3 flux estimates based on measured $(NH_3)_g$ and $[NH_x]_s$ and the Moguntia/HAMOCC3 and MSC-W models. $[NH_x]_s$ are in units of nM, $(NH_3)_g$ are in nmol m^{-3}, and fluxes are in μmol m^{-2} d^{-1}.

Region	$[NH_x]_s$ Measured	$[NH_x]_s$ MOG	$[NH_x]_s$ MSC-W	$(NH_3)_g$ Measured	$(NH_3)_g$ MOG	$(NH_3)_g$ MSC-W	Flux Measured	Flux MOG[a]	Flux MSC-W[b]
Caribbean	70–960[1]	20–60		220[5]	3–25		−150 to −110	−4.0 to 1.3	
Sargasso	10–160[2]	10–180		110–500[5]	4–10		−340 to −76		
Bermuda	10–200[3]	10–100		20[6]	2–5		−13 to −5.8	4.0	
North Sea	300–4300[4]	10–270	1000	2–140[4]	1–20	2–100	−75 to 43[4]	−1.3 to 1.3	−69 to 20
NE Atlantic 40N, 12W		10–250	250		1–3	2–50		0 to 1.3	−1.6 to 3.3
NE Atlantic 50N, 20W		10–250	250		5	0.1		1.3 to 2.7	1.3
Coast of Ireland		10–280	250		1–25	2–10		−1.3 to 0	1.4
Coast of France & Portugal		10–140	250		3–25	2–50		0 to 1.3	−1.6 to 3.3

[a] Net monthly average flux, [b] Disaggregated fluxes
[1] Glibert & McCarthy (1984), [2] Brzezinski (1988), [3] This work, [4] Asman et al. (1994), [5] Georgii & Gravenhorst (1977), [6] Lebel et al. (1985)

by the highly variable NH_3 (g) concentration which, in turn, is a result of the proximity of the measurements to continental source regions. Lower NH_3 (g) concentrations in mid-summer yield near zero fluxes while higher values at other times of the year lead to atmosphere-to-ocean fluxes.

5c. *Flux estimates derived from modeled values*

Flux estimates from the Moguntia and MSC-W models are compared to those derived from the measured concentrations in Table 4. Net fluxes were derived from the Moguntia/HAMOCC3 models using HAMOCC3 calculated NH_x (s) concentrations, sea surface temperature, and pH and Moguntia calculated NH_3 (g). Fluxes from gridboxes partly covered by the ocean were scaled using the oceanic areal fraction. Calculated fluxes for the Caribbean Sea ranged from -4.0 to 1.3 μmol m^{-2} d^{-1}. Around Bermuda, the flux consistently was from the ocean to the atmosphere with a value near 4.0 μmol m^{-2} d^{-1}. Over the northeast Atlantic, fluxes were positive and were between 0 and 1.3 μmol m^{-2} d^{-1} at 40° N and 12° W and 1.3 and 2.7 μmol m^{-2} d^{-1} at 50° N and 20° W. Fluxes from the North Sea were both slightly negative and slightly positive such that, on the average, they were near zero. Near the coasts of Ireland and France, fluxes were both negative and positive with values of -1.3 to 0 and 0 to 1.3 μmol m^{-2} d^{-1}, respectively. Over the open ocean, fluxes were on the order of 1 to 2 μmol m^{-2} d^{-1}. These agree well with the lower range of previously reported flux estimates for the central Pacific Ocean (Quinn et al. 1990).

For the flux estimates derived from the MSC-W model, reported seawater concentrations during periods of plankton blooms were used in order to estimate potential maximum fluxes. These concentrations included North Sea values of 1000 to 4000 nM (Brockman et al. 1990; Asman et al. 1994) and North Atlantic values of 250 nM (Harrison et al. 1993). All coastal waters (mixed land/sea grid squares) were given a concentration of 500 nM. Sea surface temperatures were obtained from the Numerical Weather Prediction (NWP) model of the Norwegian Meteorological Institute. Input salinity values for the northeast Atlantic and North Sea were 35 and 34 per mil, respectively (Taylor & Stevens 1980). A surface seawater pH of 8.1 was assumed for all regions. Using these input values, both net fluxes and disaggregated fluxes, i.e, fluxes separated into the upward and downward direction, were estimated for the month of August 1990. The upward flux of oceanic NH_3 is shown in Figure 5a. Emissions ranged from 1.0 to 20 μmol m^{-2} d^{-1} for the North Sea. These are comparable with the summertime values derived from measurements of Asman et al. (1994). For the same month, deposition fluxes between -0.70 and -69 μmol m^{-2} d^{-1} were estimated (Figure 5b). In both cases, the fluxes decrease from the south to the north. The net flux is smaller than the disaggregated fluxes, ranging between -1.03 to 13 μmol m^{-2} d^{-1}, and is a result of the reversals in direction during the course of the month. Over the northeast Atlantic, both emission and deposition fluxes are much smaller than those over the North Sea. Off the coasts of Ireland and France, emission values were 1.4 and 3.3 μmol m^{-2} d^{-1}, respectively (1.3 and 1.7 μmol m^{-2} d^{-1} net). These are only slightly higher than in the mid North Atlantic at distances of 1000 km from the continent where values range from 1.3 to 2.1 μmol m^{-2} d^{-1} with no deposition flux.

Results from the MSC-W model allow for an assessment of the importance of marine ammonia sources to each European country relative to agricultural and industrial sources. These preliminary calculations, assuming summertime conditions, indicate that marine emissions may be one of the larger external sources of deposited reduced nitrogen for some countries on the western margins of the continent. For Ireland, marine emissions are the second largest source of reduced nitrogen, for the United Kingdom they are the third, for Norway the fifth, and for Finland the sixth. The inclusion of a marine source of reduced nitrogen for the European domain as a whole, however, only increases total emissions by 1.5%. Winter emission fluxes would be much lower due to colder temperatures and smaller seawater NH_x (s) concentrations.

When comparing fluxes from the Moguntia/HAMOCC3 and MSC-W models, net fluxes versus disaggregated fluxes are involved. Yet, in general, the direction of the fluxes predicted by the two models is the same and the magnitudes fall within the same range (Table 4). Flux estimates derived from the measured data are larger in both the negative and positive direction than the model derived values. It is unclear whether this discrepancy is due to inaccurate chemistry and transport schemes in the models, sampling artifacts in the measurements, or factors that make a direct comparison of the measurement- and model-derived results difficult. The measurement-derived fluxes are discrete in time while the model estimates are time averaged. A first step in resolving the disagreement would be to make accurate, simultaneous measurements of $(NH_3)_g$ and $[NH_x]_s$ and then to compare these to the model derived values over equivalent time periods. Such measurements should be made in several regions of the North Atlantic that have different sea surface temperatures and NH_x (s) concentrations as well as atmospheric NH_3 (g) concentrations. Representative regions with large surface areas such as 0 to 20° N or the Sargasso Sea would be most relevant for a basin scale assessment.

5d. *Model geography of oceanic NH_3 (g) emission versus deposition*

To determine the regions of the North Atlantic that serve as a source of atmospheric NH_3, annually averaged net emissions and depositions of NH_3

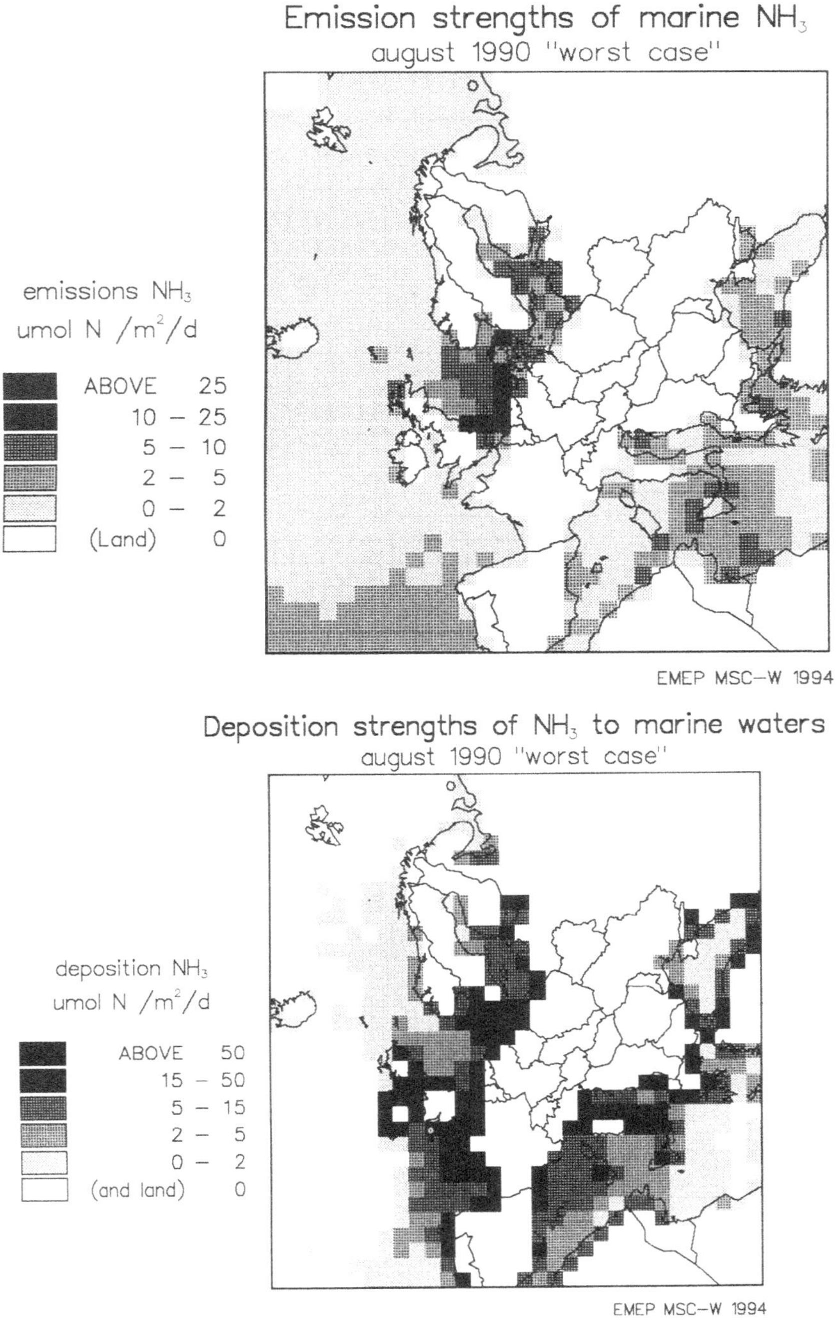

Figure 5. Disaggregated fluxes of ammonia in the a) upward (emission) direction and b) downward (deposition) direction calculated by the MSC-W model for August, 1990. Fluxes are in units of μmol m^{-2} d^{-1}.

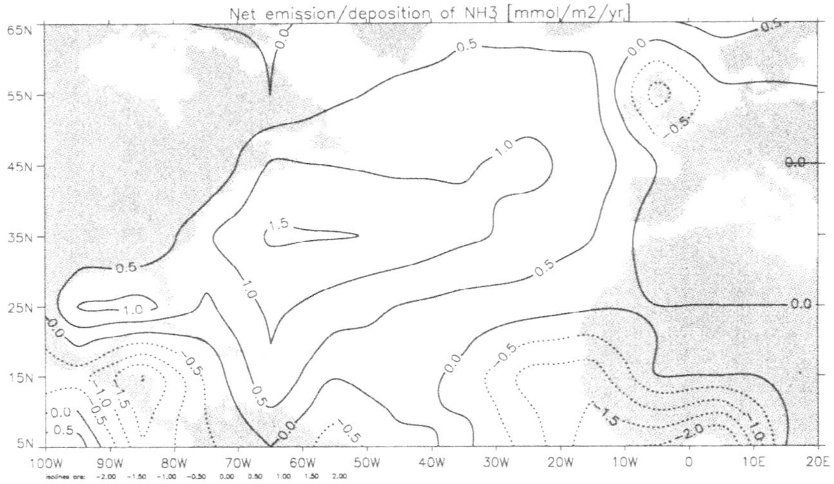

Figure 6. Net annually averaged oceanic ammonia emissions (continuous lines) and depositions (dashed lines) calculated with the Moguntia/HAMOCC3 global models. Fluxes are in mmol m^{-2} yr^{-1} (1 mmol m^{-2} yr^{-1} = 2.7 μmol m^{-2} d^{-1}).

from and to the ocean surface were calculated with the Moguntia/HAMOCC3 global models (Figure 6). Emissions and depositions over continental areas, which are an order of magnitude larger than those in oceanic regions, are not shown in this figure. Net NH_3 deposition occurs in ocean areas that are downwind of continental source regions such as Central America and eastern Africa. In these continentally-derived air masses, the model calculations indicate that ammonia emissions frequently dominate sulfur emissions. Therefore, gas phase NH_3 does not react immediately with acidic sulfate aerosol so that atmospheric residence times are long enough to allow transport to the oceans. Relatively small net fluxes are calculated over the North Sea due to the magnitude of the upward and downward disaggregated fluxes. The presence of significant NH_x (s) concentrations gives rise to large but temporary emission fluxes which are balanced by deposition due to high atmospheric concentrations at other times. The net flux results from the MSC-W model (Figure 7) also suggest a zoning in emission versus deposition during the month considered (August 1990). While the oceanic waters have a uniform emission characteristic, deposition is often dominant in coastal regions that are under the influence of terrestrial sources such as the French coast and the European shelf. Where localized marine concentrations are sufficient, however, net emission can occur such as for the central North Sea.

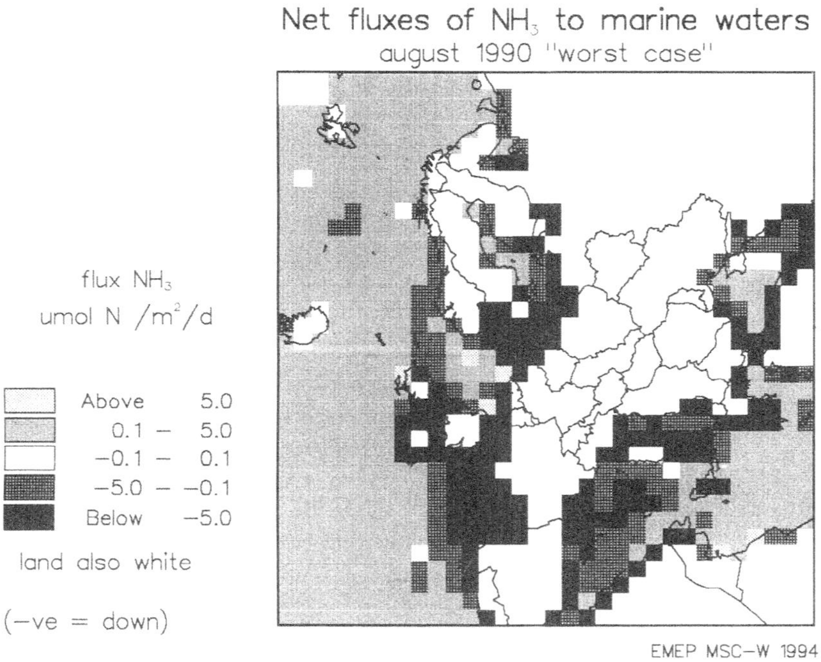

Figure 7. Net fluxes of ammonia for the northeast Atlantic Basin for August 1990 as calculated by the MSC-W model. Fluxes are in units of μmol m^{-2} d^{-1}.

6. Sensitivity of the flux calculation to input parameters

A sensitivity study was done to quantify the effect that changes in surface seawater temperature, salinity, pH, NH_x (s) concentration, the atmospheric NH_3 (g) concentration, and V_e have on the calculated flux. The input parameters for the sensitivity calculations were set for a base case of surface seawater temperature equal to 19 °C, salinity equal to 35.75 per mil, pH equal to 8.2, $[NH_x]_s$ equal to 100 nM, $(NH_3)_g$ equal to 1 nmol m^{-3}, and V_e equal to 0.8 cm sec^{-1}. One parameter was varied at a time while the others were kept at the base case values. Input parameters were varied over the range of values expected for the North Atlantic Basin: temperature from 7 to 31 °C, salinity from 34 to 37.5 per mil, pH from 7.8 to 8.5, $[NH_x]_s$ from 10 to 5000 nM, $(NH_3)_g$ from 1 to 500 nmol m^{-3}, and V_e from 0.8 to 2.0 cm sec^{-1}. As shown in Figure 8, for the range of expected values of the input parameters, changes in $[NH_x]_s$ and $(NH_3)_g$ have the largest effect on the absolute magnitude of the flux. Changes in seawater temperature also have a significant effect while changes in salinity, seawater pH, and V_e have minimal effects.

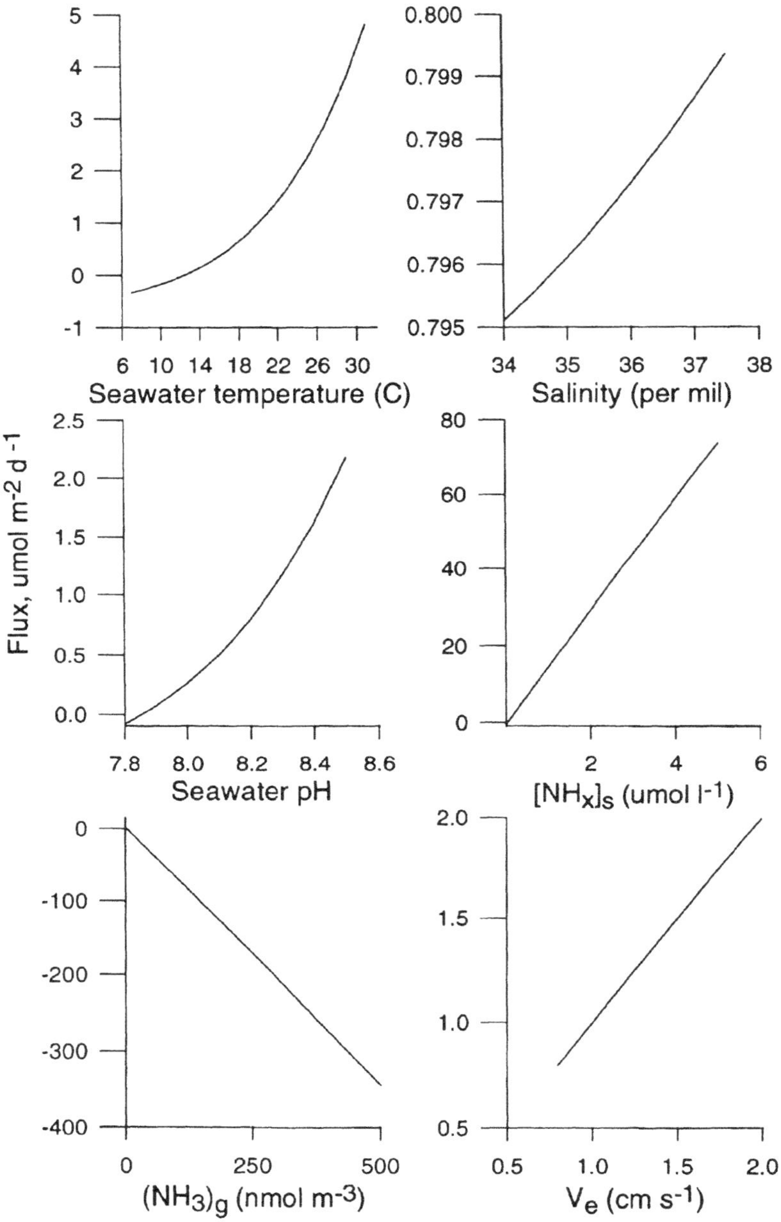

Figure 8. The sensitivity of the calculated flux to changes in surface seawater temperature, salinity, pH, $[NH_x]_s$, atmospheric $(NH_3)_g$, and the exchange velocity, V_e.

7. Conclusions

Presented here is a first attempt to determine the direction and magnitude of the air/sea ammonia flux for different regions of the North Atlantic Basin. Reported seawater and atmospheric ammonia concentrations have been compiled and the measured data have been compared to Moguntia/HAMOCC3 and MSC-W model output. In addition, air/sea NH_3 fluxes have been derived from the measured and modeled seawater and atmospheric concentration data.

Measured seawater NH_x (s) concentrations cover two orders of magnitude ranging from less than 30 to 4300 nM. The highest concentrations are reported for the Caribbean Sea, North Sea, and the Belgium coast. Very few open ocean values have been reported. The HAMOCC3 global model was able to reproduce the measured open ocean concentrations but tended to underestimate coastal values. This most likely is a result of the large scale of the model and an incomplete description within the model of the biological processes controlling the seawater ammonia concentrations. Measured atmospheric NH_3 (g) concentrations have been reported for only four North Atlantic field experiments. The concentrations range from 2 to 500 nmol m^{-3} with the largest values reported for the North Sea, the Caribbean Sea, and the Sargasso Sea. Again, there is a lack of open ocean measurements. In general, measured NH_3 (g) concentrations were larger than those calculated from the MSC-W and Moguntia models. The systematically higher measured values could be due to sample contamination and/or inadequate emission and transport schemes in the models. In addition, Moguntia concentrations are representative of the lowest 0 to 400 m of the atmosphere while the measurements are made at the surface. The difficulty in comparing point measurements with wider-scale models may also contribute to the disagreement.

Air/sea fluxes of ammonia calculated from the measured seawater NH_x (s) and atmospheric NH_3 (g) concentrations are meant to be a first approximation as, in most cases, the seawater and atmospheric concentrations were not measured simultaneously. The North Sea was found to have both atmosphere-to-ocean and ocean-to-atmosphere net fluxes. Fluxes for the Caribbean, Bermuda, and the Sargasso Sea were consistently from the atmosphere to the ocean although those from the Sargasso Sea were about a factor of 15 to 20 larger than those from Bermuda. The lack of agreement between the Bermuda and Sargasso Sea fluxes is disturbing as they are in the same region. It is likely that the NH_3 (g) concentrations measured over the Sargasso Sea (Georgii & Gravenhorst 1977) were contaminated leading to artificially large atmosphere-to-ocean fluxes. The only fluxes calculated from simultaneously measured NH_3 (g) and NH_x (s) concentrations were for the North Sea (Asman et al. 1994) where values ranged from -75 to 43 μmol m^{-2} d^{-1}. The net direction of the flux depends primarily on the highly variable $(NH_3)_g$ as for other regions under continental influence. Fluxes derived from the measured concentration data were either more negative or more positive than those derived from the modeled concentrations. This difference could be due to sampling artifacts, inadequacies in the models, or difficulties in comparing discrete and time-averaged fluxes.

Sensitivity tests indicate that the flux estimates depend strongly on changes in the seawater NH_x and atmospheric NH_3 concentrations. Therefore, to define the role of oceanic ammonia emissions in the nitrogen budget of the North Atlantic Basin, simultaneous and accurate measurements of $(NH_3)_g$ and $[NH_x]_s$ in several regions are needed. These should be compared to model-derived values over equivalent spatial and temporal scales for validation of the models. The eventual goal is to develop models which are able to simulate accurately the nitrogen dynamics of the North Atlantic Basin. Such models will help to determine the magnitude of oceanic NH_3 emissions on a regional basis, the influence of continental NH_3 emissions on the North Atlantic Basin, and the regional effects of NH_3 on aerosol chemical, physical, and optical properties in the North Atlantic atmosphere.

Acknowledgments

We would like to acknowledge W. Asman for supplying the compensation point calculation method used in the MSC-W model and W. Asman and T. Bates for comments on the manuscript. We thank the organizers of the Block Island workshop (R. Howarth and J. Galloway) for providing the forum that inspired this joint effort. We also appreciate the funding provided by UNEP and the Mellon Foundation to support the efforts of the SCOPE Nitrogen Project and by the World Meteorological Organization to investigate the interaction between the N cycles of the atmosphere and the North Atlantic Ocean. Work done by P. K. Quinn was funded by the NOAA Climate and Global Change Program. Work done by K. J. Barrett was supported by the Department of the Environment of the United Kingdom and was a voluntary contribution-in-kind to EMEP. Work done by K. D. Six was supported by EC under Contract EV5V-CT92-0124. Work done by F. Lipschultz was supported by NSF grant # OCE90-16069. This is PMEL contribution number 1609 and BBSR contribution number 1406.

References

Asman WAM, Drukker B & Janssen AJ, (1988) Modelled historical concentrations and depositions of ammonia and ammonium in Europe. Atm. Environ. 22: 725–735

Asman WAM, Harrison RM & Ottley CJ (1994) Estimation of the net air-sea flux of ammonia over the Southern Bight of the North Sea. Atm. Environ. 28: 3647–3654

Bates RG & Pinching GD (1950) Dissociation constants of aqueous ammonia at 0 to 50 C from e.m.f. studies of the ammonium salt of a weak acid. J. Am. Chem. Soc. 72: 1393–1396

Brzezinski MA (1988) Vertical distribution of ammonium in stratified oligotrophic waters. Limnol Oceanogr. 33: 1176–1182

Brockmann UH, Laane RWPM & Postma H (1990) Cycling of nutrient elements in the North Sea. Netherlands J. of Sea Res. 26: 239–264

Buijsman E, Maas HFM & Asman WAH (1987) Anthropogenic NH_3 emission in Europe. Atm. Environ. 21: 1009–1022

Carpenter EJ (1973) Nitrogen fixation by *Oscillatoria (Trichodesmium) theibautii* in the southwestern Sargasso Sea. Deep-Sea Res. 20: 285–288

Carpenter EJ & McCarthy JJ (1975) Nitrogen fixation and uptake of combined nitrogenous nutrients by *Oscillatoria (Trichodesmium) theibautii* in the western Sargasso Sea. Limnol. Oceanogr. 20: 389–401

Clarke AD & Porter JN (1993) Pacific marine aerosol 2. Equatorial gradients in chlorophyll, ammonium, and excess sulfate during SAGA–3. J. Geophys. Res. 98: 16997–17010

Cochlan (1986) Seasonal study of uptake and regeneration of nitrogen on the Scotian Shelf. Cont. Shelf Res. 5: 555–577

Dasgupta PK & Dong S (1986) Solubility of ammonia in liquid water and generation of trace levels of standard gaseous ammonia. Atm. Environ. 20: 565–570

D'Elia CF (1983) Nitrogen determination in seawater. In: Carpenter EJ & Capone DG (Eds) Nitrogen in the Marine Environment (pp 731–762). Academic Press, New York

Dentener FJ & Crutzen PJ (1994) A three dimensional model of the global ammonia cycle. J. Atm. Chem. 19: 331–369

Duce RA, Liss PS, Merrill JT, Atlas EL, Buat-Menard P, Hicks BB, Miller JM, Prospero JM, Arimoto R, Church TM, Ellis W, Galloway JN, Hansen L, Jickells TD, Knap AH, Reinhardt KH, Schneider B, Soudine A, Tokos JJ, Tsunogai S, Wollast R & Zhou M (1991) The atmospheric input of trace species to the world ocean. Global Biogeochem. Cycles 5: 193–260

Garrels RM & Christ CL (1965) Solutions, Minerals, and Equilibria. Harper and Row, New York

Garside C & Garside JC (1993) The "f-ratio" on 20W during the North Atlantic Bloom Experiment. Deep-Sea Res. 40: 75–90

Georgii HW & Gravenhorst G (1977) The ocean as a source or sink of reactive trace gases. Pageoph. 115: 503–511

Glibert PM, Dennett MR & Caron DA (1988) Nitrogen uptake and ammonium regeneration by pelagic microplankton and marine snow from the North Atlantic. J. Mar. Res. 46: 837–852

Glibert PM & McCarthy JJ (1984) Uptake and assimilation of ammonium and nitrate by phytoplankton: Indices of nutritional status for natural assemblages. J. Plankton Res. 6: 677–697

Hanson RB & Robertson CY (1988) Spring recycling rates of ammonium in turbid continental shelf waters off the southeastern United States. Cont. Shelf Res. 8: 49–68

Harrison WG (1983) The time-course of uptake of inorganic and organic nitrogen compounds by phytoplankton from the Eastern Canadian Arctic: A comparison with temperate and tropical populations. Limnol. Oceanogr. 28: 1237–1242

Harrison WG, Head EJH, Horne EPW, Irwin B, Li WKW, Longhurst AR, Paranjape MA & Platt T (1993) The western North Atlantic Bloom Experiment. Deep-Sea Res. 40: 279–305

Jones RD (1991) An improved fluorescence method for the determination of nanomolar concentrations of ammonium in natural seawaters. Limnol. Oceanogr. 36: 814–819

Kurz KD (1993) Zur saisonalen Variation des ozeanischen Kohlendioxidpartialdrucks, Examensarbeit Nr. 18. Max-Planck-Institut fuer Meteorologie

Kurz KD & Dentener FJ (1994) Exchange of dimethylsulfide and ammonia between the ocean and the atmosphere calculated using a global ocean and atmospheric model. (in preparation)

Lancelot C, Mathot S & Owens NJP (1986) Modelling protein synthesis, a step to an accurate estimate of net primary production: *Phaeocystis pouchetii* colonies in Belgium coastal waters. Marine Ecol. Prog. Ser. 32: 193–202

LeBel PJ, Hoell JM, Levine JS & Vay SA (1985) Aircraft measurements of ammonia and nitric acid in the lower troposphere. Geophys. Res. Lett. 12: 401–404

Le Bouteiller A (1986) Environmental control of nitrate and ammonium uptake by phytoplankton in the equatorial Atlantic Ocean. Marine Ecol. Prog. Ser. 30: 1647–1679

Liss PS & Galloway JN (1993) Air-sea exchange of sulfur and nitrogen and their interaction in the marine atmosphere. In: Wollast R, Mackenzie FT & Chou L (Eds) Interactions of C, N, P, and S Biogeochemical Cycles and Global Change (pp 259–281). Springer-Verlag, Berlin

Longhurst AR, Bedo A, Harrison WG, Head EJH, Horne EP, Irwin B & Morales C (1989) NFLUX: A test of vertical nitrogen flux by diel migrant biota. Deep-Sea Res. 36: 1705–1721

Lyman J & Fleming RH (1940) Composition of seawater. J. Mar. Resources 3: 134–146

Maier-Reimer E (1993) Geochemical cycles in an ocean general circulation model: Preindustrial tracer distributions. Global Biogeochem. Cycles 7: 645–677

Pickard GL & Emery WJ (1982) Descriptive Physical Oceanography, An Introduction. Pergamon Press, New York, 44 p

Quinn PK, Charlson RJ & Zoller WH (1987) Ammonia, the dominant base in the remote marine troposphere: A review. Tellus 39B: 413–425

Quinn PK, Charlson RJ & Bates TS (1988) Simultaneous observations of ammonia in the atmosphere and ocean. Nature 335: 336–338

Quinn PK, Bates TS, Johnson JE, Covert DS & Charlson RJ (1990) Interactions between the sulfur and reduced nitrogen cycles over the central Pacific Ocean. J. Geophys. Res. 95: 16405–16416

Quinn PK, Asher WE & Charlson RJ (1992) Equilibria of the marine multiphase ammonia system. J. Atmos. Chem. 14: 11–30

Seinfeld JH (1986) Atmospheric Chemistry and the Physics of Air Pollution. Wiley, New York

Slawyk G & MacIsaac JJ (1972) Comparison of two automated ammonium methods in a region of coastal upwelling. Deep-Sea Res. 19: 521–524

Taylor AH & Stevens JA (1980) Seasonal and year-to-year variations in surface salinity at the nine North Atlantic Ocean weather stations. Ocean. Acta 3: 421–429

Tuovinen JP, Barrett KJ & Styve H (1994) Transboundary acid pollution in Europe: Calculated fields and budgets 1985–1993, EMEP/MSC-W Report 1/94. Norwegian Meteorological Institute, Oslo

Warneck P (1988) Chemistry of the Natural Atmosphere. Academic Press, San Diego, CA, pp 426–441

Williams EJ, Sandholm ST, Bradshaw JD, Schendel JS, Langford AO, Quinn PK, LeBel PJ, Vay SA, Roberts PD, Norton RB, Watkins BA, Buhr MP, Parrish DD, Calvert JG & Fehsenfeld FC (1992) An intercomparison of five ammonia measurement techniques. J. Geophys. Res. 97: 11591–11611

Zhuang L & Huebert B (1994) personal communication.

GPSR Compliance

The European Union's (EU) General Product Safety Regulation (GPSR) is a set of rules that requires consumer products to be safe and our obligations to ensure this.

If you have any concerns about our products, you can contact us on ProductSafety@springernature.com

In case Publisher is established outside the EU, the EU authorized representative is:

Springer Nature Customer Service Center GmbH
Europaplatz 3
69115 Heidelberg, Germany

Batch number: 10816149

Printed by Printforce, the Netherlands